D0850853

Systems & Control: Foundations & Applications

Panagiotis D. Christofides

Nonlinear and Robust Control of PDE Systems

Methods and Applications to Transport-Reaction Processes

With 80 Figures

Birkhäuser
Boston • Basel • Berlin

Panagiotis D. Christofides
Department of Chemical Engineering
University of California
Los Angeles, CA 90095-1592
USA

Library of Congress Cataloging-in-Publication Data
Christofides, Panagiotis D.
Nonlinear and robust control of PDE systems : methods and applications to transport-reaction processes / Panagiotis D. Christofides.
p. cm.—(Systems and control)
Includes bibliographical references and index.
ISBN 0-8176-4156-4 (alk. paper)
1. Nonlinear control theory. 2. Differential equations, Nonlinear. I. Title. II. Systems & control.
QA402.35 .C48 2000
629.8′312—dc21 00-044511
CIP

Printed on acid-free paper.

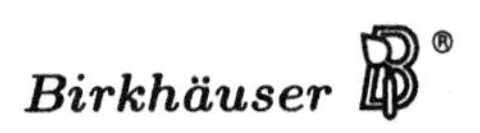

ISBN 0-8176-4156-4
ISBN 3-7643-4156-4 SPIN 10746894

Production managed by Louise Farkas; manufacturing supervised by Jeffrey Taub.
Typeset by TechBooks, Fairfax, VA.
Printed and bound by Edwards Brothers, Inc., Ann Arbor, MI.
Printed in the United States of America.

9 8 7 6 5 4 3 2 1

Contents

Preface

The interest in control of nonlinear partial differential equation (PDE) systems has been triggered by the need to achieve tight distributed control of transport-reaction processes that exhibit highly nonlinear behavior and strong spatial variations. Drawing from recent advances in dynamics of PDE systems and nonlinear control theory, control of nonlinear PDEs has evolved into a very active research area of systems and control. This book—the first of its kind—presents general methods for the synthesis of nonlinear and robust feedback controllers for broad classes of nonlinear PDE systems and illustrates their applications to transport-reaction processes of industrial interest.

Specifically, our attention focuses on quasi-linear hyperbolic and parabolic PDE systems for which the manipulated inputs and measured and controlled outputs are distributed in space and bounded. We use geometric and Lyapunov-based control techniques to synthesize nonlinear and robust controllers that use a finite number of measurement sensors and control actuators to achieve stabilization of the closed-loop system, output tracking, and attenuation of the effect of model uncertainty. The controllers are successfully applied to numerous convection-reaction and diffusion-reaction processes, including a rapid thermal chemical vapor deposition reactor and a Czochralski crystal growth process. The book includes comparisons of the proposed nonlinear and robust control methods with other approaches and discussions of practical implementation issues.

The book assumes a basic knowledge about PDE systems and nonlinear control and is intended for researchers, graduate students, and process control engineers interested in distributed parameter systems theory, nonlinear control, and process control applications.

Most of the research that led to the first half of this book was carried out when I was a doctoral student in the Department of Chemical Engineering at the University of Minnesota working under the supervision of Professor Prodromos Daoutidis. I would like to thank him for his guidance and support. The research that led to the second half of the book was performed at the Department of Chemical Engineering at the University of California at Los Angeles. In addition to my work, my doctoral students Antonios Armaou, James Baker, Charalambos Antoniades, and Nael El-Farra contributed greatly to the research results included in the book and in the preparation of the final manuscript. I would like to thank them for their hard work and contributions.

I would also like to thank all the other people who contributed in some way to this project. In particular, I would like to thank my faculty colleagues at UCLA for creating a pleasant working environment, the staff of Birkhäuser for their excellent cooperation, and the Petroleum Research Fund, the United States National Science Foundation, and the Air Force Office of Scientific Research for financial support. Last, but not least, I would like to express my deepest gratitude to my wife and my parents for their dedication, encouragement, and support over the course of this project. I dedicate this book to them.

Los Angeles, California
September 2000

Panagiotis D. Christofides

List of Figures

List of Tables

Chapter 1

Introduction

1.1 Motivation

Transport-reaction processes are characterized by the coupling of chemical reaction with significant convection, diffusion, and dispersion phenomena, and are essential in making many high-value industrial products. Examples include the plug-flow and packed-bed reactors used to produce specialty chemicals, the Czochralski crystallization of high-purity crystals, and the chemical vapor deposition of thin films for microelectronics manufacturing, as well as the solidification of liquid solution coatings for photographic films.

Traditional approaches for controlling transport-reaction processes are based on the simplifying assumption that the manipulated and to-be-controlled variables are spatially uniform. Yet, many industrial control problems in transport-reaction processes involve regulation of variables which are distributed in space (e.g., temperature profile across a wafer), and cannot be effectively solved with the traditional approaches. Moreover, future transport-reaction processes will have to meet increasingly stringent environmental and safety regulations, and tighter product quality and energy specifications. In other words, future transport-reaction processes must not only maximize product quality but also minimize by-product formation and energy use. In addition, they must be flexible enough to meet the product demands of a rapidly changing world market, utilize a wide range of feedstocks, and minimize the use of potentially dangerous chemicals. Therefore, our ability to operate highly-efficient and environmentally-benign transport-reaction processes will be critical for the future economic success of the chemical industry and for the protection of the environment.

These technological, economic and environmental needs, together with the recent advances in the development of fundamental mathematical models that accurately predict the behavior of transport-reaction processes, provide a strong motivation for developing a general framework for nonlinear feedback control of these processes based on detailed models. The philosophy behind this "model-based" approach is that the design of efficient processes and control systems should exploit the ability of a mathematical model to predict the behavior of a process and the fundamental knowledge of the underlying physico-chemical phenomena that the model contains. To provide an idea of the mathematical models arising in the description of transport-reaction processes, we consider, in the next subsection, two examples of transport-reaction processes and present their models.

1.2 Examples of Transport-Reaction Processes

1.2.1 A plug-flow reactor

Consider a plug-flow aerosol reactor, shown in Figure 1.1, used to produce NH_4Cl particles [90]. The following chemical reaction takes place close to the inlet of the reactor, $NH_3 + HCl \rightarrow NH_4Cl$, where NH_3, HCl are the reactant species and NH_4Cl is the monomer species (molecule of condensable species). The monomer molecules grow by condensation and form stable aerosol particles. These particles travel through the reactor and grow further by condensation and coagulation. The residence time in the reactor is usually very small (on the order of few seconds) owing to very large flow rate, and thus, the diffusive phenomena are negligible compared to the convective ones (plug flow). Under standard assumptions, the mathematical model, which accounts for chemical reaction, particle formation, and growth and convective transport, and describes the spatiotemporal evolution of average properties of the aerosol product (see [90] for detailed population balance modeling of the evolution of the entire aerosol size distribution), together with the evolution of monomer (NH_4Cl) and reactant (NH_3, HCl) concentrations and reactor temperature, takes the form:

$$\begin{aligned}
\frac{\partial N}{\partial t} &= -v\frac{\partial N}{\partial z} + I - \xi N^2 \\
\frac{\partial V}{\partial t} &= -v\frac{\partial V}{\partial z} + Ik^* + \eta(S-1)N \\
\frac{\partial V_2}{\partial t} &= -v\frac{\partial V_2}{\partial z} + Ik^{*2} + 2\epsilon(S-1)V + 2\zeta V^2
\end{aligned}$$

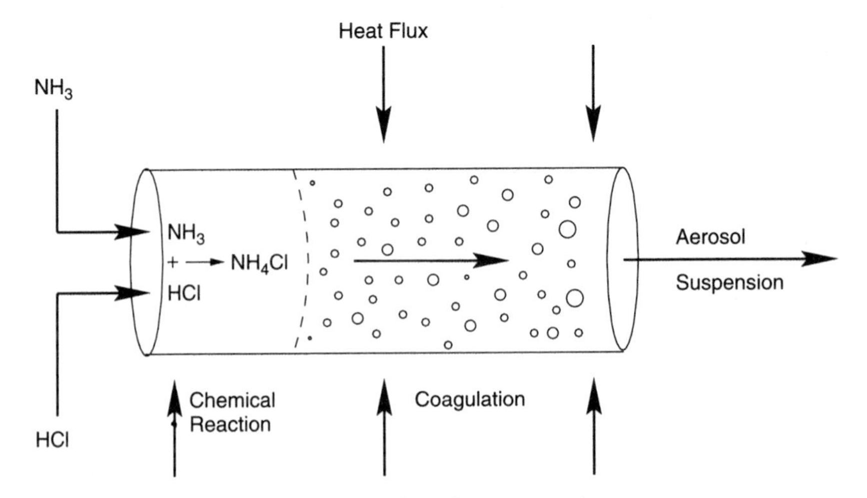

FIGURE 1.1. A plug-flow aerosol reactor.

$$\begin{aligned}
\frac{\partial S}{\partial t} &= -v\frac{\partial S}{\partial z} + CC_1C_2 - Ik^* - \eta(S-1)N \\
\frac{\partial C_1}{\partial t} &= -v\frac{\partial C_1}{\partial z} - A_1C_1C_2 \\
\frac{\partial C_2}{\partial t} &= -v\frac{\partial C_2}{\partial z} - A_2C_1C_2 \\
\frac{\partial T}{\partial t} &= -v\frac{\partial T}{\partial z} + BC_1C_2T + ET(T_w - T)
\end{aligned} \tag{1.1}$$

where N is the dimensionless total aerosol concentration, V is the dimensionless total aerosol volume, V_2 is the dimensionless second moment of the aerosol size distribution, S is the monomer concentration, C_1 and C_2 are the dimensionless concentrations of NH_3 and HCl, respectively, T and T_w are the dimensionless reactor and wall temperatures, respectively, t is the time, z is the axial coordinate, v is a dimensionless flow velocity, and I, ξ, η, ϵ, A_1, A_2, B, C, E are dimensionless quantities (I, ξ, η, ϵ depend in a nonlinear fashion on the states of the system; see [90] for details).

The process model of Eq. 1.1 consists of a system of seven first-order *hyperbolic* partial differential equations (PDEs). A typical control problem for this process is to regulate the geometric average particle volume, $v_g = V^2/(N^{1.5}V_2^{1.5})$, in the outlet of the reactor by manipulating the reactor wall temperature T_w.

1.2.2 Rapid thermal chemical vapor deposition

We consider a low-pressure rapid thermal chemical vapor deposition (RTCVD) process shown in Figure 1.2 [93] (see also [134]). The process consists of a quartz chamber, three banks of tungsten heating lamps which are used to heat the wafer, and a fan which is located at the bottom of the reactor and is used to cool the chamber. The furnace is designed so that the top lamp bank A and the bottom lamp bank C heat the total area of the wafer, while the lamp bank B, which surrounds the reactor, is used to heat the wafer edge in order to compensate for heat loss that occurs from the edge (radiative cooling between wafer edge and quartz chamber). The wafer is rotated while heated for azimuthal temperature uniformity. A small opening exists on the top of the quartz chamber which is used to feed the reacting gases. The objective of the process is to deposit uniformly a $0.5\mu m$ film of polycrystalline silicon on a 6-inch wafer in 40 seconds. To achieve this objective, the reactor is fed with 10% SiH_4 in Ar at 5 *Torr* pressure and the heating lamps are used to heat the wafer from room temperature to $1200K$ (this is the temperature where the deposition reactions take place) at a heating rate of $180\,K/sec$.

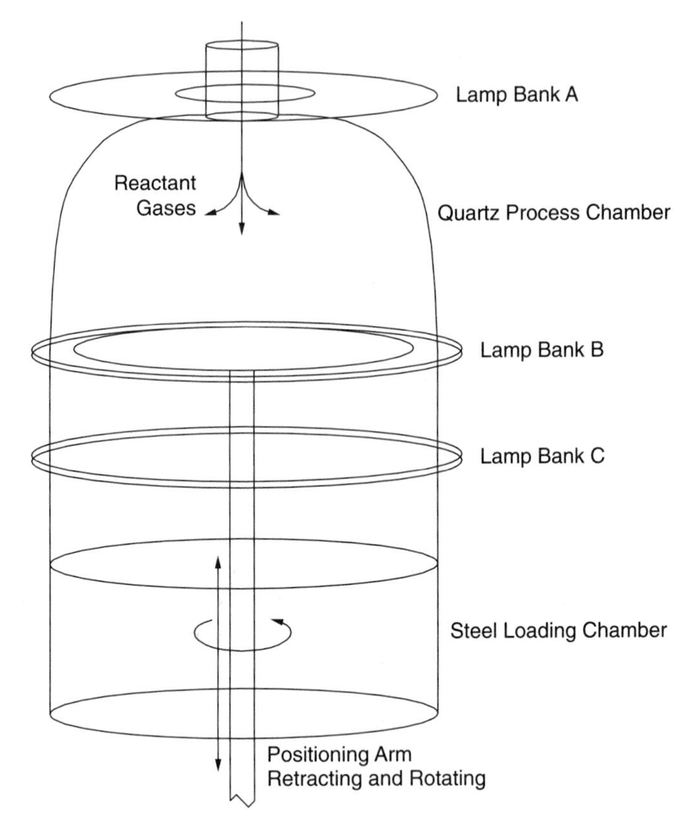

FIGURE 1.2. Rapid thermal chemical vapor deposition.

A mathematical model of the RTCVD process has been developed in [134] and consists of an energy balance on the wafer:

$$\rho_w T_{amb} \frac{\partial}{\partial t}(C_{p_w}(T)T) = \frac{T_{amb}}{R_w^2} \frac{1}{r} \frac{\partial}{\partial r}\left(\kappa(T) r \frac{\partial T}{\partial r}\right) - \frac{q_{rad}(T, r)}{\delta z} \tag{1.2}$$

subject to the boundary conditions:

$$\left.\frac{\partial T}{\partial r}\right|_{r=0} = 0 \tag{1.3}$$

$$\left(\frac{T_{amb}}{R_w} \kappa(T) \frac{\partial T}{\partial r}\right)_{r=1} = -\sigma \epsilon_w T_{amb}^4 (T^4 - T_c^4) + q_{edge} u_b \tag{1.4}$$

an energy balance on the quartz chamber:

$$\begin{aligned} T_{amb} M_c \frac{dT_c}{dt} &= \epsilon_c Q_{lamps} \cdot u - A_{hem} q_h - A_{cyl} q_c - Q_{convect} \\ &\quad -\sigma \epsilon_c A_c T_{amb}^4 (T_c^4 - 1) \end{aligned} \tag{1.5}$$

and mass balances for silane and hydrogen inside the chamber:

$$\begin{aligned}
\frac{dX_{SiH_4}}{dt} &= -\alpha \int_{A_w} R_s(T, X_{SiH4}, X_{H2})\, dA_w + \frac{1}{\tau}(X_{SiH_4}^{in} - X_{SiH_4}) \\
\frac{dX_{H_2}}{dt} &= 2\alpha \int_{A_w} R_s(T, X_{SiH4}, X_{H2})\, dA_w - \frac{1}{\tau} X'_{H_2}.
\end{aligned} \tag{1.6}$$

In the above equations, T_{amb} is the ambient temperature, T is the dimensionless wafer temperature, T_c is the chamber temperature, X_{SiH_4} is the mole fraction of silane, and X_{H_2} is the mole fraction of hydrogen. The assumptions involved in the derivation of the above model and the rest of the notation used in the modeling equations are discussed in Chapter 7.

The spatiotemporal evolution of the wafer temperature is described by Eq. 1.2, which is a nonlinear *parabolic* PDE. The control objective is to use measurements of wafer temperature at four locations to manipulate the power of the top lamps in order to achieve uniform wafer temperature, and thus, uniform deposition of the thin film on the wafer over the entire process cycle.

1.3 Background on Control of PDE Systems

The objective of this subsection is to provide a review of results on control of PDE systems; this review is not intended to be exhaustive, rather it provides the necessary background for the results presented in the book. Excellent surveys of theoretical, as well as application, papers on this topic can be found in [53, 14, 140, 98, 114].

The conventional approach for the control of PDE systems is based on the spatial discretization (typically with finite-difference or finite-element methods) of the PDE system followed by the controller design on the basis of the resulting (linear or nonlinear) ordinary differential equation (ODE) system (see, for example, [129, 61, 108, 77]). However, there are certain well-known disadvantages associated with this approach. For example, fundamental control-theoretic properties, like controllability and observability, which should depend only on the location of sensors and actuators, may also depend on the discretization method and the number and location of discretization points [115]. Moreover, neglecting the infinite dimensional nature of the original system may lead to erroneous conclusions concerning the stability properties of the open-loop and/or the closed-loop system. Furthermore, in processes where the spatially distributed nature is very strong, due to the underlying convection and diffusion phenomena, such an approach limits the controller performance and may lead to unacceptable control quality.

Motivated by the above considerations, significant research efforts have focused on the development of control methods for PDE systems that directly account for their spatially distributed nature. In this direction, the well-known classification of PDE systems into hyperbolic and parabolic [127], according to the properties of the spatial differential operator, essentially determines the approach followed for the solution of the control problem.

Specifically, for first-order hyperbolic PDE systems (e.g., convection-reaction processes), the fact that the eigenvalues of the spatial differential operator cluster along vertical or nearly vertical asymptotes in the complex plane motivates formulating and solving the control problem on the basis of the PDE system. Following this approach, optimal control methods (e.g., [142, 100, 101, 16, 115]) and Lyapunov's direct method [141, 2] have been used for the design of controllers for first-order hyperbolic PDE systems. More recently, a methodology for the design of distributed state feedback controllers, based on a combination of the method of characteristics and sliding mode techniques was proposed in [123]. This method was further developed in [79] to account for possible discontinuous behavior of the control action and was applied to a heat exchanger.

On the other hand, parabolic PDE systems (e.g., diffusion-reaction processes) typically involve spatial differential operators whose eigenspectrum can be partitioned into a finite-dimensional slow one and an infinite-dimensional stable fast complement [64, 12]. This implies that the dynamic behavior of such systems can be approximately described by finite-dimensional systems. Motivated by this, the standard approach to the control of parabolic PDEs involves the application of Galerkin's method to the PDE system to derive ODE systems that describe the dynamics of the dominant (slow) modes of the PDE system, which are subsequently used as the basis for the synthesis of finite-dimensional controllers (e.g., [69, 12, 115, 49, 15, 35]). The main advantage of this approach is that the exponential stability of the fast eigenmodes ensures that a controller that exponentially stabilizes the closed-loop ODE system, also stabilizes the closed-loop parabolic PDE system. On the other hand, the main disadvantage of this approach, especially in the context of nonlinear parabolic PDEs, is that the number of modes that should be retained to derive an ODE system that yields the desired degree of approximation may be very large [18, 1], leading to complex controller design and high dimensionality of the resulting controllers.

Motivated by the above, recent research efforts on control of parabolic PDE systems have focused on the problem of synthesizing low-dimensional output feedback controllers. In [68] (see also [34]), a method that uses the singular functions of the differential operator instead of the eigenfunctions in the series expansion of the solution was proposed to address this problem for linear parabolic PDEs. For nonlinear parabolic PDEs, a natural framework to address the problem of deriving low-dimensional ODE systems that accurately reproduce their solutions is based on the concept of inertial manifold (IM) (see [133] and the references therein). An IM is a positively invariant, finite-dimensional Lipschitz manifold, which attracts every trajectory exponentially. The dynamics of a parabolic PDE system restricted on the inertial manifold are described by a set of ODEs called the inertial form. Hence, stability and bifurcation studies and controller design for the infinite-dimensional PDE system can be readily performed on the basis of the finite-dimensional inertial form [133]. However, the explicit derivation of the inertial form requires the computation of the analytic form of the IM. Unfortunately, IMs have been proven to exist only for certain classes of PDEs (for example, Kuramoto–Sivashinsky equation and some diffusion-reaction equations [133]), and even then it is almost impossible to derive their analytic form. In order to overcome the problems associated with the existence and construction of IMs, the concept of approximate inertial manifold (AIM) has been introduced (see, for example, [63, 62, 137, 90]) and used for the derivation of ODE systems whose dynamic behavior approximates the one of the inertial form.

In the area of control of nonlinear parabolic PDE systems, the concept of inertial manifold has been used: (a) in [31, 30], to determine the extent to which linear boundary proportional control influences the dynamic and steady-state closed-loop response of a nonlinear parabolic PDE system, and (b) in [121, 97, 120], to address the problem of stabilization of a parabolic PDE with boundary low-order linear output feedback control; a standard observer-based controller augmented with a residual mode filter [17] was used to induce an inertial manifold in the closed-loop system.

In addition to being highly nonlinear and infinite dimensional, the PDE models of most transport-reaction processes are uncertain. Typical sources of model uncertainty include unknown or partially known time-varying process parameters, exogenous disturbances, and unmodeled dynamics. It is well known that the presence of uncertain variables and unmodeled dynamics, if not taken into account in the controller design, may lead to severe deterioration of the nominal closed-loop performance or even to closed-loop instability.

Research on robust control of PDEs with uncertain variables has been limited to linear systems. For linear parabolic PDEs, the problem of complete elimination of the effect of uncertain variables on the output via distributed state feedback (known as disturbance decoupling) was solved in [50, 51]. Research on the problem of robust stabilization of linear PDE systems with uncertain variables led to the development of H^∞ control methods in the frequency domain (e.g., [52, 106, 67]). The derivation of concrete relations between frequency-domain and state-space concepts for a wide class of PDE systems in [87] motivated research on the development of the state-space counterparts of the H^∞ results for linear PDE systems (see [140, 26], for example). Other results on robust control of PDE systems with time-varying uncertainty include Lyapunov-based control of parabolic PDEs [149, 148], and Lyapunov-based [91] and H^∞ (e.g., [27, 94]) control of Navier–Stokes equations. Within a state-space framework, an alternative approach for the design of controllers for linear PDE systems, that deals explicitly with time-invariant uncertain variables, involves the use of adaptive control methods [32, 146, 81, 55, 18]. On the other hand, the problem of robustness of control methods for PDE systems with respect to unmodeled dynamics is typically studied within the singular perturbation framework (e.g., [146]). This approach was employed in [146] to establish robustness of a class of finite-dimensional adaptive controllers, which asymptotically stabilize a linear PDE system with time-invariant uncertain variables, with respect to unmodeled dynamics, provided that they are stable and sufficiently fast.

1.4 Objectives and Organization of the Book

Motivated by the fact that many industrially important transport-reaction processes are naturally described by nonlinear and uncertain hyperbolic and/or parabolic PDE systems, and the lack of general nonlinear and robust control methods for such systems, the objectives of the present book are

- to present practical, general nonlinear and robust control methods for hyperbolic and parabolic PDE systems, and
- to illustrate the application of the proposed methods to transport-reaction processes of industrial interest and document their effectiveness and advantages with respect to traditional control methods for PDE systems.

Throughout the book, we will assume that all the PDE systems under consideration (with and without feedback control) possess a unique solution which is also sufficiently smooth (i.e., all the spatial and time derivatives in the PDE systems are smooth functions of space and time); the reader may refer to the books [109, 22, 23, 54, 150] for techniques and results for studying the mathematically delicate questions of existence, uniqueness, and regularity of solutions for various classes of PDE systems. In addition, we focus our attention on quasi-linear hyperbolic and parabolic PDE systems for which the manipulated inputs and measured and controlled outputs are bounded. From a practical point of view, this means that we do not deal with control problems that involve boundary actuation, measurements, and control objectives, even though several problems of this kind may be cast within the distributed control framework and can be addressed by the proposed methods. Linear infinite dimensional systems with unbounded manipulated inputs, measurements, and control objectives have been studied extensively (see, for example, [49, 112, 99, 140, 22, 23]).

The rest of the book is structured as follows. Chapter 2 focuses on nonlinear control of quasi-linear first-order hyperbolic PDE systems, for which the manipulated, controlled, and measured variables are distributed in space.

Motivated by the spectral properties of these systems, a general method, based on geometric control concepts, is presented for the synthesis of distributed output feedback controllers that enforce stability and output tracking in the closed-loop system. Theoretical analogies between the proposed approach and available feedback control methods for the stabilization of linear hyperbolic PDEs are pointed out. The developed control method is applied to the nonisothermal plug-flow reactor example, modeled by three first-order hyperbolic PDEs.

Chapter 3 focuses on hyperbolic PDE systems of the type studied in Chapter 2, which also include time-varying uncertain variables and unmodeled dynamics. Utilizing a combination of geometric control concepts and Lyapunov's direct method, a general method is presented for the synthesis of robust nonlinear controllers that guarantees boundedness of the state and achieves asymptotic output tracking with arbitrary degree of asymptotic attenuation of the effect of uncertain variables on the output of the closed-loop system. The controllers are shown to be robust with respect to stable, but sufficiently fast, unmodeled dynamics. The developed control method is successfully applied to a nonisothermal fixed-bed reactor, where the reactant wave propagates through the bed with significantly larger speed than the heat wave, and the heat of reaction is unknown.

In Chapter 4, we turn to nonlinear control of quasi-linear parabolic PDE systems whose dynamics can be separated into slow and fast ones, for which the manipulated, controlled, and measured variables are also distributed in space. Taking advantage of the low-dimensional nature of the dominant dynamics of such systems, we develop a procedure, based on a combination of Galerkin's method with approximate inertial manifolds, for the construction of low-order ODE systems that reproduce the solutions of the PDE system with desired accuracy. These ODE systems are used as the basis for the synthesis of nonlinear low-order output feedback controllers that guarantee stability and enforce output tracking in the closed-loop system. The methodology is successfully employed to control the temperature profiles of a catalytic rod and a nonisothermal tubular reactor with recycle around unstable steady-states.

Chapter 5 focuses on robust control of quasi-linear parabolic PDE systems of the type introduced in Chapter 4, which also include time-varying uncertain variables. Initially, Galerkin's method is used to derive an ODE system of dimension equal to the number of slow modes, which is subsequently used to synthesize robust state feedback controllers via Lyapunov's direct method. Singular perturbation methods are employed to establish that the degree of asymptotic attenuation of the effect of uncertain variables on the output, enforced by these controllers, is proportional to the degree of separation of the fast and slow modes of the spatial differential operator. For processes for which such a degree of uncertainty attenuation is not sufficient, a sequential procedure, based on the concept of approximate inertial manifold, is developed for the synthesis of robust controllers that achieve arbitrary degree of asymptotic uncertainty attenuation in the closed-loop parabolic PDE system. Then, under the assumption that the number of measurements is equal to the number of slow modes, we propose a procedure for obtaining estimates for the states of the approximate ODE model from the measurements. We show that the use of these estimates in the robust state feedback controllers leads to robust output feedback controllers, which enforce the desired properties in the closed-loop system, provided that the separation between the slow and fast modes is sufficiently large. The controllers are successfully applied to a catalytic rod with uncertainty.

Chapter 6 presents general methods for the synthesis of nonlinear and robust time-varying output feedback controllers for systems of quasi-linear parabolic PDEs with time-dependent spatial domains, whose dynamics can be separated into slow and fast ones. Initially, a nonlinear model reduction procedure, based on a combination of Galerkin's method with the concept of approximate inertial manifold, is employed for the derivation of ODE systems that yield solutions which are close, up to a desired accuracy, to the ones of the PDE system. Then, these ODE systems are used as the basis for the explicit construction of nonlinear and robust output feedback controllers via geometric control methods and Lyapunov techniques, respectively. Differences in the nature of the model reduction and control problems between parabolic PDE systems with fixed and moving spatial domains are identified and discussed. The control methods are applied to a catalytic rod with time-dependent spatial domain and uncertainty.

Finally, Chapter 7 presents applications of the nonlinear control methods for parabolic PDE systems with fixed and moving spatial domains presented in Chapters 4 and 6 to the rapid thermal chemical vapor deposition process introduced in subsection 1.2.2 and a Czochralski crystal growth process, respectively. Nonlinear low-dimensional output feedback controllers are synthesized and implemented on detailed fundamental models of these processes. The performance of the controllers is successfully tested through simulations.

The proofs of all the results and a brief presentation of the Karhunen–Loève expansion, a method for computing empirical eigenfunctions, are given in the appendix.

Chapter 2

Feedback Control of Hyperbolic PDE Systems

2.1 Introduction

Transport-reaction processes in which the diffusive and dispersive phenomena are negligible compared to the convective phenomena can be adequately described by systems of first-order hyperbolic PDEs. Representative chemical processes modeled by such systems include heat exchangers [115], plug-flow reactors [115], fixed-bed reactors [130], pressure swing adsorption processes [119], and so forth.

The conventional approach to the control of first-order hyperbolic PDEs is based on the direct spatial discretization (typically using finite-difference method) of the PDE model followed by the controller design on the basis of the resulting (linear or nonlinear) ordinary differential equation (ODE) model (see e.g., [129, 61, 77]). However, there are certain well-known disadvantages associated with this approach. For example, fundamental control-theoretic properties, like controllability and observability, which should depend only on the location of control actuators, specification of control objectives, and measurement sensors, may also depend on the discretization method used and the number and location of discretization points [109]. Moreover, neglecting the infinite dimensional nature of the original system may lead to erroneous conclusions concerning the stability properties of the open-loop and/or the closed-loop system.

The distinct properties of first-order hyperbolic PDE systems are: (a) the eigenvalues of the spatial differential operator cluster along vertical or nearly vertical asymptotes in the complex plane, and (b) the speed of propagation of a perturbation throughout the spatial domain is finite, which is due to the absence of diffusion terms. The first property implies that a large number of modes is required to accurately capture the dynamic behavior of first-order hyperbolic PDEs and prohibits the use of modal (spectral) decomposition techniques to derive low-order ODE models that approximately describe the dynamics of the PDE system. The second property implies that the formulation of a meaningful control problem (i.e., one in which the dynamics between a control actuator and a controlled output do not involve dead-time) depends on the location of the actuator and the specification of the controlled output, and suggests formulating and solving the control problem on the basis of the PDE system in order to avoid errors introduced by the spatial discretization.

In this chapter, we address the feedback control problem for systems described by quasi-linear first-order hyperbolic PDEs, for which the manipulated input, the controlled output, and the measured output are distributed in space. For such systems, our objective is to synthesize nonlinear distributed output feedback controllers on the basis of the hyperbolic PDE system that enforce output tracking and guarantee stability in the closed-loop system.

The chapter is structured as follows: after reviewing the necessary preliminaries, we introduce a concept of characteristic index between the controlled output and the manipulated input and use it for the synthesis of distributed state feedback controllers that induce output tracking in the closed-loop system. A notion of zero-output constraint dynamics for first-order hyperbolic PDEs is introduced and used to derive precise conditions that guarantee the stability of the closed-loop system. Then, output feedback controllers are synthesized through combination of appropriate distributed state observers with the developed state feedback controllers. Theoretical analogies between the proposed approach and available feedback control methods for the stabilization of linear hyperbolic PDEs are pointed out. Controller implementation issues are also discussed. Finally, the application of the developed control method is illustrated through a nonisothermal plug-flow reactor example modeled by a system of three quasi-linear hyperbolic PDEs. The results of this chapter were first presented in [39].

2.2 First-Order Hyperbolic PDE Systems

2.2.1 Preliminaries

We consider systems of quasi-linear first-order partial differential equations in one spatial dimension with the following state-space representation:

$$\begin{aligned}\frac{\partial x}{\partial t} &= A\frac{\partial x}{\partial z} + f(x) + g(x)\bar{u} \\ \bar{y} = h(x),& \quad \bar{q} = p(x)\end{aligned} \tag{2.1}$$

subject to the boundary condition:

$$C_1 x(\alpha, t) + C_2 x(\beta, t) = R(t) \tag{2.2}$$

and the initial condition:

$$x(z, 0) = x_0(z) \tag{2.3}$$

where $x(z,t) = [x_1(z,t) \; \dots \; x_n(z,t)]^T$ denotes the vector of state variables, $x(z,t) \in \mathcal{H}^n[(\alpha,\beta), \mathbb{R}^n]$, with $\mathcal{H}^n[(\alpha,\beta)\mathbb{R}^n]$ being the infinite-dimensional Hilbert space of n-dimensional-like vector functions defined on the interval $[\alpha, \beta]$ whose spatial derivatives up to n-th order are square integrable, $z \in [\alpha, \beta] \subset \mathbb{R}$ and $t \in [0, \infty)$ denote position and time, respectively, $\bar{u}(z,t)$ denotes the manipulated variable, $\bar{y}(z,t)$ denotes the controlled variable, and $\bar{q}(z,t)$ denotes the measured variable. A is a matrix, $f(x)$ and $g(x)$ are sufficiently smooth vector functions, $h(x)$, $p(x)$ are sufficiently smooth scalar functions, $R(t)$ is a column vector which is assumed to be a sufficiently smooth function of time, C_1, C_2 are constant matrices, and $x_0(z) \in \mathcal{H}[(\alpha,\beta),\mathbb{R}^n]$, with $\mathcal{H}[(\alpha,\beta),\mathbb{R}^n]$ being the Hilbert space of n-dimensional vector functions defined on the interval $[\alpha,\beta]$ which are square integrable.

The model of Eq. 2.1 describes the majority of convection-reaction processes arising in chemical engineering [117] and constitutes a natural generalization of *linear* PDE models (see Eq. 2.9 below) considered in [110, 115] in the context of linear distributed state estimation and control. The distributed and affine appearance of the manipulated variable u is typical in most practical applications (see, for example, [110, 68, 115]), where the jacket temperature is usually selected as the manipulated variable (see subsection 2.7.4 below for a detailed discussion on how the jacket temperature is manipulated in practice). Furthermore, the possibility that the system of Eq. 2.1 may admit boundary conditions at two separate points (for example, in the case of counter-current processes) is captured by the boundary condition of Eq. 2.2. Models of the form of Eq. 2.1 include as special cases the models considered in [123, 79].

Depending on the location of the eigenvalues of the matrix A, the system of Eq. 2.1 can be hyperbolic, parabolic, or elliptic [127]. Assumption 2.1 that follows ensures that the system of Eq. 2.1 has a well-defined solution and specifies the class of systems considered in Chapters 2 and 3.

Assumption 2.1 *The matrix A is real symmetric and its eigenvalues satisfy:*

$$\lambda_1 \leq \cdots \leq \lambda_k < 0 < \lambda_{k+1} \leq \cdots \leq \lambda_n. \tag{2.4}$$

Typical examples where Assumption 2.1 is satisfied include heat exchangers and plug-flow reactors where the matrix A is diagonal (see example in section 2.7, as well as other applications presented in [117]). Systems of the form of Eq. 2.1, for which the eigenvalues of the matrix A are real and distinct, are said to be *hyperbolic*, while systems for which some of the eigenvalues of the matrix A are equal are said to be *weakly hyperbolic* [127].

2.2.2 Specification of the control problem

Consider the system of quasi-linear PDEs of the form of Eq. 2.1, for which the manipulated variable $\bar{u}(z,t)$, the measured variable $\bar{q}(z,t)$, and the controlled variable $\bar{y}(z,t)$ are distributed in space. Let's assume that for the control of the variable $\bar{y}$, there exists a finite number of control actuators, l, and the same number of measurement sensors; clearly, it is not possible to control the variable $\bar{y}(z,t)$ at all positions. Therefore, it is meaningful

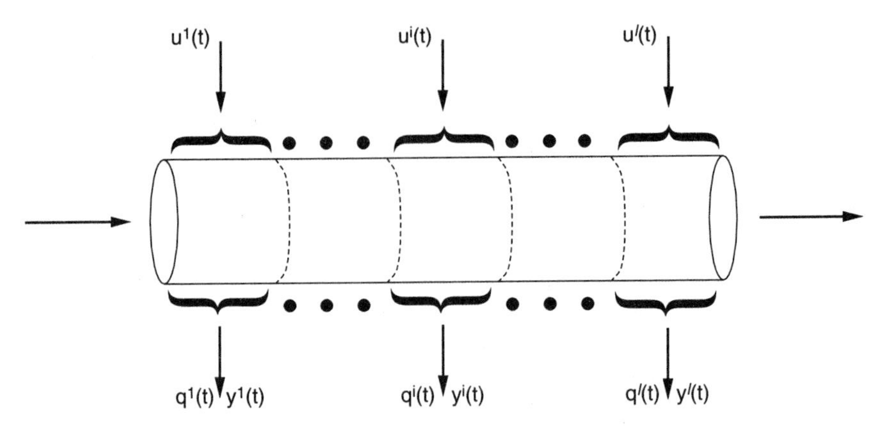

FIGURE 2.1. Control problem specification in the case of a prototype example.

to formulate the control problem as the one of controlling $\bar{y}(z,t)$ at a finite number of spatial intervals. In particular, referring to the single spatial interval $[z_i, z_{i+1}]$, we suppose that the manipulated input is $u^i(t)$, with $u^i \in \mathbb{R}$, and the measured output is $q^i(t)$, with $q^i \in \mathbb{R}$, while the controlled output is $y^i(t)$, with $y^i \in \mathbb{R}$, such that the following relations hold:

$$\begin{gathered}\bar{u}(z,t) = \sum_{i=1}^{l} b^i(z)u^i(t), \quad y^i(t) = \mathcal{C}^i\bar{y}(z,t), \\ q^i(t) = \mathcal{Q}^i\bar{q}(z,t), \quad z_i \le z \le z_{i+1}\end{gathered} \tag{2.5}$$

where $b^i(z)$ is a known smooth function of z, and $\mathcal{C}^i$, $\mathcal{Q}^i$ are bounded linear operators, mapping $\mathcal{H}^n$ into $\mathbb{R}$. Figure 2.1 shows a pictorial representation of this formulation in the case of a prototype example. From a practical point of view, the function $b^i(z)$ describes how the control action $u^i(t)$ is distributed in the spatial interval $[z_i, z_{i+1}]$, while the operator $\mathcal{Q}^i$ determines the structure of the sensor in the same spatial interval. Whenever the control action enters the system at a single point z_0 (e.g., lateral flow injections), with $z_0 \in [z_i, z_{i+1}]$ (i.e., point actuation), the function $b^i(z)$ is taken to be nonzero in a finite spatial interval of the form $[z_0 - \epsilon, z_0 + \epsilon]$, where ϵ is a small positive real number, and zero elsewhere in $[z_i, z_{i+1}]$. Similarly, in the case of a point sensor acting at z_0, the operator $\mathcal{Q}^i$ is assumed to act in $[z_0 - \epsilon, z_0 + \epsilon]$ and considered to be zero elsewhere. In mathematical terms, we consider here only hyperbolic PDE systems for which the manipulated inputs and measured and controlled outputs are bounded; the reader may refer to (e.g., [49, 112, 99, 140, 22, 23]) for results on analysis and control of infinite dimensional systems with unbounded manipulated inputs and measurements. The operator $\mathcal{C}^i$ depends on the desired performance specifications and in the majority of practical applications (see e.g.,

[115, 78]), is of the following form:

$$y^i(t) = \mathcal{C}^i h(x) = \int_{z_i}^{z_{i+1}} c^i(z) h(x(z,t))dz \tag{2.6}$$

where $c^i(z)$ is a known smooth function of z. The choice of $c^i(z)$ and the location of the control actuators so that the resulting PDE system is controllable will be addressed in section 2.3 below.

For simplicity, the functions $b^i(z)$, $c^i(z)$, $i = 1, \ldots, l$ will be assumed to be normalized in the interval $[\alpha, \beta]$, i.e.,

$$\sum_{i=1}^{l} \int_{z_i}^{z_{i+1}} b^i(z)dz = \sum_{i=1}^{l} \int_{z_i}^{z_{i+1}} c^i(z)dz = 1.$$

Using the relations of Eqs. 2.5–2.6, the system of Eq. 2.1 takes the form:

$$\begin{aligned} \frac{\partial x}{\partial t} &= A\frac{\partial x}{\partial z} + f(x) + g(x)b(z)u \\ y &= \mathcal{C}h(x), \quad q = \mathcal{Q}p(x) \\ C_1 x(\alpha, t) &+ C_2 x(\beta, t) = R(t) \end{aligned} \tag{2.7}$$

where

$$\begin{aligned} u &= [u^1 \quad \cdots \quad u^l]^T \\ y &= [y^1 \quad \cdots \quad y^l]^T \\ b(z) &= [(H(z-z_1) - H(z-z_2))\, b^1(z) \quad \cdots \quad (H(z-z_l) \\ &\quad - H(z-z_{l+1}))\, b^l(z)] \\ \mathcal{C} &= [(H(z-z_1) - H(z-z_2))\, \mathcal{C}^1 \quad \cdots \quad (H(z-z_l) - H(z-z_{l+1}))\, \mathcal{C}^l]^T \\ \mathcal{Q} &= [(H(z-z_1) - H(z-z_2))\, \mathcal{Q}^1 \quad \cdots \quad (H(z-z_l) - H(z-z_{l+1}))\, \mathcal{Q}^l]^T \end{aligned} \tag{2.8}$$

with $H(\cdot)$ being the standard Heaviside function.

Referring to the system of Eq. 2.7, we note that by setting $f(x) = Bx$, $g(x) = w$, $h(x) = kx$, $p(x) = px$ where A, B are matrices and w, k, p are vectors of appropriate dimensions, it reduces to the following system of linear first-order hyperbolic PDEs:

$$\begin{aligned} \frac{\partial x}{\partial t} &= A\frac{\partial x}{\partial z} + Bx + wb(z)u \\ y &= \mathcal{C}kx, \quad q = \mathcal{Q}px \end{aligned} \tag{2.9}$$

subject to the boundary condition of Eq. 2.2 and the initial condition of Eq. 2.3.

The following example will be used throughout the chapter to illustrate the various aspects of our methodology.

Example 2.1 Consider a steam-jacketed tubular heat exchanger [115]. The dynamic model of the process is of the form:

$$\begin{aligned} \frac{\partial T}{\partial t} &= -v_l \frac{\partial T}{\partial z} - a\,T + a\,T_j \\ T(0,t) &= R(t) \end{aligned} \tag{2.10}$$

where $T(z,t)$ denotes the temperature of the reactor, $z \in [0,1]$, $T_j(z,t)$ denotes the jacket temperature, v_l denotes the fluid velocity in the exchanger, and a is a positive constant. Considering T_j as the manipulated variable, and T as the controlled and measured variable, the above model can be put in the form of Eq. 2.1:

$$\begin{aligned} \frac{\partial x}{\partial t} &= -v_l \frac{\partial x}{\partial z} - ax + a\bar{u} \\ \bar{y} &= x, \quad \bar{q} = x \\ x(0,t) &= R(t). \end{aligned} \tag{2.11}$$

Consider the case where there exists one actuator with distribution function $b(z) = 1$, the controlled output is assumed to be $y(t) = \int_0^1 x(z,t)dz$, and there is a point sensor located at $z = 0.5$. Utilizing these relations, the system of Eq. 2.11 takes the form:

$$\begin{aligned} \frac{\partial x}{\partial t} &= -v_l \frac{\partial x}{\partial z} - ax + au(t) \\ y(t) &= \int_0^1 x(z,t)dz, \quad q = \int_{0.5-\epsilon}^{0.5+\epsilon} x(z,t)dz. \end{aligned} \tag{2.12}$$

△

2.2.3 Review of system-theoretic properties

The objective of this subsection is to review basic system-theoretic properties of systems of first-order hyperbolic PDEs, which will be used in the subsequent sections. For more details on these properties, the reader may refer to [53, 118]. We will start with the definitions of the inner product and the norm, with respect to which the notion of exponential stability for the systems under consideration will be defined.

- Let ω_1, ω_2 be two elements of $\mathcal{H}([\alpha,\beta];\mathbb{R}^n)$. Then, the inner product and the norm, in $\mathcal{H}([\alpha,\beta];\mathbb{R}^n)$, are defined as follows:

$$\begin{aligned} (\omega_1, \omega_2) &= \int_\alpha^\beta \left(\omega_1(z), \omega_2(z)\right)_{\mathbb{R}^n} dz \\ ||\omega_1||_2 &= (\omega_1, \omega_1)^{\frac{1}{2}} \end{aligned} \tag{2.13}$$

  where the notation $(\cdot,\cdot)_{\mathbb{R}^n}$ denotes the standard inner product in the Euclidean space $\mathbb{R}^n$.

Referring to the linear system of Eq. 2.9, for which assumption 2.1 holds, the operator

$$\mathcal{L}x = A\frac{\partial x}{\partial z} + Bx \tag{2.14}$$

defined on the domain in $\mathcal{H}[(\alpha,\beta);\mathbb{R}^n]$ consisting of functions $x \in \mathcal{H}^1[(\alpha,\beta);\mathbb{R}^n]$ which satisfy the boundary condition of Eq. 2.2, generates [118] a strongly continuous semigroup $U(t)$ (this is the analogue of the concept of state transition matrix used for ODE systems in the case of infinite-dimensional systems) of bounded linear operators on $\mathcal{H}[(\alpha,\beta);\mathbb{R}^n]$. This fact implies the existence, uniqueness, and continuity of solutions for the system of Eq. 2.9. In particular, the generalized solution of this system is given by:

$$x = U(t)x_0 + \int_0^t U(t-\tau)wbu(\tau)d\tau + C(t)R \tag{2.15}$$

where $C(t)$ is a bounded linear operator for each t mapping $\mathcal{H}[(0,t);\mathbb{R}^n]$ into $\mathcal{H}[(\alpha,\beta);\mathbb{R}^n]$. As can be easily seen from Eq. 2.15, $U(t)$ evolves the initial condition x_0 forward in time. From general semigroup theory [64], it is known that $U(t)$ satisfies the following growth property:

$$||U(t)||_2 \leq Ke^{at}, \quad t \geq 0 \tag{2.16}$$

where $K \geq 1$, a is the largest real part of the eigenvalues of the operator $\mathcal{L}$, and an estimate of K, a can be obtained utilizing the Hille–Yoshida theorem [64]. Whenever the parameter a is strictly negative, we will say that the operator of Eq. 2.14 generates an exponentially stable semigroup $U(t)$. We note that although there exist many stability concepts for PDEs (e.g., weak (asymptotic) stability [64, 127]), we will focus, throughout the chapter, on exponential stability, because of its robustness to bounded perturbations, which is required in most practical applications, where there is always some uncertainty associated with the process model. The aforementioned concepts allow stating precisely a standard detectability requirement (see also [110, 16]) for the system of Eq. 2.9, which will be exploited in section 2.6 for the design of distributed state observers.

Assumption 2.2 *The pair $[\mathcal{Q}p\ \mathcal{L}]$ is detectable; i.e., there exists a bounded linear operator $\mathcal{P}$, mapping $\mathbb{R}^l$ into $\mathcal{H}^n$, such that the linear operator $\mathcal{L}_o = \mathcal{L} - \mathcal{P}\mathcal{Q}p$ generates an exponentially stable semigroup.*

The above detectability assumption does not impose any restrictions on the form of the operator $\mathcal{Q}$, and thus, on the structure of the sensors (e.g., distributed, point sensors).

In closing this subsection, motivated by the lack of general stability results for systems of quasi-linear PDEs, we will review a result that will allow characterizing the local stability properties of the quasi-linear system of Eq. 2.7 on the basis of its corresponding linearized system. To this end, consider the linearization of the quasi-linear system of Eq. 2.7:

$$\begin{aligned}\frac{\partial x}{\partial t} &= A\frac{\partial x}{\partial z} + B(z)x + w(z)b(z)u \\ y &= \mathcal{C}k(z)x, \quad q = \mathcal{Q}p(z)x\end{aligned} \tag{2.17}$$

where

$$B(z) = \left(\frac{\partial f(x)}{\partial x}\right)_{x=x_s(z)}, \quad w(z) = g(x_s(z)), k(z) = \left(\frac{\partial h(x)}{\partial x}\right)_{x=x_s(z)},$$
$$p(z) = \left(\frac{\partial p(x)}{\partial x}\right)_{x=x_s(z)},$$

and $x_s(z)$ denotes some steady-state profile.

Proposition 2.1 ([127], p. 121) *The system of Eq. 2.7 (with $u = 0$) for which assumption 2.1 holds, subject to the boundary condition of Eq. 2.2, is locally exponentially stable if the operator of the linearized system of Eq. 2.17:*

$$\bar{\mathcal{L}}x = A\frac{\partial x}{\partial z} + B(z)x \tag{2.18}$$

generates an exponentially stable semigroup.

The following remark provides conditions, which can be easily verified in practice, that guarantee the open-loop stability of hyperbolic PDE systems of the form of Eq. 2.9 (and thus local stability of the system of Eq. 2.7).

Remark 2.1 Consider the system of quasi-linear first-order PDEs of Eq. 2.7 for which $u = 0$, and assume that the following conditions hold:

- $\lambda_1 \leq \lambda_2 \leq \cdots \leq \lambda_n < 0$.
- $C_2 = 0$

For such a system, it can be shown [112] that the eigenvalues of the operator of Eq. 2.18 (Eq. 2.14) are of the following form:

$$a_\mu = -\infty + \mu\pi i, \quad \mu = -\infty, \ldots, \infty. \tag{2.19}$$

Thus, first-order PDE systems that satisfy the above conditions (physical examples include plug-flow reactors, co-current heat exchangers, etc.) possess eigenvalues that lie on a vertical line crossing the real axis at $s = -\infty$, which, according to Eq. 2.16, implies that they are exponentially stable.

2.2.4 Methodology for control

Motivated by the fact that control methods for quasi-linear hyperbolic PPE systems should explicitly account for their nonlinear and spatially varying nature, we employ a methodology, that involves the following two steps.

1. Synthesize distributed nonlinear state feedback controllers that enforce output tracking and derive conditions that guarantee the exponential stability of the closed-loop system.
2. Synthesize distributed nonlinear output feedback controllers through combination of the developed state feedback controllers with appropriate distributed state observers.

Owing to the mathematical properties of these systems, the state feedback control problem is solved on the basis of the original PDE model by following an approach conceptually similar to the one used for the synthesis of inversion-based controllers for ODE systems. In order to motivate the approach followed for the quasi-linear case and identify theoretical analogies between our approach and available results on feedback stabilization of linear hyperbolic PDEs, we will also present the development for the case of systems of linear PDEs of the form of Eq. 2.9. The development for the case of quasi-linear systems will be performed by essentially generalizing the results developed for the linear case in a nonlinear context.

2.3 Characteristic Index

In this section, we will introduce a concept of characteristic index between the output y and the input u for systems of the form of Eq. 2.9, that will allow us to formulate and solve the state feedback control problem. To reveal the origin of this concept and illustrate its role, we consider the operation of differentiation of the output y^i of the system of Eq. 2.9 with respect to time, which yields:

$$\begin{aligned} y^i &= C^i k x \\ \frac{dy^i}{dt} &= \frac{d}{dt} C^i k x = C^i k \frac{\partial x}{\partial t} \\ &= C^i k \left(A \frac{\partial}{\partial z} + B \right) x + C^i k w b^i(z) u^i. \end{aligned} \tag{2.20}$$

Now, if the scalar $C^i k w b^i(z)$ is nonzero, we will say that the characteristic index of y^i with respect to u^i, denoted by σ^i, is equal to one. If $C^i k w b^i(z) = 0$, the characteristic index is greater than one, and from Eq. 2.20 we have that:

$$\frac{dy^i}{dt} = C^i k \left(A \frac{\partial}{\partial z} + B \right) x. \tag{2.21}$$

Performing one more time differentiation, we obtain:

$$\frac{d^2 y^i}{dt^2} = \mathcal{C}^i k \left(A\frac{\partial}{\partial z} + B\right)^2 x + \mathcal{C}^i k \left(A\frac{\partial}{\partial z} + B\right) wb^i(z)u^i. \qquad (2.22)$$

In analogy with the above, if the scalar $\mathcal{C}^i k(A\frac{\partial}{\partial z} + B)wb^i(z)$ is nonzero, the characteristic index is equal to two, while if $\mathcal{C}^i k(A\frac{\partial}{\partial z} + B)wb^i(z) = 0$, the characteristic index is greater than two.

Generalizing the above development, one can give the definition of the characteristic index for systems of the form of Eq. 2.9.

Definition 2.1 *Referring to the system of linear first-order partial differential equations of the form of Eq. 2.9, we define the characteristic index of the output y^i with respect to the input u^i as the smallest integer σ^i for which*

$$\mathcal{C}^i k \left(A\frac{\partial}{\partial z} + B\right)^{\sigma^i - 1} wb^i(z) \neq 0 \qquad (2.23)$$

or $\sigma^i = \infty$ if such an integer does not exist.

Remark 2.2 According to definition 2.1, the characteristic index is the smallest order time derivative of the output y^i which explicitly depends on the manipulated input u^i. In this sense, it can be thought of as a natural generalization of the concept of relative order used for nonlinear finite-dimensional systems [86] for the systems under consideration. However, note that the characteristic index allows a precise characterization of the spatiotemporal interactions between controlled outputs and manipulated inputs in hyperbolic PDE systems. For the case of linear time-invariant ODE systems, the relative order can be interpreted as the difference in the degree of the denominator polynomial and numerator polynomial. Such an interpretation cannot be given for the concept of characteristic index because the frequency domain representation of the system of Eq. 2.9 typically gives rise to transfer functions that involve complicated transcendental forms.

In analogy with the linear case, the following concept of characteristic index will be introduced for the quasi-linear PDE system of Eq. 2.7. To this end, we will need to use the Lie derivative notation [86]: $L_f h(x)$ denotes the Lie derivative of a scalar field $h(x)$ with respect to the vector field $f(x)$, defined as $L_f h(x) = \left[\frac{\partial h}{\partial x_1} \cdots \frac{\partial h}{\partial x_n}\right] f(x)$, $L_f^k h(x)$ denotes the k-th order Lie derivative, defined as $L_f^k h(x) = L_f(L_f^{k-1} h(x))$, and $L_g L_f^{k-1} h(x)$ denotes the mixed Lie derivative.

Definition 2.2 *Referring to the system of quasi-linear first-order partial differential equations of the form of Eq. 2.7, we define the characteristic* *index of the output y^i with respect to the input u^i as the smallest integer σ^i for which*

$$C^i L_g \left(\sum_{j=1}^{n} \frac{\partial x_j}{\partial z} L_{a_j} + L_f \right)^{\sigma^i - 1} h(x) b^i(z) \not\equiv 0 \tag{2.24}$$

where a_j denotes the j-th column vector of the matrix A or $\sigma^i = \infty$ if such an integer does not exist.

We assume that Eq. 2.24 holds for all $x \in \mathcal{H}^n$, $z \in [\alpha, \beta]$.

From definitions 2.1 and 2.2, one can immediately see that the characteristic index σ^i depends on the structural properties of the process (the matrices A, B and the vectors w, k for the linear case, or the matrix A and the functions $f(x), g(x), h(x)$ for the quasi-linear case), as well as on the selection of the control system and objectives (the functions $b^i(z)$ and the output operators $\mathcal{C}^i$). Note that in the control problem specification of subsection 2.2.2, we have implicitly assumed that $\mathcal{C}^i$ and $b^i(z)$ are chosen to act in the same spatial interval (collocated); in the case where $\mathcal{C}^i$ and $b^i(z)$ are chosen to act in different spatial intervals (noncollocated), it follows directly from Eqs. 2.23 and 2.24 that the characteristic index $\sigma^i = \infty$, which implies that this selection leads to loss of controllability of the output y^i from the input u^i.

In most practical applications, the selection of $(b^i(z), \mathcal{C}^i)$ is typically consistent for all pairs (y^i, u^i), in a sense which is made precise in the following assumption.

Assumption 2.3 *Referring to the system of first-order PDEs of Eq. 2.9 (Eq. 2.7), $\sigma^1 = \sigma^2 = \cdots = \sigma^l = \sigma$.*

Given the above assumption, σ can be also thought of as the characteristic index between the output vector y and the input vector u.

2.4 State Feedback Control

2.4.1 Linear systems

In this subsection, we focus on systems of linear first-order PDEs of the form of Eq. 2.9 and address the problem of synthesizing a distributed state feedback controller that forces the output of the closed-loop system to track a reference input in a prespecified manner. More specifically, we consider distributed state feedback laws of the form:

$$u = \mathcal{S}x + sv \tag{2.25}$$

where $\mathcal{S}$ is a linear operator mapping $\mathcal{H}^n$ into $\mathbb{R}^l$, s is an invertible diagonal matrix of functionals, and $v \in \mathbb{R}^l$ is the vector of reference inputs.

The structure of the control law of Eq. 2.25 is motivated by available results on stabilization of linear PDEs systems via distributed state feedback (e.g., [142, 16]) and the requirement of output tracking. Substituting the distributed state feedback law of Eq. 2.25 into the system of Eq. 2.9, the following closed-loop system is obtained:

$$\begin{aligned} \frac{\partial x}{\partial t} &= A\frac{\partial x}{\partial z} + Bx + wb(z)\mathcal{S}x + wb(z)sv. \\ y &= \mathcal{C}kx \end{aligned} \tag{2.26}$$

It is clear that feedback laws of the form of Eq. 2.25 preserve the linearity with respect to the reference input vector v. We also note that the evolution of the linear PDE system of Eq. 2.26 is governed by a strongly continuous semigroup of bounded linear operators, because $\mathcal{L}$ generates a strongly continuous semigroup and $b(z)\mathcal{S}x$, $b(z)sv$ are bounded, finite dimensional perturbations [64], ensuring that the closed-loop system has a well-defined solution (see subsection 2.2.3). Proposition 2.2 allows specifying the order of the input/output response in the closed-loop system.

Proposition 2.2 *Consider the system of first-order PDEs of Eq. 2.9 subject to the boundary condition of Eq. 2.2, for which assumptions 2.1 and 2.3 hold. Then, a distributed state feedback control law of the form of Eq. 2.25 preserves the characteristic index σ, in the sense that the characteristic index of y with respect to v in the closed-loop system of Eq. 2.26 is equal to σ.*

The fact that the characteristic index between the output y and the reference input v is equal to σ suggests requesting the following input/output response for the closed-loop system:

$$\gamma_\sigma \frac{d^\sigma y}{dt^\sigma} + \cdots + \gamma_1 \frac{dy}{dt} + y = v \tag{2.27}$$

where $\gamma_1, \gamma_2, \ldots, \gamma_\sigma$ are adjustable parameters. These parameters can be chosen to guarantee input/output stability and enforce desired performance specifications in the closed-loop system. In relation to Eq. 2.27, note that, motivated by physical arguments, we request, for each pair (y^i, v^i), $i = 1, \ldots, l$, an input/output response of order σ with the same transient characteristics (i.e., the parameters γ_k are chosen to be the same for each pair (y^i, v^i)). This requirement can be readily relaxed if necessary to impose responses with different transient characteristics for the various pairs (y^i, v^i).

We are now in a position to state the main result of this subsection in the form of a theorem (the proof can be found in appendix A).

Theorem 2.1 *Consider the system of linear first-order PDEs of Eq. 2.9 subject to the boundary condition of Eq. 2.2, for which assumptions 2.1 and*

2.3 hold. Then, the distributed state feedback law:

$$u = \left[\gamma_\sigma Ck\left(A\frac{\partial}{\partial z} + B\right)^{\sigma-1} wb(z)\right]^{-1} \times \left\{v - Ckx - \sum_{\nu=1}^{\sigma} \gamma_\nu Ck\left(A\frac{\partial}{\partial z} + B\right)^{\nu} x\right\} \tag{2.28}$$

enforces the input/output response of Eq. 2.27 in the closed-loop system.

Remark 2.3 Referring to the controller of Eq. 2.28, it is clear that the calculation of the control action requires algebraic manipulations, as well as differentiations and integrations in space, which is expected because of the distributed nature of the controller.

Remark 2.4 The distributed state feedback controller of Eq. 2.28 was derived following an approach conceptually similar to the one employed for the synthesis of inversion-based controllers for ODE systems. We note that this is possible because, for the system of Eq. 2.9: (a) the solution is well defined (i.e., the evolution of the state is locally governed by a strongly continuous semigroup of bounded linear operators), (b) the input/output spaces are finite dimensional, and (c) the manipulated input and the controlled output are distributed in space. These three requirements are standard in most control theories for PDE systems (e.g., [16, 17]), and only the third one poses some practical limitations excluding processes where the manipulated input appears in the boundary.

Remark 2.5 The class of distributed state feedback laws of Eq. 2.25 is a generalization of control laws of the form

$$u = \mathcal{F}x \tag{2.29}$$

where $\mathcal{F}$ is a bounded linear operator mapping $\mathcal{H}^n$ into $\mathbb{R}^l$, which are used for the stabilization of linear PDEs. The usual approach followed for the design of the gain operator $\mathcal{F}$ utilizes optimal control methods (e.g., [101, 115]).

Example 2.1 (Cont'd) In the case of the heat exchanger example introduced earlier, one can easily verify that the characteristic index of the system of Eq. 2.12 is equal to one. Therefore, a first-order input/output response is requested in the closed-loop system:

$$\gamma_1 \frac{dy}{dt} + y = v. \tag{2.30}$$

Using the result of theorem 2.1, the appropriate control law that enforces this response is:

$$u = \frac{1}{\gamma_1}\left\{v - \int_0^1 x(z,t)dz - \gamma_1 \int_0^1 \left(-v_l\frac{\partial x}{\partial z}(z,t) - ax(z,t)\right) dz\right\}. \tag{2.31}$$

2.4.2 Quasi-linear systems

In this subsection, we consider systems of quasi-linear first order PDEs of the form of Eq. 2.7 and control laws of the form:

$$u = \bar{\mathcal{S}}(x) + \bar{s}(x)v \tag{2.32}$$

where $\bar{\mathcal{S}}(x)$ is a nonlinear operator mapping $\mathcal{H}^n$ into $\mathbb{R}^l$, $\bar{s}(x)$ is an invertible diagonal matrix of functionals, and $v \in \mathbb{R}^l$ is the vector of reference inputs. The class of control laws of Eq. 2.32 is a natural generalization of the class of control laws considered for the case of linear systems (Eq. 2.25). Under the control law of Eq. 2.32, the closed-loop system takes the form:

$$\begin{aligned} \frac{\partial x}{\partial t} &= A\frac{\partial x}{\partial z} + f(x) + g(x)b(z)\bar{\mathcal{S}}(x) + g(x)b(z)\bar{s}(x)v. \\ y &= \mathcal{C}h(x) \end{aligned} \tag{2.33}$$

It is straightforward to show that the above system has locally a well-defined solution, and the counterpart of proposition 2.2 also holds, that is, the characteristic index of the output y with respect to v in the closed-loop system of Eq. 2.33 is equal to σ, which suggests seeking a linear input/output response of the form of Eq. 2.27 in the closed-loop system. Theorem 2.2, states the controller synthesis result for this case.

Theorem 2.2 *Consider the system of quasi-linear first-order partial differential equations of Eq. 2.7 subject to the boundary condition of Eq. 2.2, for which assumptions 2.1 and 2.3 hold. Then, the distributed state feedback law:*

$$\begin{aligned} u = &\left[\gamma_\sigma \mathcal{C} L_g \left(\sum_{j=1}^{n} \frac{\partial x_j}{\partial z} L_{a_j} + L_f\right)^{\sigma-1} h(x)b(z)\right]^{-1} \\ &\times \left\{v - \mathcal{C}h(x) - \sum_{\nu=1}^{\sigma} \gamma_\nu \mathcal{C}\left(\sum_{j=1}^{n} \frac{\partial x_j}{\partial z} L_{a_j} + L_f\right)^{\nu} h(x)\right\} \end{aligned} \tag{2.34}$$

enforces the input/output response of Eq. 2.27 in the closed-loop system.

Remark 2.6 Theorem 2.2 provides an analytical formula for a distributed nonlinear state feedback controller that enforces a linear input/output response in the closed-loop system. In this sense, the controller of Eq. 2.34

can be viewed as the counterpart of input/output linearizing control laws for nonlinear ODE systems (see [96] and the references therein), in the case of infinite dimensional systems of the form of Eq. 2.7.

2.5 Closed-Loop Stability

The goal of this section is to define a concept of zero dynamics and the associated notion of minimum phase for systems of first-order hyperbolic PDEs of the form of Eq. 2.9 (Eq. 2.7), subject to the boundary condition of Eq. 2.2; this will allow us to state conditions that guarantee exponential stability of the closed-loop system. We will initially define the concept of zero dynamics for the case of linear systems (the definition for the case of quasi-linear systems is completely similar and will be omitted for brevity). Our definition is analogous to the one given in [29] (see also [111]) for the case of linear parabolic PDE systems with boundary feedback control.

Definition 2.3 *The zero dynamics associated with the system of linear first-order PDEs of Eq. 2.9 is the system obtained by constraining the output to zero, that is, the system:*

$$\begin{aligned}\frac{\partial x}{\partial t} &= A\frac{\partial x}{\partial z} + Bx - wb(z)\left[\mathcal{C}k\left(A\frac{\partial}{\partial z}+B\right)^{\sigma-1}wb(z)\right]^{-1}\\ &\quad\times\left\{\mathcal{C}k\left(A\frac{\partial}{\partial z}+B\right)^{\sigma}x\right\}\\ &\qquad\qquad \mathcal{C}kx \equiv 0\\ &\quad C_1x(\alpha,t)+C_2x(\beta,t) = R(t).\end{aligned} \tag{2.35}$$

From definition 2.3, it is clear that the dynamical system that describes the zero dynamics is an infinite-dimensional one. The concept of zero dynamics allows us to define a notion of minimum phase for systems of the form of Eq. 2.9. More specifically, if the zero dynamics are exponentially stable, the system of Eq. 2.9 is said to be minimum phase, while if the zero dynamics are unstable, the system of Eq. 2.9 is said to be nonminimum phase.

We have now introduced the necessary elements that will allow us to address the issue of closed-loop stability. Proposition 2.3 provides conditions that guarantee the exponential stability of the closed-loop system (the proof can be found in appendix A).

Proposition 2.3 *Consider the system of linear first-order PDEs of Eq. 2.9 for which assumptions 2.1 and 2.3 hold, under the controller of Eq. 2.28. Then, the closed-loop system is exponentially stable (i.e., the differential* *operator of the closed-loop system generates an exponentially stable semi-group) if the following conditions are satisfied:*

1. *The roots of the equation*

$$1 + \gamma_1 s + \cdots + \gamma_\sigma s^\sigma = 0 \tag{2.36}$$

   lie in the open left half of the complex plane.

2. *The system of Eq. 2.35 is exponentially stable.*

Remark 2.7 Referring to the above proposition, we note that the first condition addresses the input/output stability of the closed-loop system and the second condition addresses its internal stability. Note also that the first condition is associated with the stability of a *finite* number of poles, while the second condition concerns the stability of an *infinite* number of poles. This is expected since the input/output spaces are finite dimensional, while the state of the system evolves in infinite dimensions.

Remark 2.8 From the result of proposition 2.3, it follows that the controller of theorem 2.1 places a finite number of poles of the open-loop infinite-dimensional system of Eq. 2.9 at prespecified (depending on the choice of parameters γ_k) locations, by essentially canceling an infinite number of poles, those included in the zero dynamics. Furthermore, the closed-loop system is exponentially stable if the zero dynamics of the original system are exponentially stable (condition 2 of proposition 2.3). This result is analogous to available results on stabilization of systems of linear PDEs of the form of Eq. 2.9 with feedback laws of the form of Eq. 2.29. Specifically, it is well known (e.g., [118, 16]) that control laws of the form of Eq. 2.29 allow placing a finite number of open-loop poles at prespecified locations, while in addition guaranteeing the exponential stability of the closed-loop system, if the pair $[\mathcal{L}\ wb(z)]$ is stabilizable (i.e., the remaining infinite uncontrolled poles are in the open left half of the complex plane).

Remark 2.9 From the result of proposition 2.3 and the discussion of remark 2.8, it is clear that the derivation of exponential stability results for the closed-loop system under the control law of Eq. 2.25 (or the control law of Eq. 2.29) requires that the open-loop system be minimum phase (or stabilizable). The a-priori verification of these properties can in principle be performed by utilizing spectral theory for operators in infinite dimensions [64, 111]. However, these calculations are difficult to perform in the majority of practical applications. In practice, the stabilizability and minimum-phase properties can be checked through simulations.

In closing this section, we address the issue of closed-loop stability for systems of quasi-linear PDEs. Proposition 2.4 provides the counterpart of the result of proposition 2.3 for the case of quasi-linear systems.

Proposition 2.4 *Consider the system of quasi-linear first-order PDEs of Eq. 2.7 for which assumptions 2.1 and 2.3 hold, under the controller of Eq. 2.34. Then, the closed-loop system is locally exponentially stable (i.e., the differential operator of the linearized closed-loop system generates an exponentially stable semigroup) if the following conditions are satisfied:*

1. *The roots of the equation*

$$1 + \gamma_1 s + \cdots + \gamma_\sigma s^\sigma = 0 \tag{2.37}$$

   lie in the open left half of the complex plane.

2. *The zero dynamics of the system of Eq. 2.7 are locally exponentially stable.*

Remark 2.10 The exponential stability of the closed-loop system guarantees, in both the linear and the quasi-linear case, that in the presence of small modeling errors, the states of the closed-loop system will be bounded. Furthermore, since the input/output spaces of the closed-loop system are finite dimensional, and the controller of Eq. 2.34 enforces a linear input/output dynamics between y and v, it is possible to implement a linear error feedback controller with integral action around the $(y - v)$ loop to ensure asymptotic offsetless output tracking in the closed-loop system, in the presence of small errors in model parameters and constant disturbances.

2.6 Output Feedback Control

In this section, we will consider the synthesis of distributed output feedback controllers for systems of the form of Eq. 2.9 (Eq. 2.7). The requisite controllers will be synthesized employing a combination of the developed distributed state feedback controllers with distributed state observers. Analysis of the resulting closed-loop system will allow deriving precise conditions which guarantee that the requirements of exponential stability and output tracking are enforced in the closed-loop system.

The conventional approach followed for the design of state estimators for linear PDE systems is to discretize the system equations and then apply results from estimation theory for ODE systems (e.g., [129]). It has been shown, however, that methods for state estimation that treat the full distributed parameter system lead to state observers that yield significantly superior performance [115, 47]. In this direction, available results on state estimation for systems of first-order hyperbolic PDEs concern mainly the use of Kalman filtering theory for the design of distributed state observers [110, 150] (see also [20, 21] for results on state estimation of distributed parameter systems).

2.6.1 Linear systems

We consider state observers with the following general state-space description [104]:

$$\frac{\partial \eta}{\partial t} = A\frac{\partial \eta}{\partial z} + B\eta + wb(z)u + \mathcal{P}(q - \mathcal{Q}p\eta) \tag{2.38}$$

where $\mathcal{P}$ is a bounded linear operator, mapping $\mathbb{R}^l$ into $\mathcal{H}^n$, that has to be designed so that the operator $\mathcal{L}_o = \mathcal{L} - \mathcal{P}\mathcal{Q}p$ generates an exponentially stable semigroup (note that this is possible by assumption 2.2). The system of Eq. 2.38 consists of a replica of the process system and the term $\mathcal{P}(q - \mathcal{Q}p\eta)$ used to enforce a fast decay of the discrepancy between the estimated and the actual values of the states of the system. In practice, the design of the operator $\mathcal{P}$ can be performed via (a) simple pole placement in cases where the output measurements are not corrupted by noise, or (b) Kalman filtering theory, in cases where the output measurements are noisy.

Theorem 2.3 provides a state-space realization of the output feedback controller resulting from the combination of the state observer of Eq. 2.38 with the state feedback controller of Eq. 2.28 (the proof is given in appendix A).

Theorem 2.3 *Consider the system of linear first-order PDEs of Eq. 2.9 subject to the boundary condition of Eq. 2.2, for which assumptions 2.1, 2.2, 2.3 and the conditions of proposition 2.3 hold. Consider also the linear bounded operator $\mathcal{P}$ designed such that the operator $\mathcal{L}_o = \mathcal{L} - \mathcal{P}\mathcal{Q}p$ generates an exponentially stable semigroup. Then, the distributed output feedback controller:*

$$\begin{aligned}
\frac{\partial \eta}{\partial t} &= A\frac{\partial \eta}{\partial z} + B\eta + wb(z)\left[\gamma_\sigma Ck\left(A\frac{\partial}{\partial z} + B\right)^{\sigma-1} wb(z)\right]^{-1} \\
&\quad \times \left\{v - Ck\eta - \sum_{\nu=1}^{\sigma}\gamma_\nu Ck\left(A\frac{\partial}{\partial z} + B\right)^{\nu}\eta\right\} + \mathcal{P}(q - \mathcal{Q}p\eta) \\
u &= \left[\gamma_\sigma Ck\left(A\frac{\partial}{\partial z} + B\right)^{\sigma-1} wb(z)\right]^{-1} \\
&\quad \times \left\{v - Ck\eta - \sum_{\nu=1}^{\sigma}\gamma_\nu Ck\left(A\frac{\partial}{\partial z} + B\right)^{\nu}\eta\right\}
\end{aligned} \tag{2.39}$$

a) guarantees exponential stability of the closed-loop system,

b) enforces the input/output response of Eq. 2.27 in the closed-loop system if $x(z,0) = \eta(z,0)$.

Remark 2.11 In the case of open-loop stable systems, a more convenient way to reconstruct the state of the system is to consider the observer of Eq. 2.38 with the operator $\mathcal{P}$ set identically equal to zero. This is motivated by the fact that the open-loop stability of the system guarantees the convergence of the estimated values to the actual ones with transient behavior depending on the location of the spectrum of the operator of Eq. 2.14. The robustness of such an open-loop observer with respect to small errors in model parameters and constant disturbances can be achieved by implementing a linear error feedback controller with integral action around the $(y - v)$ loop to ensure asymptotic offsetless output tracking in the closed-loop system.

Remark 2.12 Available results on stabilization of systems of linear hyperbolic PDEs via distributed output feedback (e.g., [115, 16]) concern the design of controllers with the following general state space description:

$$
\begin{aligned}
\frac{\partial \eta}{\partial t} &= A\frac{\partial \eta}{\partial z} + [B + wb(z)\mathcal{F}]\eta + \mathcal{R}(q - \mathcal{Q}p\eta) \\
u &= \mathcal{F}\eta
\end{aligned}
\tag{2.40}
$$

where $\mathcal{R}$ is a bounded linear operator, mapping $\mathbb{R}^l$ into $\mathcal{H}^n$. We note that the main similarity between a controller of the form of Eq. 2.40 and the controller of Eq. 2.39 is that both are infinite dimensional (because of the state observers utilized), while their main difference lies in the fact that the controller of Eq. 2.39 guarantees exponential stability of the closed-loop system, if the open-loop system is minimum-phase and detectable, while a controller of the form of Eq. 2.40 will exponentially stabilize the closed-loop system, if the open-loop system is jointly stabilizable/detectable [16].

Example 2.1 (Cont'd) Referring to the linear PDE system of Eq. 2.12, we note that $\lambda = -v_l < 0$, while $C_2 = 0$, and thus, the system is open-loop stable according to the result of remark 2.1. The open-loop stability of the system allows using a feedback controller that consists of the distributed state feedback controller coupled with an open-loop observer. The appropriate controller takes the form:

$$
\begin{aligned}
\frac{\partial \eta}{\partial t} &= -v_l\frac{\partial \eta}{\partial z} - a\eta \\
&\quad + a\frac{1}{\gamma_1}\left\{v - \int_0^1 \eta(z,t)dz - \gamma_1\int_0^1\left(-v_l\frac{\partial \eta}{\partial z}(z,t) - a\eta(z,t)\right)dz\right\} \\
u &= \frac{1}{\gamma_1}\left\{v - \int_0^1 \eta(z,t)dz - \gamma_1\int_0^1\left(-v_l\frac{\partial \eta}{\partial z}(z,t) - a\eta(z,t)\right)dz\right\}.
\end{aligned}
\tag{2.41}
$$

2.6.2 Quasi-linear systems

In this subsection, we consider the synthesis of distributed output feedback controllers for systems of the form of Eq. 2.7. Given the lack of general available results on state estimation of such systems, we will proceed with the design of a nonlinear state observer that guarantees local exponential convergence of the state estimates to the actual state values. In particular, the following state observer will be used to estimate the state vector of the system in space and time:

$$\frac{\partial \eta}{\partial t} = A\frac{\partial \eta}{\partial z} + f(\eta) + g(\eta)b(z)u + \bar{\mathcal{P}}(q - \mathcal{Q}p(\eta)) \tag{2.42}$$

where η denotes the observer state vector and $\bar{\mathcal{P}}$ is a linear operator, mapping $\mathbb{R}^l$ into $\mathcal{H}^n$, designed on the basis of the linearization of the system of Eq. 2.42 so that the eigenvalues of the operator $\bar{\mathcal{L}}_o = \bar{\mathcal{L}} - \mathcal{P}\mathcal{Q}p(z)$ lie in the left-half plane.

The state observer of Eq. 2.42 can be coupled with the state feedback controller of Eq. 2.34 to derive an output feedback controller that guarantees output tracking and closed-loop stability. The resulting controller is given in Theorem 2.4 (the proof is given in Appendix A).

Theorem 2.4 *Consider the system of quasi-linear first-order PDEs of Eq. 2.7 subject to the boundary condition of Eq. 2.2, for which assumptions 2.1, 2.2, 2.3 and the conditions of proposition 2.4 hold. Consider also the bounded operator $\bar{\mathcal{P}}$ designed such that the operator $\bar{\mathcal{L}}_o = \bar{\mathcal{L}} - \bar{\mathcal{P}}\mathcal{Q}p(z)$ generates an exponentially stable semigroup. Then, the distributed output feedback controller:*

$$\begin{aligned}
\frac{\partial \eta}{\partial t} &= A\frac{\partial \eta}{\partial z} + f(\eta) + g(\eta)b(z)\left[\gamma_\sigma C L_g\left(\sum_{j=1}^{n}\frac{\partial \eta_j}{\partial z}L_{a_j} + L_f\right)^{\sigma-1} h(\eta)b(z)\right]^{-1} \\
&\quad \times\left\{v - Ch(\eta) - \sum_{\nu=1}^{\sigma}\gamma_\nu C\left(\sum_{j=1}^{n}\frac{\partial \eta_j}{\partial z}L_{a_j} + L_f\right)^{\nu} h(\eta)\right\} + \bar{\mathcal{P}}(q - \mathcal{Q}p(\eta)) \\
u &= \left[\gamma_\sigma C L_g\left(\sum_{j=1}^{n}\frac{\partial \eta_j}{\partial z}L_{a_j} + L_f\right)^{\sigma-1} h(\eta)b(z)\right]^{-1} \\
&\quad \times\left\{v - Ch(\eta) - \sum_{\nu=1}^{\sigma}\gamma_\nu C\left(\sum_{j=1}^{n}\frac{\partial \eta_j}{\partial z}L_{a_j} + L_f\right)^{\nu} h(\eta)\right\}
\end{aligned} \tag{2.43}$$

(a) guarantees local exponential stability of the closed-loop system,

(b) enforces the input/output response of Eq. 2.27 in the closed-loop system if $x(z,0) = \eta(z,0)$.

In analogy with the linear case, for open-loop stable systems, the operator $\bar{\mathcal{P}}$ can be taken to be identically equal to zero since the local exponential stability of the open-loop system guarantees the local convergence of the estimated values to the actual values.

Remark 2.13 Note that in the case of imperfect initialization of the observer states (i.e., $\eta(z,0) \neq x(z,0)$), although a slight deterioration of the performance may occur, (i.e., the input/output response of Eq. 2.27 will not be exactly imposed in the closed-loop system), the output feedback controllers of theorems 2.3 and 2.4 guarantee exponential stability and asymptotic output tracking in the closed-loop system.

Remark 2.14 The nonlinear distributed output feedback of Eq. 2.43 is an infinite dimensional one, due to the infinite dimensional nature of the observer of Eq. 2.42. Therefore, a finite-dimensional approximation of the controller has to be derived for on-line implementation. This task can be performed utilizing standard discretization techniques such as finite differences. It is expected that some performance deterioration will occur in this case, depending on the discretization method used and the number and location of discretization points (see the chemical reactor application presented in the next section). We note that in the case of finite-difference discretization, as the number of discretization points increases, the closed-loop system resulting from the PDE model plus an approximate finite-dimensional controller converges to the closed-loop system resulting from the PDE model plus the infinite-dimensional controller, thereby guaranteeing the well-posedness of the approximate finite-dimensional controller (see also [16] for approximation results for a class of linear infinite-dimensional systems).

2.7 Application to a Nonisothermal Plug-Flow Reactor

2.7.1 Process description and modeling

Consider the nonisothermal plug-flow reactor shown in Figure 2.2 where two first-order reactions in series take place:

$$A \xrightarrow{k_1} B \xrightarrow{k_2} C$$

where A is the reactant species, B is the desired product, and C is an

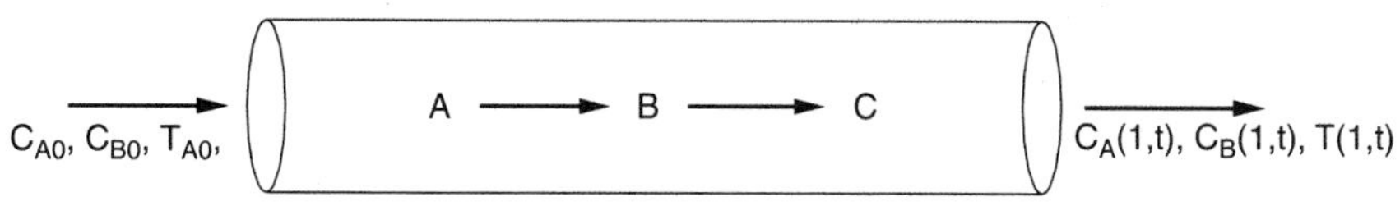

FIGURE 2.2. A nonisothermal plug-flow reactor.

undesired product. The inlet stream consists of pure A at concentration C_{A0} and temperature T_{A0}. The reactions are endothermic, and a jacket is used to heat the reactor. The reaction rate expressions are assumed to be of the following form:

$$r_1 = -k_{10} \exp\left(\frac{-E_1}{R\,T_r}\right) C_A, \quad r_2 = -k_{20} \exp\left(\frac{-E_2}{R\,T_r}\right) C_B$$

where k_{10}, k_{20} E_1, E_2 denote the pre-exponential constants and the activation energies of the reactions, C_A, C_B denote the concentrations of the species A and B in the reactor, and T_r denotes the temperature of the reactor. Under the standard assumptions of perfect radial mixing in the reactor, constant density and heat capacity of the reacting liquid, and negligible diffusive and dispersive phenomena, the material and energy balances that describe the dynamical behavior of the process take the following form:

$$\begin{aligned}
\frac{\partial C_A}{\partial t} &= -v_l \frac{\partial C_A}{\partial z} - k_{10} e^{\frac{-E_1}{RT_r}} C_A \\
\frac{\partial C_B}{\partial t} &= -v_l \frac{\partial C_B}{\partial z} + k_{10} e^{\frac{-E_1}{RT_r}} C_A - k_{20} e^{\frac{-E_2}{RT_r}} C_B \\
\frac{\partial T_r}{\partial t} &= -v_l \frac{\partial T_r}{\partial z} + \frac{(-\Delta H_{r_1})}{\rho_m c_{pm}} k_{10} e^{\frac{-E_1}{RT_r}} C_A + \frac{(-\Delta H_{r_2})}{\rho_m c_{pm}} k_{20} e^{\frac{-E_2}{RT_r}} C_B \\
&\quad + \frac{U_w}{\rho_m c_{pm} V_r} (T_j - T_r)
\end{aligned} \tag{2.44}$$

subject to the following boundary conditions:

$$C_A(0,t) = C_{A0}, \quad C_B(0,t) = 0, \quad T_r(0,t) = T_{A0}$$

where ΔH_{r_1}, ΔH_{r_2} denote the enthalpies of the two reactions, ρ_m, c_{pm} denote the density and heat capacity of the fluid in the reactor, V_r denotes the volume of the reactor, U_w denotes the heat transfer coefficient, and T_j denotes the spatially uniform temperature in the jacket. Owing to the assumption of plug flow in the reactor, the process model of Eq. 2.44 consists of a system of three first-order hyperbolic PDEs.

The control objective is the regulation of the concentration of the product species B throughout the reactor by manipulating the jacket temperature T_j. We note that, in practice, T_j is usually manipulated indirectly through manipulation of the jacket inlet flow rate (this implementation issue is addressed in subsection 2.7.4). Setting:

$$u = T_j - T_{js}, \quad x_1 = C_A, \quad x_2 = C_B, \quad x_3 = T_r, \quad y = C_B \tag{2.45}$$

the process model of Eq. 2.44 can be put in the form of Eq. 2.1 with:

$$f(x) = \begin{bmatrix} f_1(x_1, x_3) \\ f_2(x_1, x_2, x_3) \\ f_3(x_1, x_2, x_3) \end{bmatrix}$$

$$= \begin{bmatrix} -k_{10} e^{\frac{-E_1}{Rx_3}} x_1 \\ k_{10} e^{\frac{-E_1}{Rx_3}} x_1 - k_{20} e^{\frac{-E_2}{Rx_3}} x_2 \\ \frac{(-\Delta H_{r_1})}{\rho_m c_{pm}} k_{10} e^{\frac{-E_1}{Rx_3}} x_1 + \frac{(-\Delta H_{r_2})}{\rho_m c_{pm}} k_{20} e^{\frac{-E_2}{Rx_3}} x_2 \\ -\frac{U_w}{\rho_m c_{pm} V_r}(x_3 - T_{js}) \end{bmatrix}$$

$$A = [a_1 \; a_2 \; a_3] = \begin{bmatrix} -v_l & 0 & 0 \\ 0 & -v_l & 0 \\ 0 & 0 & -v_l \end{bmatrix}, \tag{2.46}$$

$$g(x) = \begin{bmatrix} g_1 \\ g_2 \\ g_3 \end{bmatrix} = \begin{bmatrix} 0 \\ 0 \\ \frac{U_w}{\rho_m c_{pm} V_r} \end{bmatrix}, \quad h(x) = [x_2]. \tag{2.47}$$

It is clear that the matrix A is real symmetric and its eigenvalues satisfy Eq. 2.4. Moreover, the three eigenvalues of A are identical, which implies that the above system of quasi-linear PDEs is weakly hyperbolic.

The values used for the process parameters used in the subsequent computations are given in Table 2.1. The steady-state profiles for C_A, C_B, T_r corresponding to the values of Table 2.1 are shown in Figure 2.3.

2.7.2 Control problem formulation—Controller synthesis

In this subsection, we will proceed with the formulation and solution of the control problem. More specifically, we assume that there are available five control actuators with unity distribution function, that is, $b^i(z)u^i = u^i$ for all $i = 1, \ldots, 5$, which act over equispaced intervals as follows:

$$u(t) = \begin{cases} u^1(t), & [0.0, 0.2] \\ u^2(t), & [0.2, 0.4] \\ u^3(t), & [0.4, 0.6] \\ u^4(t), & [0.6, 0.8] \\ u^5(t), & [0.8, 1.0] \end{cases} \tag{2.48}$$

TABLE 2.1. Parameters and steady-state values for plug-flow reactor.

v_l	=	1.0	$m\ min^{-1}$
L	=	1.0	m
V_r	=	10.0	lt
E_1	=	2.0×10^4	$kcal\ kmol^{-1}$
E_2	=	5.0×10^4	$kcal\ kmol^{-1}$
k_{10}	=	5.0×10^{12}	min^{-1}
k_{20}	=	5.0×10^6	min^{-1}
R	=	1.987	$kcal\ kmol^{-1}\ K^{-1}$
ΔH_1	=	548000.1	$kcal\ kmol^{-1}$
ΔH_2	=	986000.1	$kcal\ kmol^{-1}$
c_{pm}	=	0.231	$kcal\ kg^{-1}\ K^{-1}$
ρ_m	=	0.09	$kg\ lt^{-1}$
c_{pj}	=	2.5	$kcal\ kg^{-1}\ K^{-1}$
ρ_j	=	1.44	$kg\ lt^{-1}$
U_w	=	2000.0	$kcal\ min^{-1}\ K^{-1}$
C_{A0}	=	4.0	$mol\ lt^{-1}$
C_{B0}	=	0.0	$mol\ lt^{-1}$
T_{A0}	=	320.0	K
T_{j0}	=	350.0	K
V_j^i	=	1.0	lt
γ_{jc}	=	0.001	min
F_{js}^1	=	28.87	$lt\ min^{-1}$
F_{js}^2	=	43.66	$lt\ min^{-1}$
F_{js}^3	=	59.63	$lt\ min^{-1}$
F_{js}^4	=	72.34	$lt\ min^{-1}$
F_{js}^5	=	77.23	$lt\ min^{-1}$

The desired performance requirement is to control the averaging outputs:

$$y(t) = \begin{cases} y^1(t) = \int_{0.0}^{0.2} 5.0 x_2(z,t)dz \\ y^2(t) = \int_{0.2}^{0.4} 5.0 x_2(z,t)dz \\ y^3(t) = \int_{0.4}^{0.6} 5.0 x_2(z,t)dz \\ y^4(t) = \int_{0.6}^{0.8} 5.0 x_2(z,t)dz \\ y^5(t) = \int_{0.8}^{1.0} 5.0 x_2(z,t)dz. \end{cases} \tag{2.49}$$

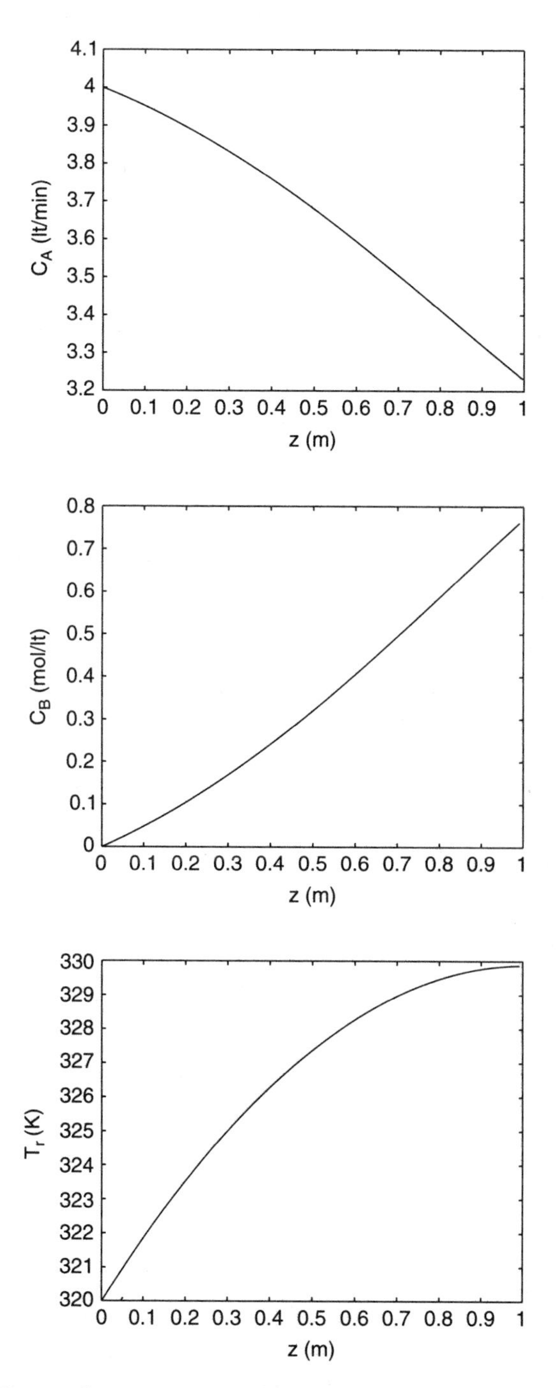

FIGURE 2.3. Steady-state profiles of reactor state variables.

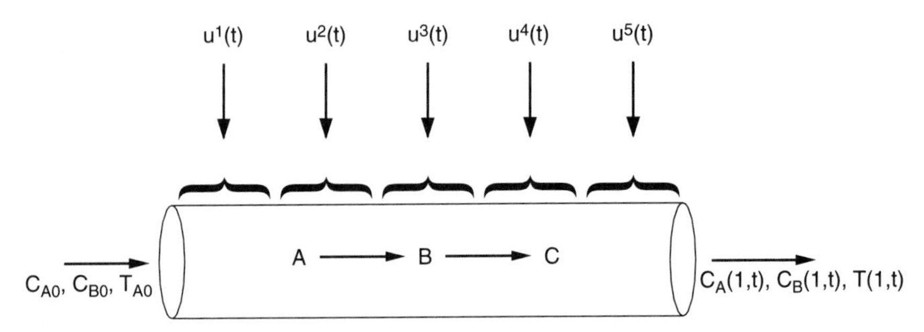

FIGURE 2.4. Specification of the control problem for the plug-flow reactor.

Using the above relations, the model that is used for the synthesis of the output feedback controller is given by:

$$\begin{aligned}\frac{\partial x}{\partial t} &= A\,\frac{\partial x}{\partial z} + f(x) + g(x)\sum_{i=1}^{5}\left(H(z-z_i) - H(z-z_{i+1})\right)u^i \\ y^i &= \int_{z_i}^{z_{i+1}} \frac{1}{z_{i+1}-z_i}\,x_2(z,t)\,dz, \quad i = 1,\ldots,5\end{aligned} \tag{2.50}$$

where the matrix A and the vector functions $f(x)$ and $g(x)$ are specified in Eq. 2.47. A schematic of the reactor along with the control system is given in Figure 2.4. Referring to the system of Eq. 2.50, its characteristic index can be calculated using definition 2.2. In particular, we have that:

$$\begin{aligned}\mathcal{C}^i L_g h(x) &= \int_{z_i}^{z_{i+1}} \frac{1}{z_{i+1}-z_i} L_g h(x)\,dz = 0, \quad \forall\ i = 1,\ldots,5 \\ \mathcal{C}^i L_g\left(\sum_{j=1}^{3}\frac{\partial x_j}{\partial z}L_{a_j} + L_f\right)h(x) &= \int_{z_i}^{z_{i+1}} \frac{1}{z_{i+1}-z_i}\frac{\partial f_2}{\partial x_3} g_3\,dz \not\equiv 0, \\ &\qquad \forall\ i = 1,\ldots,5.\end{aligned} \tag{2.51}$$

Thus, the characteristic index of the system of quasi-linear PDEs of Eq. 2.50 is equal to 2. This allows requesting the following second-order response in the closed-loop system between y^i and v^i, for all $i = 1,\ldots,5$:

$$\gamma_2\frac{d^2y^i}{dt^2} + \gamma_1\frac{dy^i}{dt} + y^i = v^i. \tag{2.52}$$

Moreover, the eigenvalues of the matrix A are negative and the boundary conditions are specified at a single point (i.e., inlet of the reactor). Thus, the result of Remark 2.1 applies directly, yielding that the system is open-loop stable. Furthermore, the process is minimum-phase as verified through simulations. Therefore, the developed control method can be applied, and the distributed output feedback controller of theorem 2.4, with $\bar{\mathcal{P}} \equiv 0$, is employed in the simulations (note that due to open-loop stability of the process the controller does not use measured process outputs). The explicit form of the controller is as follows:

$$
\begin{aligned}
\frac{\partial \eta_1}{\partial t} &= -v_l \frac{\partial \eta_1}{\partial z} - k_{10} e^{\frac{-E_1}{R\eta_3}} \eta_1 \\
\frac{\partial \eta_2}{\partial t} &= -v_l \frac{\partial \eta_2}{\partial z} + k_{10} e^{\frac{-E_1}{R\eta_3}} \eta_1 - k_{20} e^{\frac{-E_2}{R\eta_3}} \eta_2 \\
\frac{\partial \eta_3}{\partial t} &= -v_l \frac{\partial \eta_3}{\partial z} + \frac{(-\Delta H_{r_1})}{\rho_m c_{pm}} k_{10} e^{\frac{-E_1}{R\eta_3}} \eta_1 + \frac{(-\Delta H_{r_2})}{\rho_m c_{pm}} k_{20} e^{\frac{-E_2}{R\eta_3}} \eta_2 \\
&\quad + \frac{U_w}{\rho_m c_{pm} V_r} (T_{w_s} - w_3) + \frac{U_w}{\rho_m c_{pm} V_r} \sum_{i=1}^{5} (H(z - z_i) - H(z - z_{i+1})) \\
&\quad \times \left[\gamma_2 \mathcal{C}^i L_g \left(\sum_{j=1}^{3} \frac{\partial \eta_j}{\partial z} L_{a_j} + L_f \right) h(\eta) b^i(z) \right]^{-1} \\
&\quad \times \left\{ v^i - \mathcal{C}^i h(\eta) - \gamma_1 \mathcal{C}^i \left(\sum_{j=1}^{3} \frac{\partial \eta_j}{\partial z} L_{a_j} + L_f \right) h(\eta) \right. \\
&\quad \left. - \gamma_2 \mathcal{C}^i \left(\sum_{j=1}^{3} \frac{\partial \eta_j}{\partial z} L_{a_j} + L_f \right)^2 h(\eta) \right\}
\end{aligned}
$$

$$
\begin{aligned}
u^i &= \left[\gamma_2 \mathcal{C}^i L_g \left(\sum_{j=1}^{3} \frac{\partial \eta_j}{\partial z} L_{a_j} + L_f \right) h(\eta) b^i(z) \right]^{-1} \left\{ v^i - \mathcal{C}^i h(\eta) \right. \\
&\quad \left. - \gamma_1 \mathcal{C}^i \left(\sum_{j=1}^{3} \frac{\partial \eta_j}{\partial z} L_{a_j} + L_f \right) h(\eta) - \gamma_2 \mathcal{C}^i \left(\sum_{j=1}^{3} \frac{\partial \eta_j}{\partial z} L_{a_j} + L_f \right)^2 h(\eta) \right\}
\end{aligned}
\tag{2.53}
$$

where the analytical expressions of the terms included in the controller are as follows:

$$
\mathcal{C}^i L_g \left(\sum_{j=1}^{3} \frac{\partial \eta_j}{\partial z} L_{a_j} + L_f \right) h(\eta) b^i(z) = \int_{z_i}^{z_{i+1}} \frac{1}{z_{i+1} - z_i} \frac{\partial f_2}{\partial \eta_3} g_3 \, dz
$$

$$
\mathcal{C}^i \left(\sum_{j=1}^{3} \frac{\partial \eta_j}{\partial z} L_{a_j} + L_f \right) h(\eta) = \int_{z_i}^{z_{i+1}} \frac{1}{z_{i+1} - z_i} \left(-v_l \frac{\partial \eta_2}{\partial z} + f_2(\eta_1, \eta_2, \eta_3) \right) dz
$$

$$C^i L_f \left(\sum_{j=1}^{3} \frac{\partial \eta_j}{\partial z} L_{a_j} \right) h(\eta) = \int_{z_i}^{z_{i+1}} \frac{1}{z_{i+1} - z_i} (-v_l) \times \left(\frac{\partial f_2}{\partial \eta_1} \frac{\partial \eta_1}{\partial z} + \frac{\partial f_2}{\partial \eta_2} \frac{\partial \eta_2}{\partial z} + \frac{\partial f_2}{\partial \eta_3} \frac{\partial \eta_3}{\partial z} \right) dz$$

$$C^i \left(\sum_{j=1}^{3} \frac{\partial \eta_j}{\partial z} L_{a_j} \right) \left(\sum_{j=1}^{3} \frac{\partial \eta_j}{\partial z} L_{a_j} \right) h(\eta) = \int_{z_i}^{z_{i+1}} \frac{1}{z_{i+1} - z_i} v_l^2 \frac{\partial^2 \eta_2}{\partial z^2}$$

$$C^i \left(\sum_{j=1}^{3} \frac{\partial \eta_j}{\partial z} L_{a_j} \right) L_f h(\eta) = \int_{z_i}^{z_{i+1}} \frac{1}{z_{i+1} - z_i} (-v_l) \times \left(\frac{\partial f_2}{\partial \eta_1} \frac{\partial \eta_1}{\partial z} + \frac{\partial f_2}{\partial \eta_2} \frac{\partial \eta_2}{\partial z} + \frac{\partial f_2}{\partial \eta_3} \frac{\partial \eta_3}{\partial z} \right) dz$$

$$C^i L_f^2 h(\eta) = \int_{z_i}^{z_{i+1}} \frac{1}{z_{i+1} - z_i} \left(\frac{\partial f_2}{\partial \eta_1} f_1(\eta_1, \eta_3) + \frac{\partial f_2}{\partial \eta_2} f_2(\eta_1, \eta_2, \eta_3) + \frac{\partial f_2}{\partial \eta_3} f_3(\eta_1, \eta_2, \eta_3) \right) dz. \tag{2.54}$$

The controller is tuned to give an overdamped response between the output y^i and the reference input v^i. In particular, the parameters γ_1 and γ_2 are chosen to be:

$$\gamma_1 = 3.0 \ min, \quad \gamma_2 = 0.5 \ min^2$$

to achieve the following time constant and damping factor:

$$\tau = 0.707 \ min, \quad \zeta = 2.12.$$

2.7.3 Evaluation of controller performance

Several simulation runs are performed to evaluate the performance of the distributed output feedback controller of Eq. 2.53. The method of finite differences is employed to derive a finite dimensional approximation of the output feedback controller of Eq. 2.53, with a choice of 200 discretization points. In all the simulation runs, the process is initially assumed to be at steady state.

In the first simulation run, we address the reference input tracking capabilities of the controller. Initially, a 30% increase in the reference inputs v^i, $i = 1, \ldots, 5$ is imposed at time $t = 0.0 \ min$. Figure 2.5 shows the corresponding output profiles. It is clear that the controller enforces the requested input/output response in the closed-loop system and regulates the output at the new reference input values. The corresponding manipulated input profiles for each control actuator are depicted in Figure 2.6.

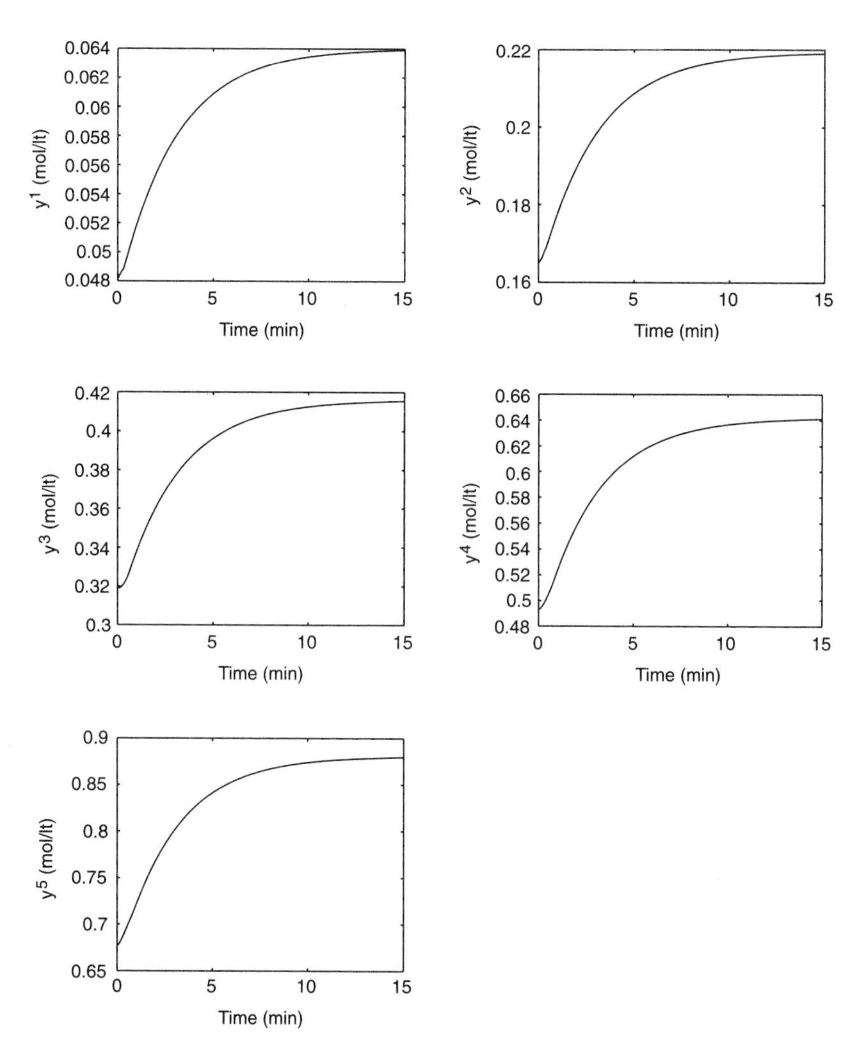

FIGURE 2.5. Closed-loop output profiles.

We observe that the control action, required by each actuator to drive the corresponding output to the new reference value, increases as we approach the outlet of the reactor. This is expected because the amount of heat required to maintain the reaction rate that yields the necessary conversion increases along the length of the reactor. Figure 2.7 shows the evolution of the concentration of the species B throughout the reactor, while Figure 2.8 shows the profile of the concentration of the species B at the outlet of the reactor. We observe that by using a finite number of control actuators, we achieve satisfactory control of the output variable C_B at all positions and times.

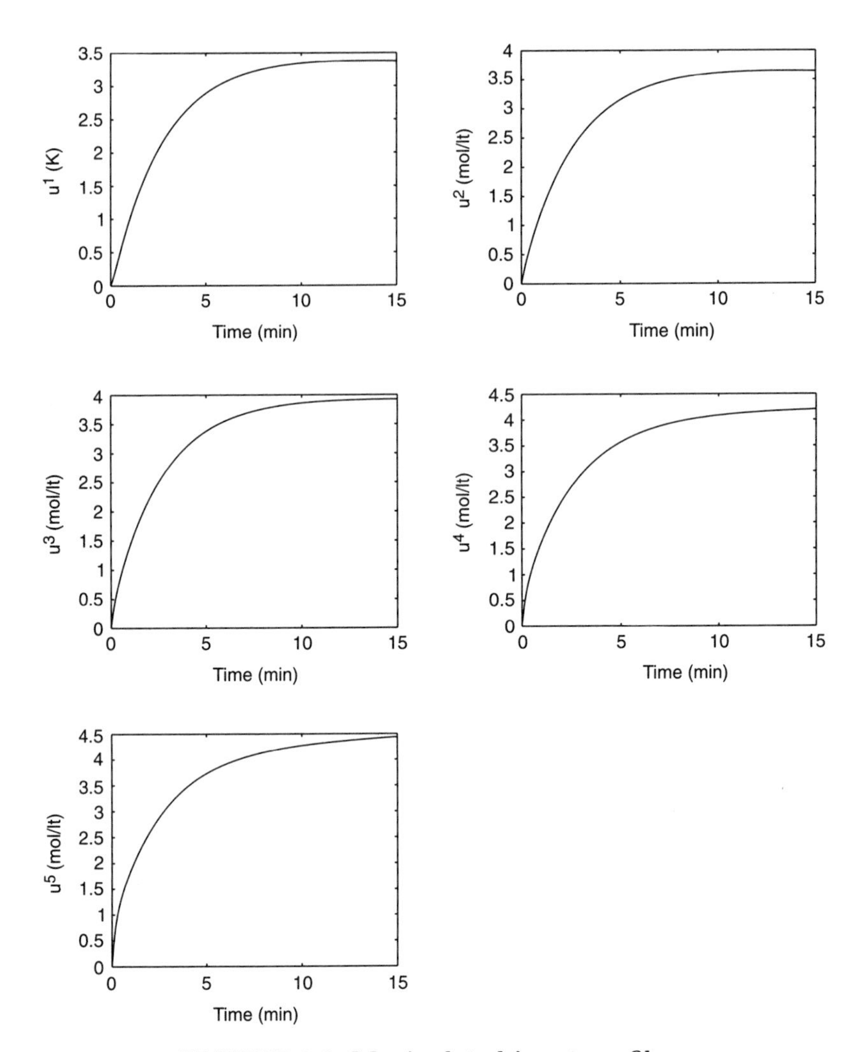

FIGURE 2.6. Manipulated input profiles.

For the sake of comparison, we also consider the control of the reactor using a controller that is designed on the basis of a model resulting from discretization of the original PDE system in space. In particular, the method of finite differences is used to discretize the original PDE model of Eq. 2.50 into a set of five (equal to the number of control actuators) ODEs in time. Subsequently, an input/output linearizing controller [90] is designed on the basis of the resulting ODE model. The corresponding output profiles are shown in Figure 2.9, while Figure 2.10 shows the profile of the concentration of the species B in the outlet of the reactor. It is clear that this controller leads to poor performance (oscillations, longer transient response

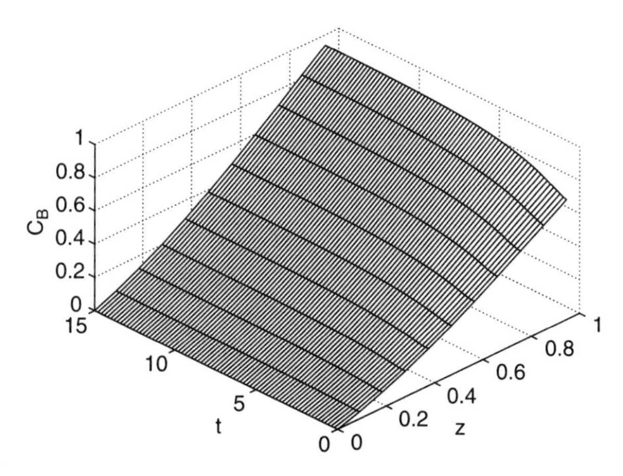

FIGURE 2.7. Profile of evolution of concentration of species B throughout the reactor.

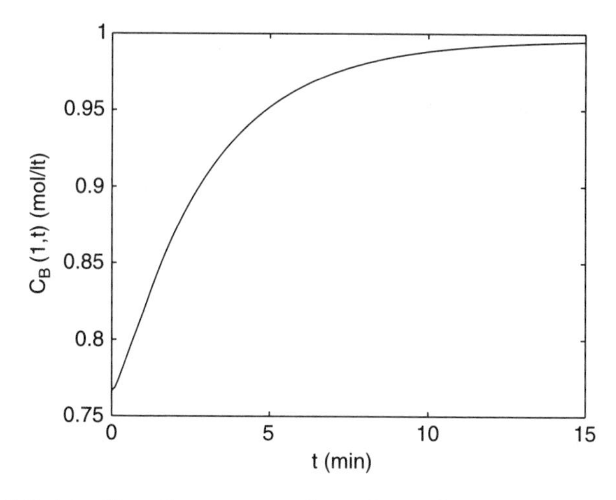

FIGURE 2.8. Profile of concentration of species B in the outlet of the reactor.

and offset) because it does not explicitly take into account the spatially varying nature of the process.

2.7.4 A practical implementation issue

The distributed output feedback controller of Eq. 2.53 assumes that the jacket temperature can be manipulated directly. In practice, the jacket temperature is usually manipulated indirectly through the jacket inlet flow rate. This can be achieved in a straightforward way by designing a controller to ensure that the jacket temperature obtains the values requested by the distributed controller of Eq. 2.53.

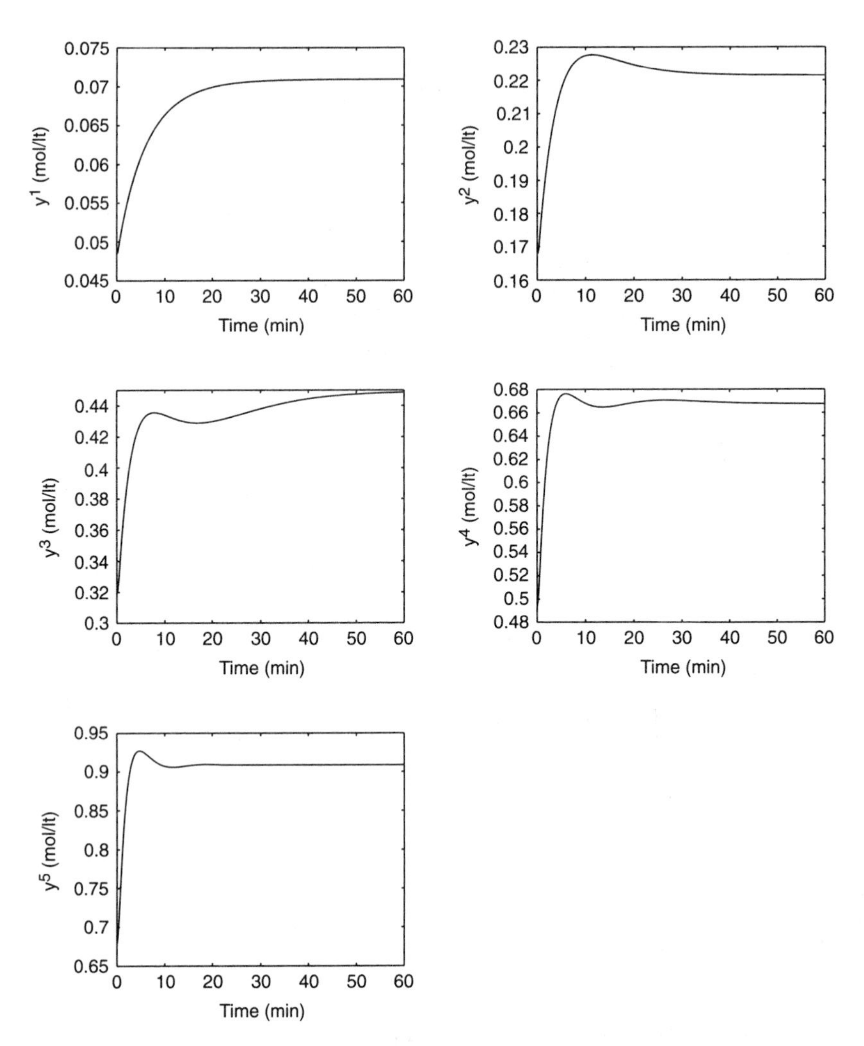

FIGURE 2.9. Closed-loop output profiles—discretization based controller.

Specifically, under the assumption of perfect mixing and constant volume, the dynamic model of the jacket takes the form:

$$\frac{d\,T_j^i}{dt} = \left(T_{j0}^i - T_j^i\right)\frac{F_{js}^i}{V_j^i} + \frac{U_w}{\rho_j c_{pj} V_j^i}\left(\int_{z_i}^{z_{i+1}} \frac{1}{z_{i+1} - z_i}\, T_r(z,t)\,dz - T_j^i\right) + \frac{\left(T_{j0}^i - T_j^i\right)}{V_j^i} u_{fl}^i, \quad i = 1, \ldots, 5 \tag{2.55}$$

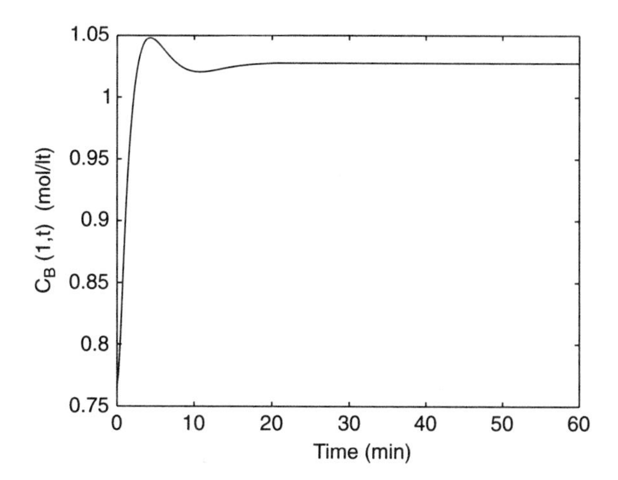

FIGURE 2.10. Profile of concentration of species B in the outlet of the reactor—discretization-based controller.

where F_{js}^i is the steady-state jacket inlet flow rate, V_j^i is the jacket volume, ρ_j, c_{pj} are the density and heat capacity of the fluid in the jacket, T_{j0}^i is the temperature of the inlet stream to a jacket, and u_{fl}^i is the jacket inlet flow rate (chosen as the new manipulated input) in deviation variable form. Requesting a first-order response of the form:

$$\gamma_{jc}\frac{d\,T_j^i}{dt} + T_j^i = u^i \tag{2.56}$$

where γ_{jc} is the time constant, the necessary controller takes the form:

$$u_{fl}^i = \frac{(u^i - T_j^i)V_j^i - \gamma_{jc}\left[(T_{j0}^i - T_j^i)F_{js}^i + \frac{U_w}{\rho_j\,c_{pj}}\left(\int_{z_i}^{z_{i+1}} \frac{1}{z_{i+1}-z_i}\,T_r(z,t)\,dz - T_j^i\right)\right]}{\gamma_{jc}\left(T_{j0}^i - T_j^i\right)}. \tag{2.57}$$

The parameter γ_{jc} should be chosen such that the response of Eq. 2.56 is sufficiently fast compared to the response of Eq. 2.52 to ensure that the jacket temperature obtains the values requested by the distributed controller of Eq. 2.53, while estimates of $T_r(z,t)$ can be obtained from the state observer of Eq. 2.53. Moreover, the implementation of the controller of Eq. 2.57 requires only measurements of T_j^i since measurements of $T_r(z,t)$ can be obtained from the state observer incorporated in the controller of Eq. 2.53.

The performance of the control scheme resulting from the combination of this controller with the distributed controller of Eq. 2.53 is evaluated through simulations on the plug-flow reactor. The values used for the

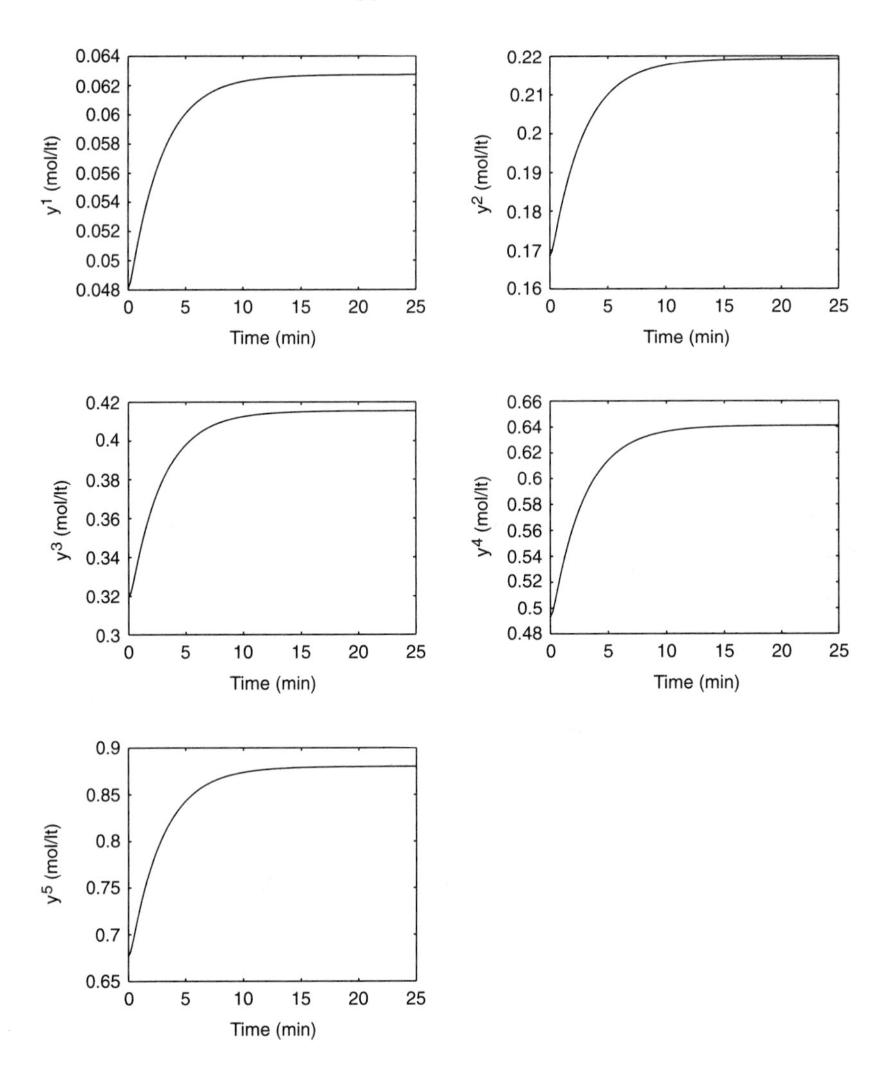

FIGURE 2.11. Closed-loop output profiles—the jacket inlet flow rates are used as the manipulated inputs.

parameters (ρ_j, c_{pj}, V_j^i, T_{j0}^i, γ_{jc}) and the steady-state values of the jacket inlet flow rate F_{js}^i, $i = 1, \ldots, 5$, are given in Table 2.1.

Figure 2.11 shows the output profiles for the same 30% increase in the value of the reference inputs as previously (the profiles for the jacket inlet flow rate are displayed in Figure 2.12).

Clearly, the performance of the control scheme is excellent, enforcing closed-loop output responses that are very close to the ones obtained by neglecting the jacket dynamics (compare Figure 2.10 with Figure 2.4).

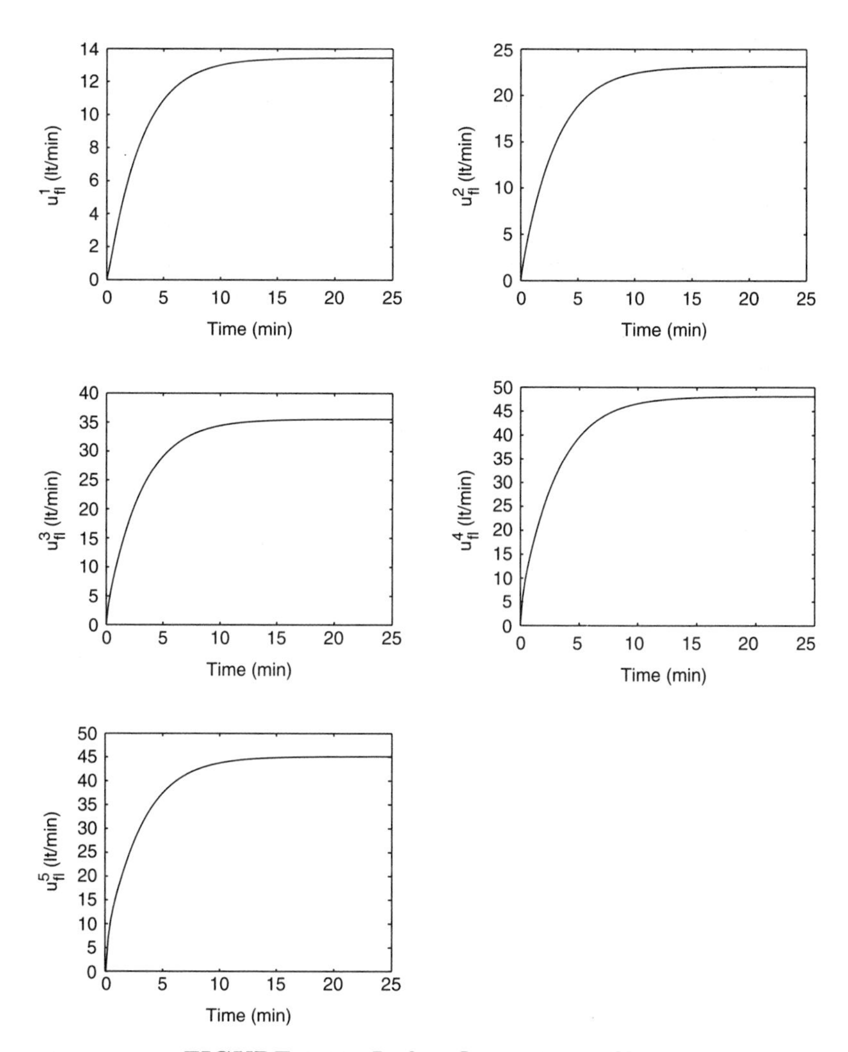

FIGURE 2.12. Jacket flow-rate profiles.

2.8 Conclusions

In this chapter, we presented an output feedback control methodology for systems described by quasi-linear first-order hyperbolic PDEs, for which the manipulated inputs, the controlled outputs, and the measured output are distributed in space. The central idea of our approach is the combination of theory of partial differential equations and concepts from geometric control. Initially, a concept of characteristic index was introduced and used for the synthesis of distributed state feedback controllers that guarantee output tracking in the closed-loop system. Conditions that ensure exponential stability of the closed-loop system were derived. Analytical formulas of distributed output feedback controllers were also derived through a combination of distributed state observers with the developed distributed state feedback controllers. The proposed control methodology was implemented, through simulations, on a nonisothermal plug-flow reactor, modeled by three quasi-linear hyperbolic PDEs. Comparisons with a control method that involves discretization in space of the original PDE model and application of standard nonlinear control methods for ODE systems established that the proposed control method yields superior performance.

Chapter 3

Robust Control of Hyperbolic PDE Systems

3.1 Introduction

In the previous chapter, we addressed the control of hyperbolic PDE systems without accounting explicitly for the presence of uncertainty (i.e., presence of mismatch between the model used for controller design and the actual process model) in the design of the controller. This chapter focuses on systems of quasi-linear first-order hyperbolic PDEs with uncertainty for which the manipulated variables and the controlled variables are distributed in space. The objective is to develop a framework for the synthesis of distributed robust controllers that handle explicitly time-varying uncertain variables and unmodeled dynamics. For systems with uncertain variables, the problem of complete elimination of the effect of uncertainty on the output via distributed feedback is initially considered; a necessary and sufficient condition for its solvability, as well as explicit controller synthesis formulas, is derived. Then, a distributed robust controller is derived that guarantees boundedness of the state and achieves asymptotic output tracking with arbitrary degree of asymptotic attenuation of the effect of uncertain variables on the output of the closed-loop system. The controller is designed constructively using Lyapunov's direct method and requires that there exist known bounding functions that capture the magnitude of the uncertain terms and a matching condition is satisfied. The problem of robustness with respect to unmodeled dynamics is then addressed within the context of control of two-time-scale systems modeled in singularly perturbed form. Initially, a robustness result of the bounded stability property of a reduced-order PDE model with respect to stable and fast dynamics is proved. This result is then used to establish that the controllers, which are synthesized on the basis of a reduced-order slow model and achieve uncertainty decoupling or uncertainty attenuation, continue to enforce these control objectives in the full-order closed-loop system, provided that the unmodeled dynamics are stable and sufficiently fast. The developed control method is tested through simulations on a nonisothermal fixed-bed reactor, where the reactant wave propagates through the bed with significantly larger speed than the heat wave and the heat of reaction is unknown and time varying. The results of this chapter were first presented in [44].

3.2 Preliminaries

We will focus on systems of quasi-linear hyperbolic first-order PDEs in one spatial variable, with the following state-space description:

$$\begin{aligned} \frac{\partial x}{\partial t} &= A\frac{\partial x}{\partial z} + f(x) + g(x)b(z)u + W(x)r(z)\theta(t) \\ y &= \mathcal{C}h(x) \end{aligned} \tag{3.1}$$

subject to the boundary condition:

$$C_1 x(\alpha, t) + C_2 x(\beta, t) = R(t) \tag{3.2}$$

and the initial condition:

$$x(z, 0) = x_0(z). \tag{3.3}$$

In the above equations, $W(x)$ is a sufficiently smooth matrix of appropriate dimensions, $\theta = [\theta_1 \ \cdots \ \theta_q] \in \mathbb{R}^q$ denotes the vector of uncertain variables, which may include uncertain process parameters or exogenous disturbances, and $r(z)$ is a known matrix whose (i, k)-*th* element is of the form $r_k^i(z)$, where the function $r_k^i(z)$ specifies the position of action of the uncertain variable θ_k on $[z_i, z_{i+1}]$. The rest of the notation used in Eqs. 3.1–3.3 is given in subsection 2.2.1. For simplicity, the function $r_k^i(z)$, $i = 1, \ldots, l$, $k = 1, \ldots, q$ is also assumed to be normalized in the interval $[\alpha, \beta]$, that is,

$$\sum_{i=1}^{l} \int_{z_i}^{z_{i+1}} b^i(z)dz = \sum_{i=1}^{l} \int_{z_i}^{z_{i+1}} c^i(z)dz = \sum_{i=1}^{l} \int_{z_i}^{z_{i+1}} r_k^i(z)dz = 1.$$

Furthermore, for any measurable (with respect to the Lebesgue measure) function $\theta : \mathbb{R}_{\geq 0} \to \mathbb{R}^m$, $\|\theta\|$ denotes the ess.sup.$|\theta(t)|$, $t \geq 0$.

Referring to the model of Eq. 3.1, the assumption of affine and separable appearance of u and θ is a standard one in uncertainty decoupling and robust control studies for linear PDEs (e.g., [50, 51]), and is satisfied in most practical applications, where the jacket temperature is usually chosen to be the manipulated input (see for example [115, 38]), and the heat of reactions, the pre-exponential constants, and the temperature and concentrations of lateral inlet streams, and so forth, are typical uncertain variables. Figure 3.1 shows the location of the manipulated inputs, controlled outputs, and uncertain variables in the case of a prototype example.

In the remainder of this section, we review a basic stability result (converse Lyapunov theorem) for systems of the form of Eq. 3.1 and introduce a concept of characteristic index between the output y^i and the uncertainty vector θ that will be used in our development.

Theorem 3.1 [141] (Converse Lyapunov theorem) *Consider the system of Eq. 3.1, with* $(|u| = |\theta| = 0)$*, and assume that the operator of*

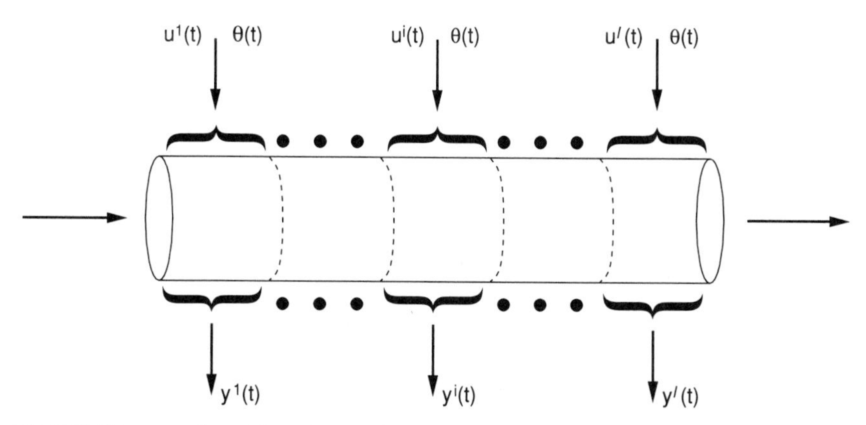

FIGURE 3.1. Specification of the control problem in a prototype example.

Eq. 2.18 generates a locally exponentially stable semigroup. Then, there exists a smooth functional $V : \mathcal{H}^n \times [\alpha, \beta] \to \mathbb{R}_{\geq 0}$ *of the form:*

$$V(t) = \int_{\alpha}^{\beta} x^T q(z) x \, dz \tag{3.4}$$

where $q(z)$ *is a known positive definite function satisfying* $\sum_{i=1}^{l} \int_{z_i}^{z_{i+1}} q(z) dz = 1$, *and a set of positive real numbers* a_1, a_2, a_3, a_4, a_5 *such that if* $\|x\|_2 \leq a_5$ *the following conditions hold:*

$$\begin{aligned} a_1 \|x\|_2^2 \leq V(t) \leq a_2 \|x\|_2^2 \\ \frac{dV}{dt} = \frac{\partial V}{\partial x} \left[A \frac{\partial x}{\partial z} + f(x) \right] \leq -a_3 \|x\|_2^2 \\ \left\| \frac{\partial V}{\partial x} \right\|_2 \leq a_4 \|x\|_2. \end{aligned} \tag{3.5}$$

Remark 3.1 Note that, for infinite-dimensional systems, stability with respect to one norm does not necessarily imply stability with respect to another norm. This difficulty is not encountered in finite-dimensional systems since all norms defined in a finite-dimensional vector space are equivalent. In our case, we choose to study stability with respect to the L_2-norm since for hyperbolic systems it represents a measure of the total energy of the system at any time, and thus, exponential stability with respect to this norm implies that the system's total energy tends to zero as $t \to \infty$ (i.e., since $V(t)$ is a quadratic function: $V(t) \to 0$ as $t \to \infty \Longrightarrow$ exponential stability).

We now define a concept of characteristic index between the output y^i and the uncertainty vector θ, for systems of the form of Eq. 3.1 that will be used to express the solvability condition of the uncertainty decoupling problem via distributed state feedback and the matching condition in the robust uncertainty attenuation problem.

Definition 3.1 *Referring to the system of Eq. 3.1, we define the characteristic index of the output y^i with respect to the vector of uncertainties θ as the smallest integer δ^i for which there exists $k \in [1, q]$ such that*

$$C^i L_{W_k} \left(\sum_{j=1}^{n} \frac{\partial x_j}{\partial z} L_{a_j} + L_f \right)^{\delta^i - 1} h(x) r_k^i(z) \neq 0 \tag{3.6}$$

where W_k denotes the k-th column vector of the matrix W, or $\delta^i = \infty$ if such an integer does not exist.

From the above definition, it follows that δ^i depends on the structure of the process (matrices A, $W(x)$ and functions $f(x)$, $h(x)$), as well as on the position of action of the uncertain variables θ (functions $r_k^i(z)$) and the function $c^i(z)$. In order to simplify the statement of our results, we define the characteristic index of the output vector y with respect to the vector of uncertainties θ as $\delta = \min\{\delta^1, \delta^2, \ldots, \delta^l\}$.

3.3 Uncertainty Decoupling

In this section, we consider systems of the form of Eq. 3.1 and address the problem of synthesizing a distributed state feedback controller that stabilizes the closed-loop system and forces the output to track the external reference input in a prespecified manner for all times, independently of the uncertain variables. More specifically, we consider control laws of the form:

$$u = \mathcal{S}(x) + s(x)v \tag{3.7}$$

where $\mathcal{S}(x)$ is a smooth nonlinear operator mapping $\mathcal{H}^n$ into $\mathbb{R}^l$, $s(x)$ is an invertible matrix of smooth functionals, and $v \in \mathbb{R}^l$ is the vector of external reference inputs. Under the control law of Eq. 3.7, the closed-loop system takes the form:

$$\begin{aligned} \frac{\partial x}{\partial t} &= A\frac{\partial x}{\partial z} + f(x) + g(x)b(z)\mathcal{S}(x) + g(x)b(z)s(x)v + W(x)r(z)\theta(t) \\ y &= \mathcal{C}h(x). \end{aligned} \tag{3.8}$$

It is clear that feedback laws of the form of Eq. 3.7 preserve the linearity with respect to the external reference input v. We also note that the evolution of the closed-loop system of Eq. 3.8 is locally governed by a strongly continuous semigroup of bounded linear operators, because $b(z)\mathcal{S}(x)$, $b(z)s(x)$, $r(z)\theta(t)$ are bounded, finite-dimensional perturbations, ensuring that the solution of the system of Eq. 3.8 is well defined. Proposition 3.1 allows specifying the order of the input/output response in the nominal closed-loop system (the proof can be found in Appendix B).

Proposition 3.1 *Consider the system of Eq. 3.1, for which assumptions 2.1 and 2.3 hold, subject to the distributed state feedback law of Eq. 3.7. Then, referring to the closed-loop system of Eq. 3.8:*

(a) the characteristic index of y with respect to the external reference input v is equal to σ, and

(b) the characteristic index of y with respect to θ is equal to δ, if $\delta \leq \sigma$.

The fact that the characteristic index between the output y and the external reference input v is equal to σ suggests requesting the following input/output response for the closed-loop system:

$$\gamma_\sigma \frac{d^\sigma y}{dt^\sigma} + \cdots + \gamma_1 \frac{dy}{dt} + y = v \tag{3.9}$$

where $\gamma_1, \gamma_2, \ldots, \gamma_\sigma$ are adjustable parameters that can be chosen to guarantee input/output stability in the closed-loop system. Referring to Eq. 3.9, note that, motivated by physical arguments, we request, for each pair (y^i, v^i), $i = 1, \ldots, l$, an input/output response of order σ with the same transient characteristics (i.e., the parameters γ_k are chosen to be the same for each pair (y^i, v^i)). The following theorem provides the main result of this section (the proof is given in Appendix B).

Theorem 3.2 *Consider the system of Eq. 3.1 for which assumption 2.1 holds, and assume that: (i) the roots of the equation $1 + \gamma_1 s + \cdots + \gamma_\sigma s^\sigma = 0$ lie in the open left half of the complex plane, (ii) the zero dynamics (Eq. 2.35) are locally exponentially stable, and (iii) there exists a positive real number $\hat{\delta}$ such that $\max\{\|\theta\|, \|\dot{\theta}\|\} \leq \hat{\delta}$. Then, the condition $\sigma < \delta$ is necessary and sufficient in order for a distributed state feedback law of the form of Eq. 3.7 to completely eliminate the effect of θ on y in the closed-loop system. Whenever this condition is satisfied, the control law:*

$$\begin{aligned} u = & \left[\gamma_\sigma C L_g \left(\sum_{j=1}^{n} \frac{\partial x_j}{\partial z} L_{a_j} + L_f \right)^{\sigma-1} h(x) b(z) \right]^{-1} \\ & \times \left\{ v - Ch(x) - \sum_{\nu=1}^{\sigma} \gamma_\nu C \left(\sum_{j=1}^{n} \frac{\partial x_j}{\partial z} L_{a_j} + L_f \right)^{\nu} h(x) \right\} \end{aligned} \tag{3.10}$$

(a) guarantees boundedness of the state of the closed-loop system, and

(b) enforces the input/output response of Eq. 3.9 in the closed-loop system.

Remark 3.2 Referring to the solvability condition of the problem, notice that it depends not only on the structural properties of the process but also on the shape of the actuator distribution functions $b(z)$, the position of action of the vector of uncertain variables (matrix function $r(z)$), and the performance specification functions $c(z)$ since σ and δ depend on $c(z)$, $b(z)$

and $c(z), r(z)$, respectively. This implies that it is generally possible to select $(b(z), c(z))$ to allow achieving uncertainty decoupling via distributed state feedback (see illustrative example in remark 3.5 below).

Remark 3.3 The solvability condition of the problem and the distributed state feedback controller of Eq. 3.10 were derived following an approach conceptually similar to the one used to solve the counterpart of this problems in the case of ODE systems (e.g., [86]). We note that this is possible because, for the system of Eq. 3.1: (a) the spatial differential operator generates a strongly continuous semigroup, (b) the input/output spaces are finite dimensional, and (c) the input/output operators (functions $c^i(z)$, $b^i(z)$) are bounded.

Remark 3.4 We note that the distributed state feedback controller of Eq. 3.10 that solves the uncertainty decoupling problem is identical to the one given in theorem 2.2, which solves the output tracking problem via distributed state feedback, while the conditions (i) and (ii) of theorem 3.2, as proved in theorem 2.4, ensure that the controller of Eq. 3.10 exponentially stabilizes the nominal closed-loop system ($\theta(t) \equiv 0$).

Remark 3.5 (Illustrative example) Consider a plug-flow reactor with lateral feed, where a first-order reaction of the form $A \rightarrow B$ takes place. Assuming that the lateral flow rate is significantly smaller than the inlet flow rate (i.e., $v_l \gg F$, where v_l denotes the velocity of the inlet stream to the reactor and F denotes the lateral flow rate per unit volume), the dynamic model of the process is given by:

$$\begin{aligned}
\frac{\partial C_A}{\partial t} &= -v_l \frac{\partial C_A}{\partial z} - k C_A + F(C_{Al} - C_A) \\
\frac{\partial T}{\partial t} &= -v_l \frac{\partial T}{\partial z} + \frac{(-\Delta H)}{\rho_f c_{pf}} k C_A + F(T_l - T) + \frac{U_w}{\rho_f c_{pf} V_r}(T_j - T) \\
& C_A(0,t) = C_{A0}, \quad T(0,t) = T_0
\end{aligned} \tag{3.11}$$

where $C_A(z,t)$ and $T(z,t)$ denote the concentration of the species A and the temperature in the reactor, $z \in [0,1]$, $T_j(z,t)$ denotes the wall temperature, C_{Al}, T_l denote the concentration and the temperature of the lateral inlet stream, k denotes the reaction rate constant, ΔH denotes the enthalpy of the reaction, ρ_f, c_{pf} denote the density and heat capacity of the reacting liquid, U_w denotes the heat transfer coefficient, and V_r denotes the volume of the reactor. Set $x_1 = C_A$, $x_2 = T$, $u = T_j - T_{js}$, $\theta_1 = T_l - T_{ls}$, $\theta_2 = C_{Al} - C_{Als}$, $y = T$, $R_1(t) = C_{A0}(t)$ and $R_2(t) = T_0(t)$, where T_{js}, T_{ls}, C_{Als} denote steady-state values. Assume that for the control of the system there is available one control actuator with distribution function $b(z)$; let also $r_1(z)$ and $r_2(z)$ determine the position of action of θ_1 and θ_2, respectively, and $y(t) = \int_0^1 c(z) x_2(z,t) dz$; then the system of

Eq. 3.11 can be written as:

$$
\begin{aligned}
\frac{\partial x_1}{\partial t} &= -v_l \frac{\partial x_1}{\partial z} - k x_1 + F(C_{Als} - x_1) + F\theta_2(t) \\
\frac{\partial x_2}{\partial t} &= -v_l \frac{\partial x_2}{\partial z} + \frac{(-\Delta H)}{\rho_f c_{pf}} k x_1 + \frac{U_w}{\rho_f c_{pf} V_r}(T_{js} - x_2) + F(T_{ls} - x_2) \\
&+ \frac{U_w}{\rho_f c_{pf} V_r} b(z) u + F r_1(z)\theta_1(t) \\
y(t) &= \int_0^1 c(z) x_2(z,t)\, dz, \quad x_1(0,t) = R_1, \quad x_2(0,t) = R_2.
\end{aligned}
\tag{3.12}
$$

Performing a time differentiation of the output $y(t)$, we have:

$$
\begin{aligned}
\frac{dy}{dt} = \int_0^1 c(z) \Bigg(& -v_l \frac{\partial x_2}{\partial z} + \frac{(-\Delta H)}{\rho_f c_{pf}} k x_1 \\
& + \frac{U_w}{\rho_f c_{pf} V_r}(T_{js} - x_2) + F(T_{ls} - x_2) \Bigg) dz \\
& + \frac{U_w}{\rho_f c_{pf} V_r} \int_0^1 c(z) b(z) u\, dz + F \int_0^1 c(z) r_1(z) \theta_1(t)\, dz.
\end{aligned}
\tag{3.13}
$$

From the above equation and the result of theorem 3.2, it follows that if $(c(z), b(z)) = \int_0^1 c(z) b(z) dz \neq 0$ (in which case $\sigma = 1$), then the uncertainty decoupling problem via distributed state feedback is solvable as long as $\sigma < \delta$, that is, when $(c(z), r_1(z)) = \int_0^1 c(z) r_1(z) dz = 0$. Whenever this condition holds, the controller that achieves uncertainty decoupling in the closed-loop system can be derived from the synthesis formula of Eq. 3.10.

3.4 Robust Control: Uncertain Variables

In this section, we consider systems of the form of Eq. 3.1 for which the condition $\sigma < \delta$ is not necessarily satisfied and address the problem of synthesizing a distributed robust nonlinear controller of the form:

$$
u = \bar{\mathcal{S}}(x) + \bar{s}(x) v + \bar{\mathcal{R}}(x, t) \tag{3.14}
$$

where $\bar{\mathcal{S}}(x)$ is a smooth nonlinear operator mapping $\mathcal{H}^n$ into $\mathbb{R}^l$, $\bar{s}(x)$ is an invertible matrix of smooth functionals, where $\bar{\mathcal{R}}(x, t)$ is a vector of nonlinear functionals, that guarantees closed-loop stability, enforces asymptotic output tracking, and achieves arbitrary degree of asymptotic attenuation of the effect of uncertain variables on the output of the closed-loop system. The control law of Eq. 3.14 consists of two components, the component $\bar{\mathcal{S}}(x) + \bar{s}(x) v$, which is used to enforce output tracking and stability in the nominal closed-loop system, and the component $\bar{\mathcal{R}}(x,t)$, which is used to asymptotically attenuate the effect of $\theta(t)$ on the output of the closed-loop system. The design of $\bar{\mathcal{S}}(x) + \bar{s}(x)v$ will be performed by employing the geometric approach presented in the previous section, while the design of $\bar{\mathcal{R}}(x,t)$ will be performed constructively using Lyapunov's direct method. The central idea of Lyapunov-based controller design is to construct $\bar{\mathcal{R}}(x,t)$, using the knowledge of bounding functions that capture the size of the uncertain terms and assuming a certain path under which the uncertainty may affect the output, so that the ultimate discrepancy between the output of the closed-loop system and the reference input can be made arbitrarily small by a suitable selection of the controller parameters.

We will assume that there exists a known smooth (not necessarily small) vector function, which captures the magnitude of the vector of uncertain variables θ for all times. Information of this kind is usually available in practice as a result of experimental data, preliminary simulations, and so forth. Assumption 3.1 states precisely our requirement.

Assumption 3.1 *There exists a known vector* $\bar{\theta}(t) = [\bar{\theta}_1(t) \ \cdots \ \bar{\theta}_q(t)]$ *whose elements are continuous positive definite functions defined on* $t \in [0, \infty)$ *such that:*

$$|\theta_k(t)| \leq \bar{\theta}_k(t), \quad k = 1, \ldots, q. \tag{3.15}$$

Assumption 3.2, which follows, characterizes precisely the class of systems of the form of Eq. 3.1 for which our robust control methodology is applicable. It determines the path under which the vector of uncertain variables $\theta(t)$ may affect the output of the closed-loop system. More specifically, we assume that the vector of uncertain variables $\theta(t)$ enters the system in such a manner such that the expressions for the time derivatives of the output y up to order $\sigma - 1$ are independent of θ. This assumption is significantly weaker than the solvability condition required for the uncertainty decoupling problem (cf. theorem 3.2) and is satisfied by many convection-reaction processes of practical interest (including the plug-flow and fixed-bed reactor examples studied in this chapter).

Assumption 3.2 $\sigma \leq \delta$.

Theorem 3.3 provides the main result of this section (the proof is given in Appendix B).

Theorem 3.3 *Consider the system of Eq. 3.1, for which assumptions 2.1, 2.2, 3.1, and 3.2, and the conditions* (i, ii, iii) *of theorem 3.2 hold under* *the distributed state feedback law of the form:*

$$u = \left[\mathcal{C} L_g \left(\sum_{j=1}^{n} \frac{\partial x_j}{\partial z} L_{a_j} + L_f \right)^{\sigma-1} h(x) b(z) \right]^{-1}$$
$$\times \left\{ \gamma_0 \left[\frac{\gamma_1}{\gamma_\sigma} v - \sum_{\nu=1}^{\sigma-1} \frac{\gamma_\nu}{\gamma_\sigma} \mathcal{C} \left(\sum_{j=1}^{n} \frac{\partial x_j}{\partial z} L_{a_j} + L_f \right)^{\nu-1} h(x) \right] \right.$$
$$- \sum_{\nu=1}^{\sigma-1} \frac{\gamma_\nu}{\gamma_\sigma} \mathcal{C} \left(\sum_{j=1}^{n} \frac{\partial x_j}{\partial z} L_{a_j} + L_f \right)^{\nu} h(x) - 2K(t)\Delta(x, \phi)$$
$$\left. \left(\sum_{\nu=1}^{\sigma-1} \frac{\gamma_\nu}{\gamma_\sigma} \mathcal{C} \left(\sum_{j=1}^{n} \frac{\partial x_j}{\partial z} L_{a_j} + L_f \right)^{\nu-1} h(x) - \frac{\gamma_1}{\gamma_\sigma} v \right) \right\}, \tag{3.16}$$

where ϕ, γ_k are adjustable parameters, and γ_k are chosen so that $\gamma_0 > 0$ and the polynomial $s^{\sigma-1} + \frac{\gamma_{\sigma-1}}{\gamma_\sigma} s^{\sigma-2} + \cdots + \frac{\gamma_2}{\gamma_\sigma} s + \frac{\gamma_1}{\gamma_\sigma} = 0$ is Hurwitz, and $\Delta(x, \phi)$ is an $l \times l$ diagonal matrix whose (i,i)-th element is of the form

$$\Delta(x, \phi) = \left(\left| \sum_{\nu=1}^{\sigma-1} \frac{\gamma_\nu}{\gamma_\sigma} \mathcal{C}^i \left(\sum_{j=1}^{n} \frac{\partial x_j}{\partial z} L_{a_j} + L_f \right)^{\nu-1} h(x) - \frac{\gamma_1}{\gamma_\sigma} v^i \right| + \phi \right)^{-1},$$

and the time-varying gain $K(t)$ is of the form:

$$K(t) = \left[\sum_{k=1}^{q} \mathcal{C} L_{W_k} \left(\sum_{j=1}^{n} \frac{\partial x_j}{\partial z} L_{a_j} + L_f \right)^{\sigma-1} h(x) r_k(z) \bar{\theta}_k(t) \right]. \tag{3.17}$$

Then, for each positive real number d, there exists a pair of positive real numbers $(\tilde{\delta}, \phi^*)$ such that if $\max\{\|x_0\|_2, \|\theta\|, \|\dot{\theta}\|\} \le \tilde{\delta}$ and $\phi \in (0, \phi^*]$, then:

(a) the state of the closed-loop system is bounded, and

(b) the output of the closed-loop system satisfies:

$$\lim_{t \to \infty} |y^i - v^i| \le d, \quad i = 1, \ldots, l. \tag{3.18}$$

Remark 3.6 The calculation of the control action from the controllers of Eqs. 3.10–3.16 requires algebraic manipulations, as well as differentiations and integrations in space, which is expected because of their distributed nature.

Remark 3.7 The robust nonlinear controller of Eq. 3.16 guarantees an arbitrary degree of asymptotic attenuation of the effect of a large class of uncertain *time-varying* variables on the output (those that satisfy assumption 3.2). In many practical applications, there exist unknown time-invariant process parametersthat may not necessarily satisfy assumption 3.2. Since the manipulated input and controlled output spaces are finite dimensional, the asymptotic rejection (in the sense that Eq. 3.18 holds) of such constant uncertainties can be achieved by combining the controller of Eq. 3.16 with an external linear error-feedback controller with integral action.

Remark 3.8 Comparing the uncertainty decoupling result of theorem 3.2 with the result of theorem 3.3 (Eq. 3.18), we note that the latter is clearly applicable to a larger class of systems (those satisfying the condition $\sigma \leq \delta$), achieving, however, a weaker performance requirement. More specifically, in the uncertainty decoupling problem a well-characterized input/output response is enforced for all times independently of $\theta(t)$, while in the case of robust uncertainty rejection the output error $(y^i - v^i)$ is ensured to stay bounded and asymptotically approach arbitrarily close to zero (by appropriate choice of the parameter ϕ).

Remark 3.9 (Illustrative example (cont'd)) Whenever $(c(z), r_1(z)) \neq 0$ (which implies that $\delta = \sigma = 1$), the uncertainty decoupling problem via distributed state feedback is not solvable. However, since $\delta = \sigma$, we have that the matching condition of our robust control methodology is satisfied. Assuming that there exists a known upper bound $\bar{\theta}(t)$ on the size of the uncertain term $\theta(t)$, such that $|\theta| \leq \bar{\theta}(t)$, the control law of theorem 3.3 takes the form:

$$
\begin{aligned}
u = & \left[\frac{U_w}{\rho_f c_{pf} V_r}\int_0^1 c(z)b(z)dz\right]^{-1}\left\{\gamma_0\left(v - \int_0^1 x_2 dz\right)\right. \\
& - \int_0^1 c(z)\left(-v_l\frac{\partial x_2}{\partial z} + \frac{(-\Delta H)}{\rho_f c_{pf}}kx_1 + \frac{U_w}{\rho_f c_{pf} V_r}(T_{js} - x_2)\right. \\
& \left.\left. + F(T_{ls} - x_2)\right)dz - 2\bar{\theta}(t)\frac{\int_0^1 x_2(z,t)dz - v}{\left|\int_0^1 x_2(z,t)dz - v\right| + \phi}\right\}.
\end{aligned} \tag{3.19}
$$

3.5 Two-Time-Scale Hyperbolic PDE Systems

In the rest of the theoretical part of this chapter, we will analyze the problem of robustness of the controllers of Eqs. 3.10–3.16 with respect to stable unmodeled dynamics. This problem will be addressed within the broader context of control of two-time-scale hyperbolic PDE systems, modeled within the framework of singular perturbations. Such a formulation provides a natural setting for addressing robustness with respect to unmodeled dynamics. To this end, in this section we will derive two general stability results for two-time-scale hyperbolic PDE systems, which will be used in the next section to derive conditions that guarantee robustness of the controllers of Eqs. 3.10–3.16 to stable unmodeled dynamics.

3.5.1 Two-time-scale decomposition

We will focus on singularly perturbed hyperbolic PDE systems with the following state-space description:

$$\begin{aligned}\frac{\partial x}{\partial t} &= A_{11}\frac{\partial x}{\partial z} + A_{12}\frac{\partial \eta}{\partial z} + f_1(x) + Q_1(x)\eta + g_1(x)b(z)u + W_1(x)r(z)\theta(t)\\ \epsilon\frac{\partial \eta}{\partial t} &= A_{21}\frac{\partial x}{\partial z} + A_{22}\frac{\partial \eta}{\partial z} + f_2(x) + Q_2(x)\eta + g_2(x)b(z)u + W_2(x)r(z)\theta(t)\\ y &= \mathcal{C}h(x)\end{aligned} \tag{3.20}$$

subject to the boundary conditions:

$$\begin{aligned}C_{11}x(\alpha,t) + C_{12}x(\beta,t) &= R_1(t)\\ C_{21}\eta(\alpha,t) + C_{22}\eta(\beta,t) &= R_2(t)\end{aligned} \tag{3.21}$$

and the initial conditions:

$$\begin{aligned}x(z,0) &= x_0(z)\\ \eta(z,0) &= \eta_0(z)\end{aligned} \tag{3.22}$$

where $x(z,t) = [x_1(z,t) \cdots x_n(z,t)]^T$, $\eta(z,t) = [\eta_1(z,t) \cdots \eta_p(z,t)]^T$, denote vectors of state variables, $x(z,t) \in \mathcal{H}^{(n)}[(\alpha,\beta),\mathbb{R}^n]$, $\eta(z,t) \in \mathcal{H}^{(n)}$ $[(\alpha,\beta),\mathbb{R}^p]$, with $\mathcal{H}^{(i)}[(\alpha,\beta),\mathbb{R}^j]$ being the infinite-dimensional Hilbert space of j-dimensional vector functions defined on the interval $[\alpha,\beta]$ whose spatial derivatives up to i-th order are square integrable, A_{11}, A_{12}, A_{21}, A_{22} are constant matrices, $f_1(x)$, $f_2(x)$, $g_1(x)$, $g_2(x)$ are sufficiently smooth vector functions, $Q_1(x)$, $Q_2(x)$, $W_1(x)$, $W_2(x)$ are sufficiently smooth matrices, ϵ is a small parameter that quantifies the degree of coupling between the fast and slow modes of the system, and $R_1(t)$, $R_2(t)$ are time-varying column vectors. The following assumption states that the two-time-scale system of Eq. 3.20 is hyperbolic.

Assumption 3.3 *The* $(n+p)\times(n+p)$ *matrix*

$$\mathcal{A} = \begin{bmatrix} A_{11} & A_{12}\\ A_{21} & A_{22}\end{bmatrix} \tag{3.23}$$

is real symmetric, and its eigenvalues satisfy

$$\bar{\lambda}_1 \le \cdots \le \bar{\lambda}_k < 0 < \bar{\lambda}_{k+1} \le \cdots \le \bar{\lambda}_{n+p}. \tag{3.24}$$

The time-scale multiplicity of the system of Eq. 3.20 can be explicitly taken into account by decomposing it into separate reduced-order models associated with different time scales. Defining $\mathcal{L}_{i1}x = A_1\frac{\partial x}{\partial z} + f_i(x)$, $\mathcal{L}_{j2}\eta = A_{j2}\frac{\partial \eta}{\partial z} + Q_j(x)\eta$, $i,\ j = 1,2$, and setting $\epsilon = 0$, the PDE system of Eq. 3.20, reduces to a system of coupled partial and ordinary differential equations of the form:

$$\begin{aligned}\frac{\partial x}{\partial t} &= \mathcal{L}_{11}x + \mathcal{L}_{12}\eta_s + g_1(x)b(z)u + W_1(x)r(z)\theta(t)\\ 0 &= \mathcal{L}_{21}x + \mathcal{L}_{22}\eta_s + g_2(x)b(z)u + W_2(x)r(z)\theta(t)\\ y_s &= \mathcal{C}h(x)\end{aligned} \tag{3.25}$$

where the subscript s in y_s denotes that this output is associated with a slow subsystem. The solution of the ODE $\mathcal{L}_{21}x + \mathcal{L}_{22}\eta_s + g_2(x)b(z)u + W_2(x)r(z)\theta(t) = 0$ subject to the boundary conditions of Eq. 3.21 is of the form

$$\eta_s = \mathcal{L}_{22}^{-1}(\mathcal{L}_{21}x + g_2(x)b(z)u + W_2(x)r(z)\theta(t)). \tag{3.26}$$

The slow subsystem, which captures the slow dynamics of the system of Eq. 3.21, takes the form:

$$\begin{aligned}\frac{\partial x}{\partial t} &= \mathcal{L}x + G(x)b(z)u + W(x)r(z)\theta(t)\\ y_s &= \mathcal{C}h(x)\end{aligned} \tag{3.27}$$

where $\mathcal{L}x = \mathcal{L}_{11}x - \mathcal{L}_{12}\mathcal{L}_{22}^{-1}\mathcal{L}_{21}x$, $G(x) = g_1(x) - \mathcal{L}_{12}\mathcal{L}_{22}^{-1}g_2(x)$, $W(x) = W_1(x) - \mathcal{L}_{12}\mathcal{L}_{22}^{-1}W_2(x)$. Defining a fast time-scale $\tau = \frac{t}{\epsilon}$ and setting $\epsilon = 0$, the fast subsystem, which describes fast dynamics of the system of Eq. 3.20, takes the form:

$$\frac{\partial \eta}{\partial \tau} = \mathcal{L}_{21}x + \mathcal{L}_{22}\eta + g_2(x)b(z)u + W_2(x)r(z)\theta(t) \tag{3.28}$$

where x, θ can be thought of as constants.

3.5.2 Stability results

In this subsection, we will give two basic stability results for systems of the form of Eq. 3.20 which are important in their own right and will also be used in the subsequent sections to prove a robustness property of the controllers of Eqs. 3.10–3.16 with respect to exponentially stable unmodeled dynamics. We will begin with the statement of stability requirements on the fast and slow subsystems. In particular, we assume that the fast subsystem of Eq. 3.28 and the unforced ($u(t) \equiv \theta(t) \equiv 0$) slow subsystem of Eq. 3.27 are locally exponentially stable. Assumptions 3.4 and 3.5 formalize these requirements.

Assumption 3.4 *The differential operator $\mathcal{L}_{22}\eta$ generates an exponentially stable, strongly continuous semigroup $U_f(t)$ that satisfies:*

$$\| U_f(t) \|_2 \leq K_F e^{-a_f \frac{t}{\epsilon}}, \quad t \geq 0 \tag{3.29}$$

where $K_f \geq 1$ and a_f is some positive real number.

Assumption 3.5 *The differential operator $\mathcal{L}x$ generates a locally exponentially stable (i.e., for $x \in \mathcal{H}$ that satisfy $\|x\|_2 \leq \delta_x$, where δ_x is a positive real number), strongly continuous semigroup $U_s(t)$ that satisfies:*

$$\| U_s(t)\|_2 \leq K_s e^{-a_s t}, \quad t \geq 0 \tag{3.30}$$

where $K_s \geq 1$ and a_s is some positive real number.

From the above assumption and using Eq. 2.15, the smoothness of $G(x)$ and $W(x)$, and the fact that for $\|x\|_2 \leq \delta_x$ there exists a pair of positive real numbers $\bar{M}_1$, $\bar{M}_2$ such that $\| G(x)\|_2 \leq \bar{M}_1$ and $\| W(x)\|_2 \leq \bar{M}_2$, we have that the following estimate holds for the system of Eq. 3.27:

$$\begin{aligned}\|x\|_2 \leq\; & K_s\|x_0\|_2 e^{-a_s t} + K_s \int_0^t e^{-a_s(t-\tau)}\| G(x)b(z)\|_2|u(\tau)|d\tau \\ & + K_s \int_0^t e^{-a_s(t-\tau)}\| W(x)r(z)\|_2|\theta(\tau)|d\tau \\ \leq\; & K_s\|x_0\|_2 e^{-a_{st}} + M_1\|u\| + M_2\|\theta\|\end{aligned} \tag{3.31}$$

where $M_1 = \frac{K_s \bar{M}_1\|b(z)\|_2}{a_s}$, $M_2 = \frac{K_s \bar{M}_2\|r(z)\|_2}{a_s}$, provided that the initial condition ($\|x_0\|_2$) and the inputs (u, θ) are sufficiently small to satisfy $K_s\|x_0\|_2 + M_1\|u\| + M_2\|\theta\| \leq \delta_x$. Theorem 3.4 provides the main stability result of this section (the proof is given in Appendix B).

Theorem 3.4 *Consider the system of Eq. 3.20 with $u \equiv 0$, for which assumptions 3.5, 3.6, and 3.7 hold, and define $\eta_f := \eta - \eta_s$. Then, there exist positive real numbers $(\bar{K}_s, \bar{a}_s, \bar{K}_f, \bar{a}_f)$, such that for each positive real number d, there exist positive real numbers $(\tilde{\delta}, \epsilon^*)$ such that if* $\max\{\|x_0\|_2, \|\eta_{f_0}\|_2, \|\theta\|, \|\dot{\theta}\|\} \leq \tilde{\delta}$ *and $\epsilon \in (0, \epsilon^*]$, then, for all $t \geq 0$:*

$$\|x\|_2 \leq \bar{K}_s\|x_0\|_2 e^{-\bar{a}_s t} + M_2\|\theta\| + d \tag{3.32}$$

$$\|\eta_f\|_2 \leq K_f\|\eta_{f_0}\|_2 e^{-\bar{a}_f \frac{t}{\epsilon}} + d. \tag{3.33}$$

Theorem 3.4 establishes a robustness result of the boundedness property that the slow subsystem of Eq. 3.27 possesses, with respect to exponentially stable singular perturbations, provided that they are sufficiently fast. Theorem 3.5 establishes another conceptually important result, namely that the exponential stability property of a reduced system (consider the system of Eq. 3.27 with $|u(t)| \equiv |\theta(t)| \equiv 0$) is preserved in the presence of exponentially stable singular perturbations. The proof of this theorem in analogous to the one of theorem 3.4 and will be omitted for brevity.

Theorem 3.5 *Consider the system of Eq. 3.20 with $|u| = |\theta(t)| \equiv 0$, for which assumptions 3.3, 3.4, and 3.5 hold, and define $\eta_f := \eta - \eta_s$. Then, there exist positive real numbers $(\tilde{\delta}, \epsilon^*)$ such that if* $\max\{\|x_0\|_2, \|\eta_{f_0}\|_2\} \leq \tilde{\delta}$ *and $\epsilon \in (0, \epsilon^*]$, the system of Eq. 3.20 is exponentially stable (i.e., Eqs. 3.32–3.33 hold with $d = 0$, $\theta(t) \equiv 0$).*

3.6 Robustness with Respect to Unmodeled Dynamics

In this section, we utilize the general stability results of the previous section to establish robustness properties of the controllers of Eqs. 3.10–3.16 with respect to unmodeled dynamics. In particular, we use the hyperbolic singularly perturbed system of Eq. 3.20 to represent the actual process model, and we assume that the model available for controller design is the slow system of Eq. 3.27. Our objective is to establish that the controllers of Eqs. 3.10–3.16, synthesized on the basis of the slow system, are robust to unmodeled dynamics in the sense that they enforce approximate uncertainty decoupling and robust disturbance rejection in the actual model of Eq. 3.20, provided that the separation of the fast and slow modes is sufficiently large.

3.6.1 Robustness of uncertainty decoupling to unmodeled dynamics

In this subsection, we consider the uncertainty decoupling problem for the two-time-scale hyperbolic system of Eq. 3.20 whose fast dynamics are stable (assumption 3.4). Substitution of the distributed state feedback law of Eq. 3.7 into the system of Eq. 3.20 yields:

$$
\begin{aligned}
\frac{\partial x}{\partial t} &= \mathcal{L}_{11}x + \mathcal{L}_{12}\eta + g_1(x)b(z)\mathcal{S}(x) + g_1(x)b(z)s(x)v + W_1(x)r(z)\theta(t)\\
\epsilon\frac{\partial \eta}{\partial t} &= \mathcal{L}_{21}x + \mathcal{L}_{22}\eta + g_2(x)b(z)\mathcal{S}(x) + g_2(x)b(z)s(x)v + W_2(x)r(z)\theta(t)\\
y &= \mathcal{C}h(x).
\end{aligned}
\tag{3.34}
$$

From the above representation of the closed-loop system, it is clear that the control law of Eq. 3.7 preserves the two-time-scale property of the open-loop system and guarantees that the system of Eq. 3.34 has a well-defined solution. Performing a two-time-scale decomposition on the system of Eq. 3.34, the closed-loop fast subsystem takes the form:

$$
\frac{\partial \eta}{\partial \tau} = \mathcal{L}_{21}x + \mathcal{L}_{22}\eta + g_2(x)b(z)\mathcal{S}(x) + g_2(x)b(z)s(x)v + W_2(x)r(z)\theta(t)
\tag{3.35}
$$

where x, θ can be thought of as constants. From the representation of the closed-loop fast subsystem it is clear that the control law of Eq. 3.34 preserves the exponential stability property of the fast dynamics of the closed-loop system. The closed-loop slow system takes the form:

$$
\begin{aligned}
\frac{\partial x}{\partial t} &= \mathcal{L}x + G(x)b(z)\mathcal{S}(x) + G(x)b(z)s(x)v + W(x)r(z)\theta(t)\\
y_s &= \mathcal{C}h(x).
\end{aligned}
\tag{3.36}
$$

For the system of Eq. 3.36 it is straightforward to show that the result of proposition 3.1 holds, which suggests requesting an input/output response of the form of Eq. 3.9 in the closed-loop reduced system.

Theorem 3.6 establishes a robustness property of the solvability condition of the uncertainty decoupling problem ($\sigma < \delta$) and the input/output response of Eq. 3.9 with respect to stable, sufficiently fast, unmodeled dynamics (the proof of the theorem is given in Appendix B).

Theorem 3.6 *Consider the system of Eq. 3.20 for which assumptions 3.3 and 3.4 hold, under the controller of Eq. 3.10. Consider also the slow subsystem of Eq. 3.27 and assume that the condition $\sigma < \delta$ and the stability conditions of theorem 3.2 hold. Then, for each positive real number d, there exist positive real numbers $(\tilde{\delta}, \epsilon^*)$ such that if $\epsilon \in (0, \epsilon^*]$ and $\max\{\|x_0\|_2, \|\eta_{f_0}\|_2, \|\theta\|, \|\dot{\theta}\|\} \leq \tilde{\delta}$, then:*

(a) the state of the closed-loop system is bounded, and

(b) the output of the closed-loop system satisfies for all $t \geq 0$:

$$|y^i(t) - y_s^i(t)| \leq d, \quad i = 1, \ldots, l \tag{3.37}$$

where $y_s^i(t)$ are the solutions of Eq. 3.9.

Remark 3.10 Referring to the above theorem, note that we do not impose any assumptions on the way the uncertain variables θ enter the actual process model of Eq. 3.20 (i.e., the condition $\sigma < \delta$ has to be satisfied only in the slow subsystem of Eq. 3.27). This is achieved at the expense of the controller of Eq. 3.10, which is synthesized on the basis of the slow subsystem of Eq. 3.27, enforcing approximate uncertainty decoupling in the actual process model. The approximate uncertainty decoupling is taken in in the sense that the discrepancy between the output of the actual closed-loop system and the output of the closed-loop reduced system (which is independent of θ) can be made smaller than a given positive number d (possibly small) for all times, provided that ϵ is sufficiently small (Eq. 3.37).

3.6.2 Robust control: uncertain variables and unmodeled dynamics

In this subsection, we consider the robust uncertainty rejection problem for two-time-scale hyperbolic systems of the form of Eq. 3.20 with stable fast dynamics. Under a control law of the form of Eq. 3.14, the closed-loop system takes the form:

$$\begin{aligned}
\frac{\partial x}{\partial t} &= \mathcal{L}_{11}x + \mathcal{L}_{12}\eta + g_1(x)b(z)[\bar{\mathcal{S}}(x) + \bar{s}(x)v + \bar{\mathcal{R}}(x,t)] + W_1(x)r(z)\theta(t) \\
\epsilon\frac{\partial \eta}{\partial t} &= \mathcal{L}_{21}x + \mathcal{L}_{22}\eta + g_2(x)b(z)[\bar{\mathcal{S}}(x) + \bar{s}(x)v + \bar{\mathcal{R}}(x,t)] + W_2(x)r(z)\theta(t) \\
y &= \mathcal{C}h(x).
\end{aligned} \tag{3.38}$$

The fast dynamics of the above system are exponentially stable and the reduced system takes the form:

$$\begin{aligned}\frac{\partial x}{\partial t} &= \mathcal{L}x + G(x)b(z)[\bar{\mathcal{S}}(x) + \bar{s}(x)v + \bar{\mathcal{R}}(x,t)] + W(x)r(z)\theta(t) \\ y_s &= \mathcal{C}h(x).\end{aligned} \tag{3.39}$$

Theorem 3.7 that follows establishes a robustness property of the controller of Eq. 3.16 to sufficiently fast unmodeled dynamics (the proof is given in Appendix B).

Theorem 3.7 *Consider the system of Eq. 3.20 for which assumptions 3.3 and 3.4 hold, under the controller of Eq. 3.16. Consider also the slow subsystem of Eq. 3.27 and suppose that assumption 3.2 and the stability conditions of theorem 3.2 hold. Then, for each positive real number d, there exist positive real numbers* $(\tilde{\delta}, \phi^*)$ *such that if* $\max\{\|x_0\|_2, \|\eta_{f0}\|_2, \|\theta\|, \|\dot{\theta}\|\} \leq \tilde{\delta}$, $\phi \in (0, \phi^*]$, *there exists a positive real number* $\epsilon^*(\phi)$ *such that if* $\max\{\|x_0\|_2, \|\eta_{f0}\|_2, \|\theta\|, \|\dot{\theta}\|\} \leq \tilde{\delta}$, $\phi \in (0, \phi^*]$, *and* $\epsilon \in (0, \epsilon^*(\phi)]$, *then:*

(a) the state of the closed-loop system is bounded, and

(b) the output of the closed-loop system satisfies:

$$\lim_{t \to \infty} |y^i - v^i| \leq d, \quad i = 1, \ldots, l. \tag{3.40}$$

Remark 3.11 Regarding the result of the above theorem, a few observations are in order: (i) no matching condition is imposed in the actual process model of Eq. 3.20 (instead it is imposed on the slow subsystem of Eq. 3.27), (ii) the dependence of ϵ^* on ϕ is due to the presence of ϕ in the closed-loop fast subsystem, and (iii) a bound for the output error, for all times (and not only asymptotically as in Eq. 3.40) can be obtained from the proof of the theorem in terms of the initial conditions and d.

Remark 3.12 Whenever the open-loop fast subsystem of Eq. 3.28 is unstable, that is, the operator $\mathcal{L}_{22}\eta$ generates an exponentially unstable semigroup, a preliminary distributed state feedback law of the form:

$$u = \mathcal{F}\eta + \tilde{u} \tag{3.41}$$

can be used to stabilize the fast dynamics, under the assumption that the pair $[\mathcal{L}_{22}\ g_2(x)b(z)]$ is stabilizable, thus yielding a two-time-scale hyperbolic system for which assumption 3.7 holds. The design of the gain operator $\mathcal{F}$ can be performed using for example standard optimal control methods [14] (see also [42] for results on nonlinear distributed output feedback control of two-time-scale hyperbolic PDE systems, and [128] for the design of well-conditioned state estimators for two-time-scale parabolic PDE systems).

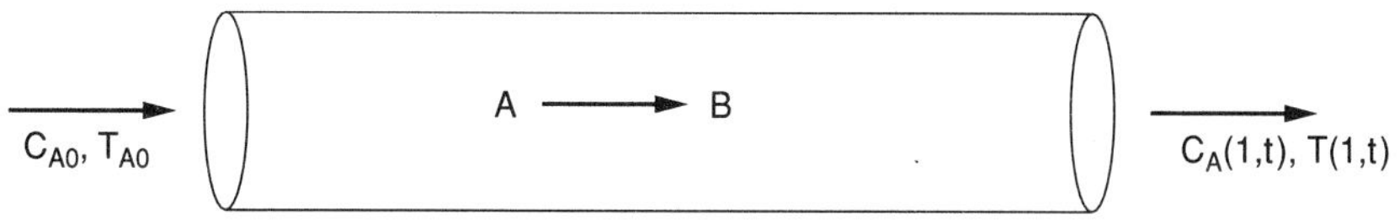

FIGURE 3.2. A nonisothermal fixed-bed reactor.

3.7 Application to a Fixed-Bed Reactor

Consider a fixed-bed reactor where an elementary reaction of the form $A \rightarrow B$ takes place, shown in Figure 3.2. The reaction is endothermic and a jacket is used to heat the reactor. Under the assumptions of perfect radial mixing, constant density and heat capacity of the reacting liquid, and negligible diffusive phenomena, a dynamic model of the process can be derived from material and energy balances and has the form:

$$\begin{aligned} \rho_b c_{pb} \frac{\partial T}{\partial t} &= -\rho_f c_{pf} v_l \frac{\partial T}{\partial z} + (-\Delta H) k_0 e^{-\frac{E}{RT}} C_A + \frac{U_w}{V_r}(T_j - T) \\ \bar{\epsilon} \frac{\partial C_A}{\partial t} &= -v_l \frac{\partial C_A}{\partial z} - k_0 e^{-\frac{E}{RT}} C_A \end{aligned} \tag{3.42}$$

subject to the initial and boundary conditions:

$$\begin{aligned} C_A(0,t) = C_{A0}, \quad & C_A(z,0) = C_{A_s}(z) \\ T(0,t) = T_{A0}, \quad & T(z,0) = T_s(z) \end{aligned}$$

where C_A denotes the concentration of the species A; T denotes the temperature in the reactor; $\bar{\epsilon}$ denotes the reactor porosity; ρ_b, c_{pb} denote the density and heat capacity of the bed; ρ_f, c_{pf} denote the density and heat capacity of the fluid phase; v_l denotes the velocity of the fluid phase; U_w denotes the heat transfer coefficient; T_j denotes the spatially uniform temperature in the jacket; V_r denotes the volume of the reactor; k_0, E, ΔH denote the pre-exponential factor, the activation energy, and the enthalpy of the reaction; C_{A0}, T_{A0} denote the concentration and temperature of the inlet stream; and $C_{A_s}(z)$, $T_s(z)$ denote the steady-state profiles for the concentration of species A and temperature in the reactor.

The main feature of fixed-bed reactors is that the reactant wave propagates through the bed with a significantly larger speed than the heat wave because the exchange of heat between the fluid and packing slows the thermal wave down [130]. Therefore, the system of Eq. 3.42 possesses an inherent two-time-scale property, that is, the concentration dynamics are much faster than the temperature dynamics (this fact was also verified through open-loop simulations). This implies that C_A is the fast variable, while T is the slow variable. In order to obtain a singularly perturbed representation of the process, where the partition to fast and slow variables is consistent with the dynamic characteristics of the process, the singular

TABLE 3.1. Parameters for fixed-bed reactor.

v_l	=	30.0	$m\ hr^{-1}$
V_r	=	1.0	m^3
$\bar{\epsilon}$	=	0.01	
L	=	1.0	m
E	=	2.0×10^4	$kcal\ kmol^{-1}$
k_0	=	5.0×10^{12}	hr^{-1}
R	=	1.987	$kcal\ kmol^{-1}\ K^{-1}$
ΔH_0	=	35480.111	$kcal\ kmol^{-1}$
c_{pf}	=	0.0231	$kcal\ kg^{-1}\ K^{-1}$
ρ_f	=	90.0	$kg\ m^{-3}$
c_{pb}	=	6.67×10^{-4}	$kcal\ kg^{-1}\ K^{-1}$
ρ_b	=	1500.0	$kg\ m^{-3}$
U_w	=	500.0	$kcal\ hr^{-1}\ K^{-1}$
C_{A0}	=	4.0	$kmol\ m^{-3}$
T_{A0}	=	320.0	K

perturbation parameter ϵ is defined as:

$$\epsilon = \frac{\bar{\epsilon}}{\rho_b c_{pb}}. \tag{3.43}$$

Setting $\bar{t} = \frac{t}{\rho_b c_{pb}}$, $x = T$, $\eta = C_A$, the system of Eq. 3.42 can be written in the following singularly perturbed form:

$$\begin{aligned} \frac{\partial x}{\partial \bar{t}} &= -\rho_f c_{pf} v \frac{\partial x}{\partial z} + (-\Delta H) k_0 e^{-\frac{E}{Rx}} + \frac{U_w}{V_r}(T_j - x) \\ \epsilon \frac{\partial \eta}{\partial \bar{t}} &= -v_l \frac{\partial \eta}{\partial z} - k_0 e^{-\frac{E}{Rx}}. \end{aligned} \tag{3.44}$$

The values of the process parameters are given in Table 3.1. The above values correspond to a stable steady state for the open-loop system. The control problem considered is the one of controlling the temperature of the reactor (which is the variable that essentially determines the dynamics of the process) by manipulating the jacket temperature. Notice that T_j is chosen to be the manipulated input with the understanding that in practice its manipulation is achieved indirectly through manipulation of the jacket inlet flow rate (for details on this issue see the discussion in subsection 2.7.4). The enthalpy of the reaction is considered to be the main uncertain variable. Assuming that there is available one control actuator with distribution function $b(z) = 1$ and defining $u = T_j - T_{js}$, $y = \int_0^1 x\, dz$, $\theta = \Delta H - \Delta H_0$, we have from Eq. 3.44:

$$\begin{aligned} \frac{\partial x}{\partial \bar{t}} &= -\rho_f c_{pf} v_l \frac{\partial x}{\partial z} + (-\Delta H_0) k_0 e^{-\frac{E}{Rx}} \eta + \frac{U_w}{V_r}(T_{js} - x) \\ &\quad + \frac{U_w}{V_r} u + k_0 e^{-\frac{E}{Rx}} \eta \theta \end{aligned}$$

$$\begin{aligned}\epsilon \frac{\partial \eta}{\partial \bar{t}} &= -v_l \frac{\partial \eta}{\partial z} - k_0 e^{-\frac{E}{Rx}} \eta \\ y &= \int_0^1 x \, dz.\end{aligned} \tag{3.45}$$

Performing a two-time-scale decomposition of the above system, the fast subsystem takes the form:

$$\frac{\partial \eta}{\partial \bar{\tau}} = -v_l \frac{\partial \eta}{\partial z} - k_0 e^{-\frac{E}{Rx}} \eta \tag{3.46}$$

where $\bar{\tau} = \frac{\bar{t}}{\epsilon}$ and x in the above system depends only on the position z. From the system of Eq. 3.46, it is clear that the fast dynamics of the system of Eq. 3.45 are exponentially stable uniformly in x, and thus they can be neglected in the controller design. Setting $\epsilon = 0$, the model of Eq. 3.45 reduces to a set of a partial and an ordinary differential equation of the form:

$$\begin{aligned}\frac{\partial x}{\partial \bar{t}} &= -\rho_f c_{pf} v_l \frac{\partial x}{\partial z} + (-\Delta H_0) k_0 e^{-\frac{E}{Rx}} \eta \\ &+ \frac{U_w}{V_r}(T_{js} - x) + \frac{U_w}{V_r} u + k_0 e^{-\frac{E}{Rx}} \eta \theta \\ 0 &= -v_l \frac{d\eta}{dz} - k_0 e^{-\frac{E}{Rx}} \eta.\end{aligned} \tag{3.47}$$

The structure of the ordinary differential equation allows an analytic derivation of its solution subject to the boundary condition $\eta(0,t) = C_{A0}(t)$, which is of the form:

$$\eta(z) = C_{A0} exp\left(\frac{k_0 \int_0^z e^{-\frac{E}{Rx}} dz}{v_l} \right) \tag{3.48}$$

Substituting Eq. 3.48 into Eq. 3.47, the following reduced system can be obtained:

$$\begin{aligned}\frac{\partial x}{\partial \bar{t}} &= -\rho_f c_{pf} v_l \frac{\partial x}{\partial z} + (-\Delta H_0) k_0 e^{-\frac{E}{Rx}} C_{A0} exp\left(\frac{k_0 \int_0^z e^{-\frac{E}{Rx}} dz}{v_l} \right) \\ &+ \frac{U_w}{V_r}(T_{js} - x) + \frac{U_w}{V_r} u + k_0 e^{-\frac{E}{Rx}} C_{A0} exp\left(\frac{k_0 \int_0^z e^{-\frac{E}{Rx}} dz}{v_l} \theta \right) \\ y_s &= \int_0^1 x \, dz.\end{aligned} \tag{3.49}$$

The above system is clearly in the form of Eq. 3.1, and a straightforward calculation of the time derivative of the output yields that the characteristic indices of the output y with respect to the manipulated input u and the uncertain variable θ are $\sigma = \delta = 1$. This implies that it is not possible to decouple the effect of θ on y via distributed state feedback, and a robust controller of the form of Eq. 3.16 should be synthesized to attenuate the effect of θ on y (this is possible because the matching condition, assumption 3.4, holds for the system of Eq. 3.49). Moreover, the zero dynamics of the system of Eq. 3.49 are locally exponentially stable as verified through simulations. The explicit form of the controller of Eq. 3.16 is:

$$u = \frac{V_r}{U_w}\left\{ \gamma_0\left(v - \int_0^1 x\,dz\right) - \int_0^1 \left(-\rho_f c_{pf} v_l \frac{\partial x}{\partial z} + (-\Delta H_0) k_0 e^{-\frac{E}{Rx}} \right.\right.$$
$$\left. \times C_{A0}\, exp\left(\frac{k_0 \int_0^z e^{-\frac{E}{Rx}}\,dz}{v_l} \right) + \frac{U_w}{V_r}(T_{js} - x) \right) dz - K(t)$$
$$\left. \times \frac{\int_0^1 x(z,t)\,dz - v}{\left| \int_0^1 x(z,t)\,dz - v \right| + \phi} \right\} \tag{3.50}$$

where

$$K(t) = 2\bar{\theta} \int_0^1 \frac{k_0 V_r}{U_w} e^{-\frac{E}{Rx}} C_{A0}\, exp\left(\frac{k_0 \int_0^z e^{-\frac{E}{Rx}}\,dz}{v_l} \right) dz.$$

A time-varying uncertainty is considered and expressed by a sinusoidal function of the form:

$$\theta = 0.5(-\Delta H_0) sin(\bar{t}). \tag{3.51}$$

The upper bound on the uncertainty is taken to be $\bar{\theta} = 0.5|(-\Delta H_0)|$. In this application, the value of the singular perturbation parameter is fixed that is, $\epsilon = 0.01$. From theorem 3.6, it is clear that for a given value of ϵ^*, there exists a lower bound on the level of asymptotic attenuation d that can be achieved. We perform a set of computer simulations (for the regulation problem) to calculate ϕ^* for certain values of d, and, in turn, the value of ϵ^* for $\phi \leq \phi^*$. The following set of parameters is found to give an $\epsilon^* \leq 0.01$ and is used in the simulations:

$$\gamma_0 = 0.2, \quad \phi = 0.1 \tag{3.52}$$

to achieve an ultimate degree of attenuation $d = 0.5$.

Two simulation runs are performed to test the regulatory, set-point tracking and uncertainty rejection capabilities of the controller. In both runs,

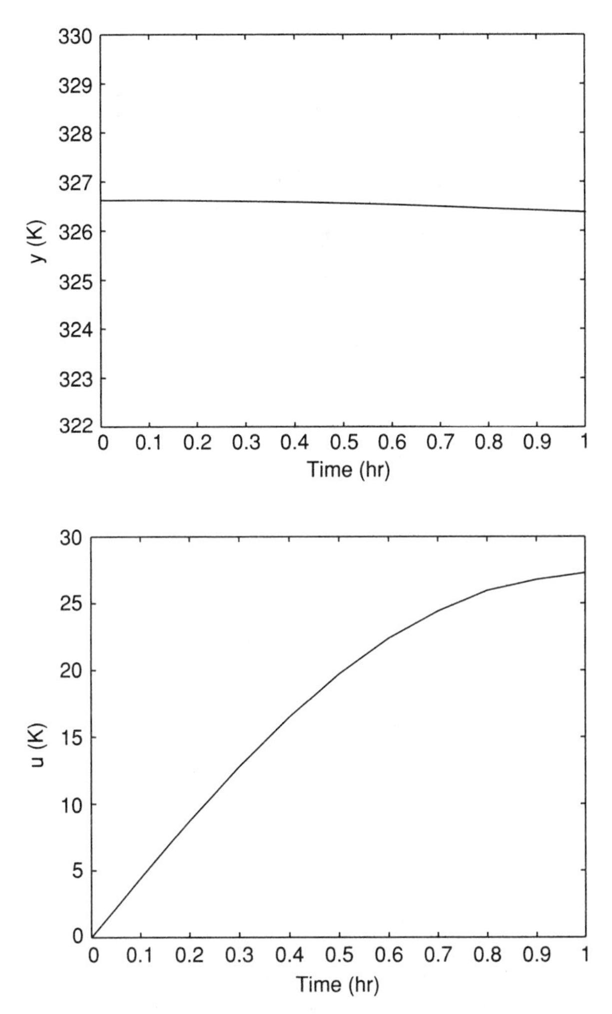

FIGURE 3.3. Closed-loop output and manipulated input profiles for regulation.

the process is initially ($t = 0.0\ hr$) assumed to be at steady state. In the first simulation run, we test the regulatory capabilities of the controller. Figure 3.3 shows the closed-loop output and manipulated input profiles, while Figure 3.4 displays the evolution of the temperature profile throughout the reactor. Clearly the controller regulates the output at the operating steady-state compensating for the effect of uncertainty and satisfying the requirement $\lim_{t\to\infty} |y - v| \leq 0.5\ K$. Note that the manipulated input changes with time to compensate for the effect of the time-varying uncertain variable. For the sake of comparison, we also implement the same controller without the term responsible for the compensation of uncertainty.

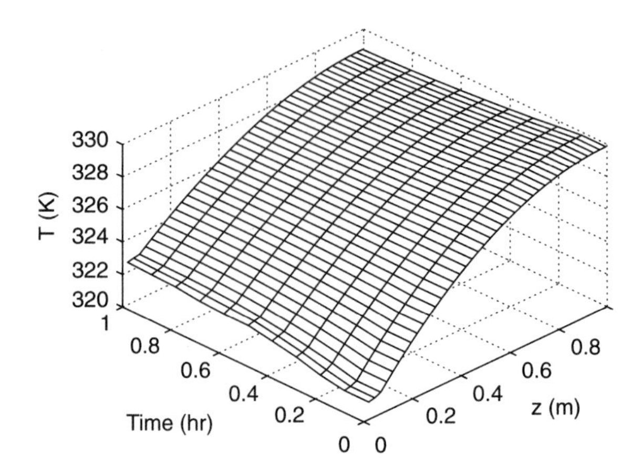

FIGURE 3.4. Profile of evolution of reactor temperature for regulation.

The output and manipulated input profiles for this simulation run are given in Figure 3.5. It is obvious that this controller cannot attenuate the effect of the uncertainty on the output of the process, leading to poor transient performance and offset. In the next simulation run, we test the output tracking capabilities of the controller. A $6.6K$ increase in the value of the reference input is imposed at time $t = 0.0\ hr$. The output and manipulated input profiles are shown in Figure 3.6. It is clear that the controller drives the output y to its new reference input value, achieving the requirement $\lim_{t\to\infty} |y - v| \leq 0.5$. From the results of the simulation study, we conclude that there is a need to compensate for the effect of uncertainty and that the proposed robust control methodology is a very efficient tool for this purpose.

Remark 3.13 The reduction in the dimensionality of the original model of the reactor using the fact that the process exhibits a two-time-scale property eliminates the need for using measurements of the concentration of the species A in the controller of Eq. 3.50, which greatly facilitates its practical implementation.

3.8 Conclusions

In this chapter, we considered systems of first-order hyperbolic PDEs with uncertainty for which the manipulated variable and the controlled variable are distributed in space. Both uncertain variables and unmodeled dynamics were considered. In the case of uncertain variables, we initially derived a necessary and sufficient condition for the solvability of the problem of complete elimination of the effect of uncertainty on the output via distributed feedback as well as an explicit controller synthesis formula. Then,

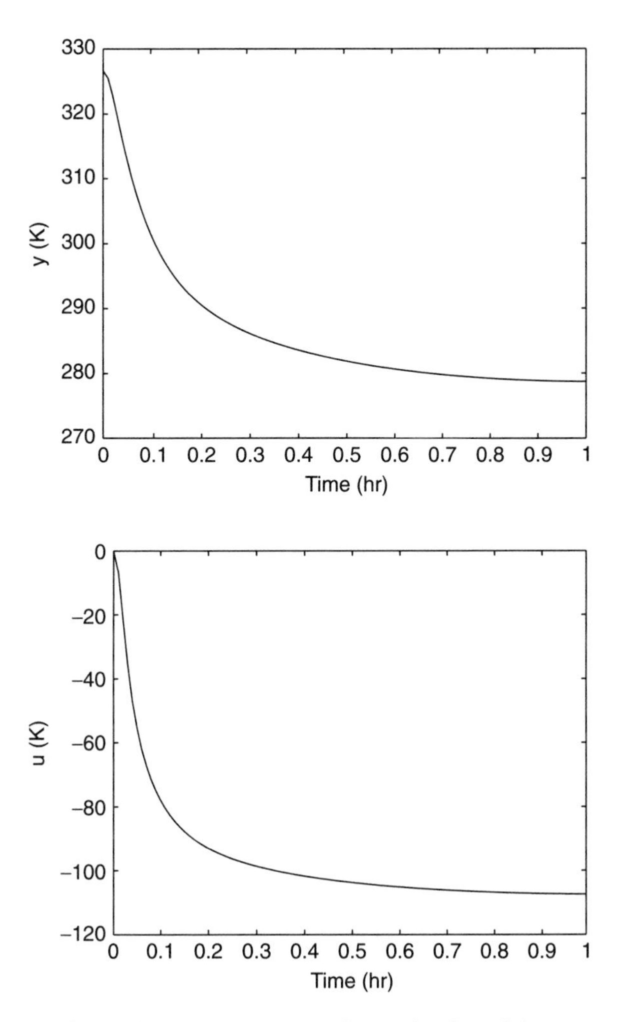

FIGURE 3.5. Closed-loop output and manipulated input profiles for regulation (no uncertainty compensation).

assuming that there exist known bounding functions that capture the magnitude of the uncertain terms and a matching condition is satisfied, we synthesized, using Lyapunov's direct method, a distributed robust controller that guarantees boundedness of the state and asymptotic output tracking with arbitrary degree of asymptotic attenuation of the effect of uncertain variables on the output of the closed-loop system. In the presence of uncertain variables and unmodeled dynamics, we established that the proposed distributed controllers enforce approximate uncertainty decoupling or uncertainty attenuation in the closed-loop system as long as the unmodeled dynamics are stable and sufficiently fast. A nonisothermal fixed-bed reactor with unknown heat of the reaction was used to illustrate the application of

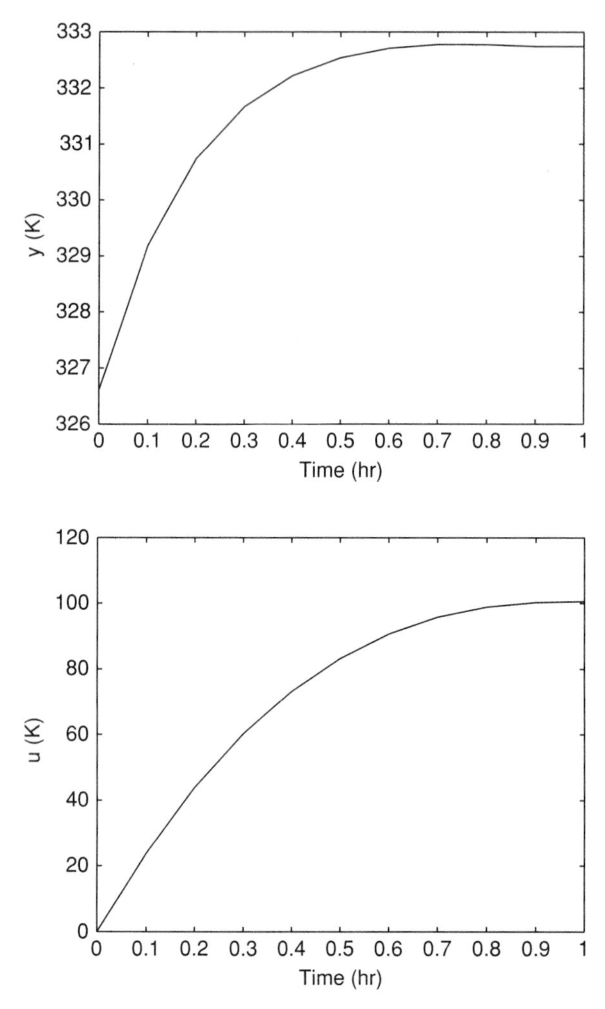

FIGURE 3.6. Closed-loop output and manipulated input profiles for reference input tracking.

the proposed control method. The fact that the reactant wave propagates through the bed faster than the heat wave was explicitly taken into account, and the original model of the process, which consists of two quasi-linear hyperbolic PDEs that describe the spatio-temporal evolution of the reactant concentration and the reactor temperature, was reduced to one quasi-linear hyperbolic PDE that describes the evolution of the reactor temperature. A distributed robust controller was then designed on the basis of the reduced-order model to enforce output tracking and compensate for the effect of the uncertainty. The performance and robustness properties of the controller were successfully tested through computer simulations.

Chapter 4

Feedback Control of Parabolic PDE Systems

4.1 Introduction

In Chapters 2 and 3, we presented nonlinear and robust control methods for systems of first-order hyperbolic PDEs. In the remainder of this book, we focus our attention on nonlinear and robust control of parabolic PDE systems. Such systems arise naturally in the modeling of transport-reaction processes with significant diffusive and dispersive mechanisms (e.g., packed-bed reactors, rapid thermal processing systems, chemical vapor deposition reactors, etc.). In contrast to hyperbolic PDEs, the main feature of parabolic PDEs is that the eigenspectrum of the spatial differential operator can be partitioned into a finite-dimensional slow one and an infinite-dimensional stable fast complement [65, 12, 133]. This motivates addressing the controller synthesis problem for parabolic PDEs on the basis of finite-dimensional systems that accurately describe their dynamic behavior.

Motivated by the above, we propose in this chapter a methodology for the synthesis of nonlinear low-dimensional output feedback controllers for quasi-linear parabolic PDE systems, for which the eigenspectrum of the spatial differential operator can be partitioned into a finite-dimensional slow one and an infinite-dimensional stable fast complement. Singular perturbation methods are initially employed to establish that the discrepancy between the solutions of an ODE system of dimension equal to the number of slow modes, obtained through Galerkin's method, and the PDE system is proportional to the degree of separation of the fast and slow modes of the spatial operator. Then, a procedure, motivated by the theory of singular perturbations, is proposed for the construction of AIMs for the PDE system. The AIMs are used for the derivation of ODE systems of dimension equal to the number of slow modes, that yield solutions which are close, up to a desired accuracy, to the ones of the PDE system, for almost all times. These ODE systems are used as the basis for the synthesis of nonlinear output feedback controllers that guarantee stability and enforce the output of the closed-loop system to follow up to a desired accuracy, a prespecified response for almost all times. The methodology is successfully employed to control the temperature profiles of a catalytic rod and a nonisothermal tubular reactor with recycle around unstable steady states. The theoretical results of this chapter were first presented in [40], and the applications in [43, 3].

4.2 Preliminaries

4.2.1 Description of parabolic PDE systems

We consider parabolic PDE systems in one spatial dimension with a state-space description of the form:

$$\begin{aligned}
\frac{\partial \bar{x}}{\partial t} &= A\frac{\partial \bar{x}}{\partial z} + B\frac{\partial^2 \bar{x}}{\partial z^2} + wb(z)u + f(\bar{x}) \\
y^i &= \int_\alpha^\beta c^i(z)k\bar{x}dz, \quad i = 1, \ldots, l \\
q^\kappa &= \int_\alpha^\beta s^\kappa(z)\omega\bar{x}dz, \quad \kappa = 1, \ldots, p
\end{aligned} \tag{4.1}$$

subject to the boundary conditions:

$$\begin{aligned}
C_1\bar{x}(\alpha, t) + D_1\frac{\partial \bar{x}}{\partial z}(\alpha, t) &= R_1 \\
C_2\bar{x}(\beta, t) + D_2\frac{\partial \bar{x}}{\partial z}(\beta, t) &= R_2
\end{aligned} \tag{4.2}$$

and the initial condition:

$$\bar{x}(z, 0) = \bar{x}_0(z) \tag{4.3}$$

where $\bar{x}(z,t) = [\bar{x}_1(z,t) \ \cdots \ \bar{x}_n(z,t)]^T$ denotes the vector of state variables; $[\alpha, \beta] \subset \mathbb{R}$ is the domain of definition of the process; $z \in [\alpha, \beta]$ is the spatial coordinate; $t \in [0, \infty)$ is the time; $u = [u^1 \ u^2 \ \cdots \ u^l]^T \in \mathbb{R}^l$ denotes the vector of manipulated inputs; $y^i \in \mathbb{R}$ denotes the i-th controlled output; $q^\kappa \in \mathbb{R}$ denotes the κ-th measured output; $\partial\bar{x}/\partial z$, $\partial^2\bar{x}/\partial z^2$ denote the first- and second-order spatial derivatives of $\bar{x}$, $f(\bar{x})$ is a nonlinear vector function; w, k are constant vectors; A, B, C_1, D_1, C_2, D_2 are constant matrices; R_1, R_2 are column vectors; and $\bar{x}_0(z)$ is the initial condition. The vector $b(z)$ is a known smooth function of z of the form $b(z) = [b^1(z) \ b^2(z) \ \cdots \ b^l(z)]$, where $b^i(z)$ describes how the control action $u^i(t)$ is distributed in the spatial interval $[\alpha, \beta]$, $c^i(z)$ is a known smooth function of z which is determined by the desired performance specifications in the interval $[\alpha, \beta]$, and s^κ is a known smooth function of z which is determined by the shape (point or distributed) of the κ-th measurement sensor. Similar to the case of hyperbolic systems, we only consider bounded manipulated inputs and controlled and measured outputs.

Referring to the system of Eq. 4.1, several remarks are in order: (a) the spatial differential operator is assumed to be linear; this assumption is valid for diffusion-convection-reaction processes where the diffusion and the conductivity coefficients can be taken to be independent of temperature and concentration, (b) the manipulated input enters the system in a linear and affine fashion; this is typically the case in many practical applications where, for example, the wall temperature is chosen as the manipulated input, (c) the nonlinearities appear in an additive fashion (e.g., complex reaction rates, Arrhenius dependence of reaction rates on temperature), and (d) the boundaries of the spatial domain of definition of the system of Eq. 4.1 are fixed; this assumption will be removed in Chapter 6 where the control of parabolic PDEs with time-dependent spatial domains is addressed.

Finally, in the remainder of the book we will use the order of magnitude notation $O(\epsilon)$. In particular, $\delta(\epsilon) = O(\epsilon)$ if there exist positive real numbers k_1 and k_2 such that: $|\delta(\epsilon)| \leq k_1|\epsilon|, \ \forall|\epsilon| < k_2$.

4.2.2 Formulation of parabolic PDE system as infinite-dimensional system—Eigenvalue problem

In this subsection, we characterize precisely the class of parabolic PDE systems of the form of Eq. 4.1 which we consider in the present and next chapters. To this end, we formulate the parabolic PDE system of Eq. 4.1 as an infinite dimensional system in the Hilbert space $\mathcal{H}([\alpha,\beta],\mathbb{R}^n)$ (this will also simplify the notation of the paper since the boundary conditions of Eq. 4.2 will be directly included in the formulation; see Eq. 4.8 below), with $\mathcal{H}$ being the space of n-dimensional vector functions defined on $[\alpha,\beta]$ that satisfy the boundary conditions of Eq. 4.2, with inner product and norm:

$$\begin{aligned} (\omega_1,\omega_2) &= \int_{\alpha}^{\beta} (\omega_1(z),\omega_2(z))_{\mathbb{R}^n}\,dz \\ ||\omega_1||_2 &= (\omega_1,\omega_1)^{\frac{1}{2}} \end{aligned} \tag{4.4}$$

where ω_1, ω_2 are two elements of $\mathcal{H}([\alpha,\beta];\mathbb{R}^n)$ and the notation $(\cdot,\cdot)_{\mathbb{R}^n}$ denotes the standard inner product in $\mathbb{R}^n$. Defining the state function x on $\mathcal{H}([\alpha,\beta],\mathbb{R}^n)$ as:

$$x(t) = \bar{x}(z,t), \quad t > 0, \quad z \in [\alpha,\beta], \tag{4.5}$$

the operator $\mathcal{A}$ in $\mathcal{H}([\alpha,\beta],\mathbb{R}^n)$ as:

$$\begin{aligned} \mathcal{A}x &= A\frac{\partial \bar{x}}{\partial z} + B\frac{\partial^2 \bar{x}}{\partial z^2}, \\ x \in D(\mathcal{A}) &= \Big\{ x \in \mathcal{H}([\alpha,\beta];\mathbb{R}^n);\ C_1\bar{x}(\alpha,t) + D_1\frac{\partial \bar{x}}{\partial z}(\alpha,t) = R_1; \\ &\qquad C_2\bar{x}(\beta,t) + D_2\frac{\partial \bar{x}}{\partial z}(\beta,t) = R_2 \Big\} \end{aligned} \tag{4.6}$$

and the input and output operators as:

$$\mathcal{B}u = wbu, \quad \mathcal{C}x = (c,kx), \quad \mathcal{Q}x = (s,\omega x), \tag{4.7}$$

where $c = [c^1 \ c^2 \ \cdots \ c^l]$ and $s = [s^1 \ s^2 \ \cdots \ s^q]$, the system of Eqs. 4.1–4.2–4.3 takes the form:

$$\begin{aligned} \dot{x} &= \mathcal{A}x + \mathcal{B}u + f(x), \quad x(0) = x_0 \\ y &= \mathcal{C}x, \quad q = \mathcal{Q}x \end{aligned} \tag{4.8}$$

where $f(x(t)) = f(\bar{x}(z,t))$ and $x_0 = \bar{x}_0(z)$. We assume that the nonlinear term $f(x)$ satisfies $f(0) = 0$ and is also locally Lipschitz continuous, i.e., there exist positive real numbers a_0, K_0 such that for any $x_1, x_2 \in \mathcal{H}$ that satisfy $\max\{||x_1||_2, ||x_2||_2\} \leq a_0$, we have that:

$$||f(x_1) - f(x_2)||_2 \leq K_0||x_1 - x_2||_2 \tag{4.9}$$

For the operator $\mathcal{A}$, the eigenvalue problem is defined as:

$$\mathcal{A}\phi_j = \lambda_j \phi_j, \quad j = 1, \ldots, \infty \tag{4.10}$$

where λ_j denotes an eigenvalue and ϕ_j denotes the corresponding eigenfunction; the eigenspectrum of $\mathcal{A}$, $\sigma(\mathcal{A})$, is defined as the set of all eigenvalues of $\mathcal{A}$, i.e., $\sigma(\mathcal{A}) = \{\lambda_1, \lambda_2, \ldots, \}$. Assumption 4.1 states that the eigenspectrum of $\mathcal{A}$ can be partitioned into a finite-dimensional part consisting of m slow eigenvalues and a stable infinite-dimensional complement containing the remaining fast eigenvalues, and that the separation between the slow and fast eigenvalues of $\mathcal{A}$ is large.

Assumption 4.1

1. *$Re\{\lambda_1\} \geq Re\{\lambda_2\} \geq \ldots \geq Re\{\lambda_j\} \geq \ldots$, where $Re\{\lambda_j\}$ denotes the real part of λ_j.*

2. *$\sigma(\mathcal{A})$ can be partitioned as $\sigma(\mathcal{A}) = \sigma_1(\mathcal{A}) + \sigma_2(\mathcal{A})$, where $\sigma_1(\mathcal{A})$ consists of the first m (with m finite) eigenvalues, i.e., $\sigma_1(\mathcal{A}) = \{\lambda_1, \ldots, \lambda_m\}$, and $\frac{|Re\{\lambda_1\}|}{|Re\{\lambda_m\}|} = O(1)$.*

3. *$Re\{\lambda_{m+1}\} < 0$ and $\frac{|Re\{\lambda_m\}|}{|Re\{\lambda_{m+1}\}|} = O(\epsilon)$, where $\epsilon < 1$ is a small positive number.*

The assumption of finite number of unstable eigenvalues is always satisfied for parabolic PDE systems [65], while the assumption of discrete eigenspectrum and the assumption of existence of only a few dominant modes that describe the dynamics of the parabolic PDE system are usually satisfied by the majority of diffusion-convection-reaction processes. We note that assumption 4.1 is not satisfied in the cases of: (a) first-order hyperbolic PDE systems (i.e., convection-reaction processes) where the eigenvalues cluster along vertical, or nearly vertical, asymptotes in the complex plane (see Chapter 2 of this book), and (b) parabolic PDE systems for which the spatial coordinate is defined in the infinite domain, where the eigenspectrum is continuous and wavelike behavior is usually exhibited (see [102] for a detailed discussion on this issue).

Whenever assumption 4.1 holds, $\mathcal{A}$ generates a strongly continuous semigroup of bounded linear operators $U(t)$, which implies that the generalized solution of the system of Eq. 4.8 is given by [64, 65, 109]:

$$x(t) = U(t)x_0 + \int_0^t U(t-\tau)[\mathcal{B}u(\tau) + f(x(\tau))]d\tau. \quad (4.11)$$

$U(t)$ satisfies the following growth property:

$$||U(t)||_2 \leq K_1 e^{a_1 t}, \quad \forall t \geq 0 \quad (4.12)$$

where K_1, a_1 are real numbers, with $K_1 \geq 1$ and $a_1 \geq Re\{\lambda_1\}$ (an estimate of K_1, a_1 can be obtained utilizing the Hille–Yoshida theorem [64]). If a_1 is strictly negative, we will say that $\mathcal{A}$ generates an exponentially stable semigroup $U(t)$. Finally, we focus, throughout the chapter, on local exponential stability, and not on weak (asymptotic) stability [64], because of its robustness to bounded perturbations, which are always present in most practical applications [17].

4.3 Examples of Processes Modeled by Parabolic PDEs

In this section, we present two examples of processes modeled by nonlinear parabolic PDE systems of the form of Eq. 4.1 and compute the eigenvalues and eigenfunctions of the corresponding spatial differential operators.

4.3.1 Catalytic rod

Consider a long, thin rod in a reactor (Figure 4.1). The reactor is fed with pure species A and a zero-th order exothermic catalytic reaction of the form $A \rightarrow B$ takes place on the rod. Since the reaction is exothermic, a cooling medium that is in contact with the rod is used for cooling. Under the assumptions of constant density and heat capacity of the rod, constant conductivity of the rod, and constant temperature at both ends of the rod, and excess of species A in the furnace, the mathematical model which describes the spatiotemporal evolution of the dimensionless rod temperature consists of the following parabolic PDE:

$$\frac{\partial \bar{x}}{\partial t} = \frac{\partial^2 \bar{x}}{\partial z^2} + \beta_T e^{-\frac{\gamma}{1+\bar{x}}} + \beta_U(b(z)u(t) - \bar{x}) - \beta_T e^{-\gamma} \quad (4.13)$$

subject to the Dirichlet boundary conditions:

$$\bar{x}(0,t) = 0, \quad \bar{x}(\pi,t) = 0 \quad (4.14)$$

and the initial condition:

$$\bar{x}(z,0) = \bar{x}_0(z) \quad (4.15)$$

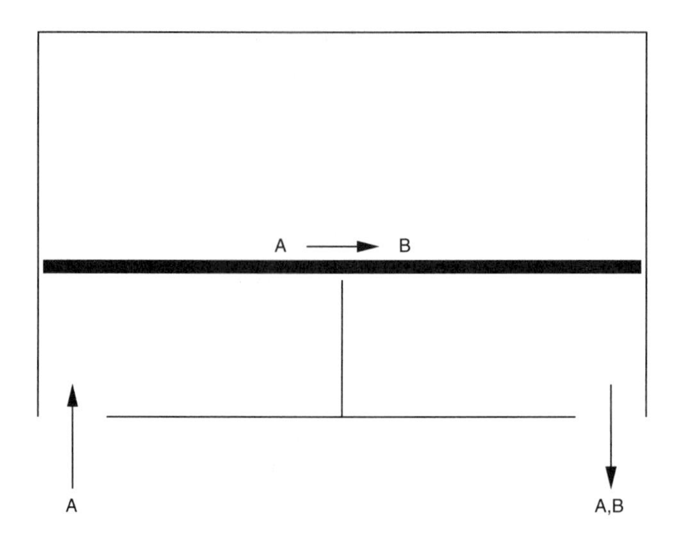

FIGURE 4.1. Catalytic rod.

where $\bar{x}$ denotes the dimensionless temperature in the reactor, β_T denotes a dimensionless heat of reaction, γ denotes a dimensionless activation energy, β_U denotes a dimensionless heat transfer coefficient, and u denotes the manipulated input (temperature of the cooling medium). The following typical values are given to the process parameters:

$$\beta_T = 50.0, \quad \beta_U = 2.0, \quad \gamma = 4.0. \tag{4.16}$$

For the above values, the operating steady state $\bar{x}(z,t) = 0$ is an unstable one (Figure 4.2 shows the evolution of open-loop rod temperature starting

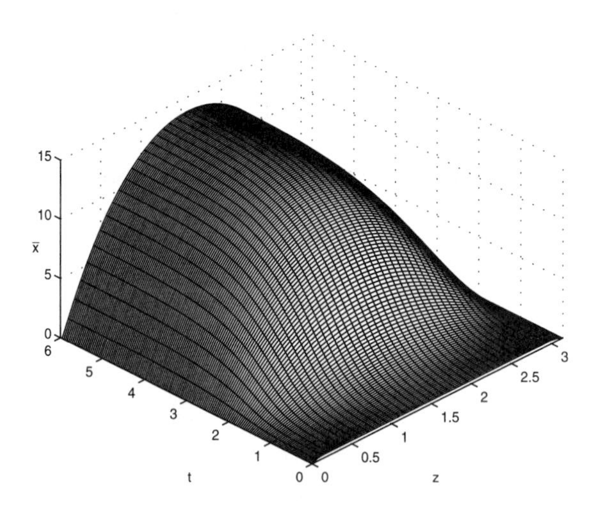

FIGURE 4.2. Profile of evolution of open-loop rod temperature.

from initial conditions close to the steady state $\bar{x}(z,t) = 0$; the process moves to another stable steady state characterized by a maximum in the temperature profile, *hot-spot*, in the middle of the rod). Therefore, the control objective is to stabilize the rod temperature profile at the unstable steady state $\bar{x}(z,t) = 0$. Since the maximum open-loop temperature occurs in the middle of the rod (Figure 4.2), the controlled output is defined as:

$$y(t) = \int_0^{\pi} \sqrt{\frac{2}{\pi}} sin(z)\bar{x}(z,t)\,dz \tag{4.17}$$

and the actuator distribution function is taken to be $b(z) = \sqrt{\frac{2}{\pi}} sin(z)$, in order to apply maximum cooling to the middle of the rod. One distributed sensor that measures

$$q(t) = \int_0^{\pi} \sqrt{\frac{2}{\pi}} sin(z)\bar{x}(z,t)\,dz \tag{4.18}$$

is assumed to be available (the control of the rod when a single point measurement is available is discussed in section 5.8). The eigenvalue problem for the spatial differential operator of the process:

$$\mathcal{A}x = \frac{\partial^2 \bar{x}}{\partial z^2}, \quad x \in D(\mathcal{A}) = \left\{x \in \mathcal{H}([0,\pi];\mathbb{R}); \quad \bar{x}(0,t) = 0, \quad \bar{x}(\pi,t) = 0\right\} \tag{4.19}$$

can be solved analytically and its solution is of the form:

$$\lambda_j = -j^2, \quad \phi_j(z) = \sqrt{\frac{2}{\pi}} sin(j\ z), \quad j = 1, \ldots, \infty. \tag{4.20}$$

4.3.2 Nonisothermal tubular reactor with recycle

We consider a nonisothermal tubular reactor shown in Figure 4.3, where an irreversible first-order reaction of the form $A \to B$ takes place. The reaction is exothermic and a cooling jacket is used to remove heat from the reactor. The outlet of the reactor is fed to a separator where the unreacted

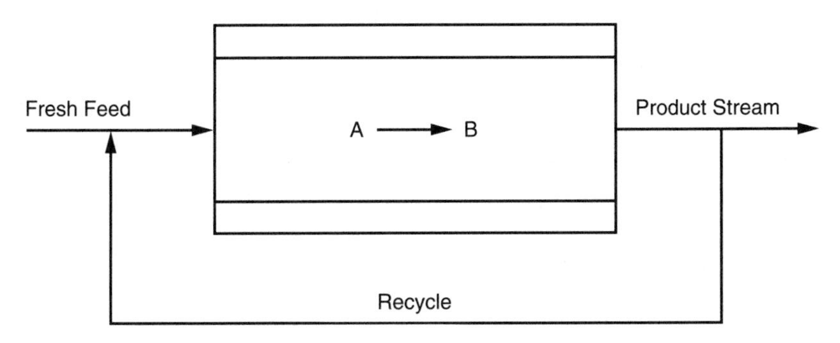

FIGURE 4.3. Nonisothermal tubular reactor.

species A is separated from the product B. The unreacted amount of species A is then fed back to the reactor through a recycle loop. Under standard modeling assumptions, the dynamic model of the process can be derived from mass and energy balances and takes the following dimensionless form:

$$\begin{aligned}\frac{\partial \bar{x}_1}{\partial t} &= -\frac{\partial \bar{x}_1}{\partial z} + \frac{1}{Pe_T}\frac{\partial^2 \bar{x}_1}{\partial z^2} + B_T B_C e^{\frac{\gamma \bar{x}_1}{1+\bar{x}_1}}(1+\bar{x}_2) + \beta_T(b(z)u(t) - \bar{x}_1) \\ \frac{\partial \bar{x}_2}{\partial t} &= -\frac{\partial \bar{x}_2}{\partial z} + \frac{1}{Pe_C}\frac{\partial^2 \bar{x}_2}{\partial z^2} - B_C e^{\frac{\gamma \bar{x}_1}{1+\bar{x}_1}}(1+\bar{x}_2)\end{aligned} \tag{4.21}$$

subject to the boundary conditions:

$$\begin{aligned}&\frac{\partial \bar{x}_1}{\partial z}(0,t) = Pe_T(\bar{x}_1(0,t) - (1-r)x_{1f} - r\bar{x}_1(1,t)), \\ &\frac{\partial \bar{x}_2}{\partial z}(0,t) = Pe_C(\bar{x}_2(0,t) - (1-r)x_{2f} - r\bar{x}_2(1,t)); \\ &\qquad \frac{\partial \bar{x}_1}{\partial z}(1,t) = 0, \quad \frac{\partial \bar{x}_2}{\partial z}(1,t) = 0\end{aligned} \tag{4.22}$$

where $\bar{x}_1$ and $\bar{x}_2$ denote dimensionless temperature and concentration of species A in the reactor, respectively; Pe_T and Pe_C are the heat and thermal Peclet numbers, respectively; B_T and B_C denote a dimensionless heat of reaction and a dimensionless pre-exponential factor, respectively; r is the recirculation coefficient (it varies from zero to one, with one corresponding to total recycle and zero fresh feed and zero corresponding to no recycle); γ is a dimensionless activation energy; β_T is a dimensionless heat transfer coefficient; u is the manipulated input (dimensionless jacket temperature); and $b(z)$ is the actuator distribution function.

In order to transform the boundary condition of Eq. 4.22 to a homogeneous one, we insert the nonhomogeneous part of the boundary condition into the differential equation and obtain the following PDE representation of the process:

$$\begin{aligned}\frac{\partial \bar{x}_1}{\partial t} &= -\frac{\partial \bar{x}_1}{\partial z} + \frac{1}{Pe_T}\frac{\partial^2 \bar{x}_1}{\partial z^2} + B_T B_C e^{\frac{\gamma \bar{x}_1}{1+\bar{x}_1}}(1+\bar{x}_2) + \beta_T(b(z)u(t) - \bar{x}_1) \\ &\quad + \delta(z-0)((1-r)x_{1f} + r\bar{x}_1(1,t)) \\ \frac{\partial \bar{x}_2}{\partial t} &= -\frac{\partial \bar{x}_2}{\partial z} + \frac{1}{Pe_C}\frac{\partial^2 \bar{x}_2}{\partial z^2} - B_C e^{\frac{\gamma \bar{x}_1}{1+\bar{x}_1}}(1+\bar{x}_2) \\ &\quad + \delta(z-0)((1-r)x_{2f} + r\bar{x}_2(1,t))\end{aligned} \tag{4.23}$$

where $\delta(\cdot)$ is the standard Dirac function, subject to the homogeneous boundary conditions:

$$
\begin{aligned}
\frac{\partial \bar{x}_1}{\partial z}(0,t) = Pe_T \bar{x}_1(0,t), \quad \frac{\partial \bar{x}_2}{\partial z}(0,t) = Pe_C \bar{x}_2(0,t); \\
\frac{\partial \bar{x}_1}{\partial z}(1,t) = 0, \quad \frac{\partial \bar{x}_2}{\partial z}(1,t) = 0
\end{aligned}
\tag{4.24}
$$

The following values for the process parameters are used in the following calculations:

$$
\begin{aligned}
Pe_T = 7.0, \quad Pe_C = 7.0, \quad B_C = 0.1, \quad B_T = 2.5, \\
\beta_T = 2.0, \quad \gamma = 10.0, \quad r = 0.5.
\end{aligned}
\tag{4.25}
$$

For the above values, the operating steady state of the open-loop system is unstable (the linearization around the steady state possesses one real unstable eigenvalue and infinitely many stable eigenvalues), thereby implying the need to operate the process under feedback control. A 400-th order Galerkin truncation of the system of Eq. 4.23–4.25 (200 equations for each PDE) is used in our simulations in order to accurately describe the process (further increase on the order of the Galerkin truncation was found to give negligible improvement on the accuracy of the results). Figure 4.4 shows the open-loop profile of $\bar{x}_1$ along the length of the reactor, which corresponds to the operating unstable steady state. We note that in the absence of recycle loop (i.e., $r = 0$), the above process parameters correspond to a unique stable steady state for the open-loop system.

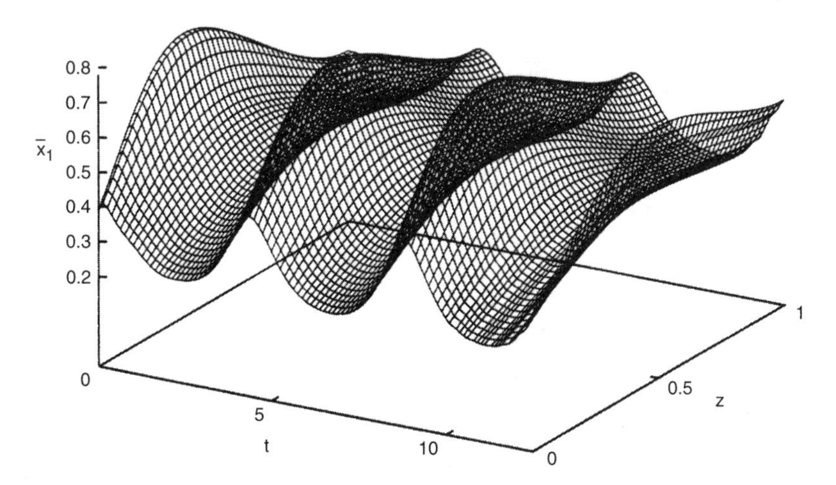

FIGURE 4.4. Profile of evolution of open-loop reactor temperature.

The spatial differential operator of the system of Eq. 4.23 is of the form:

$$\mathcal{A}x = \begin{bmatrix} \mathcal{A}_1 x_1 & 0 \\ 0 & \mathcal{A}_2 x_2 \end{bmatrix} = \begin{bmatrix} \frac{1}{Pe_T}\frac{\partial^2 \bar{x}_1}{\partial z^2} - \frac{\partial \bar{x}_1}{\partial z} & 0 \\ 0 & \frac{1}{Pe_C}\frac{\partial^2 \bar{x}_2}{\partial z^2} - \frac{\partial \bar{x}_2}{\partial z} \end{bmatrix}, \tag{4.26}$$

$$x \in D(\mathcal{A}) = \Big\{ x \in \mathcal{H}([0,1]; \mathrm{IR}^2);\ \frac{\partial \bar{x}_1}{\partial z}(0,t) = Pe_T \bar{x}_1(0,t),$$
$$\frac{\partial \bar{x}_2}{\partial z}(0,t) = Pe_C \bar{x}_2(0,t),\ \frac{\partial \bar{x}_1}{\partial z}(1,t) = 0,\ \frac{\partial \bar{x}_2}{\partial z}(1,t) = 0 \Big\}.$$

Assuming for simplicity that $Pe = Pe_T = Pe_C$, the solution of the eigenvalue problem for $\mathcal{A}_i$, $i = 1, 2$, can be obtained by utilizing standard techniques from linear operator theory and is of the form [115] (see also [113]):

$$\lambda_{ij} = \frac{\bar{a}_{ij}^2}{Pe} + \frac{Pe}{4}, \quad i = 1, 2, \quad j = 1, \ldots, \infty$$
$$\phi_{ij}(z) = B_{ij} e^{Pe\frac{z}{2}} \left(cos(\bar{a}_{ij} z) + \frac{Pe}{2\bar{a}_{ij}} sin(\bar{a}_{ij} z) \right), \quad i = 1, 2, \quad j = 1, \ldots, \infty$$
$$\bar{\phi}_{ij}(z) = e^{-Pez} \phi_{ij}(z), \quad i = 1, 2, \quad j = 1, \ldots, \infty \tag{4.27}$$

where λ_{ij}, ϕ_{ij}, $\bar{\phi}_{ij}$, denote the eigenvalues, eigenfunctions, and adjoint eigenfunctions of $\mathcal{A}_i$, respectively. $\bar{a}_{ij}$, B_{ij} can be calculated from the following formulas:

$$tan(\bar{a}_{ij}) = \frac{Pe\bar{a}_{ij}}{\bar{a}_{ij}^2 - \left(\frac{Pe}{2}\right)^2}, \quad i = 1, 2, \quad j = 1, \ldots, \infty$$
$$B_{ij} = \left\{ \int_0^1 \left(cos(\bar{a}_{ij} z) + \frac{Pe}{2\bar{a}_{ij}} sin(\bar{a}_{ij} z) \right)^2 dz \right\}^{-\frac{1}{2}}, \tag{4.28}$$
$$i = 1, 2, \quad j = 1, \ldots, \infty.$$

A direct computation of the first nine eigenvalues of $\mathcal{A}$ yields: $\lambda_{11} = \lambda_{21} = -2.36$, $\lambda_{12} = \lambda_{22} = -4.60$, $\lambda_{13} = \lambda_{23} = -9.13$, $\lambda_{14} = \lambda_{24} = -16.30$, $\lambda_{15} = \lambda_{25} = -26.22$, $\lambda_{16} = \lambda_{26} = -38.94$, $\lambda_{17} = \lambda_{27} = -54.47$, $\lambda_{18} = \lambda_{28} = -72.81$, and $\lambda_{19} = \lambda_{29} = -93.96$.

Owing to the open-loop instability, the control objective is to stabilize the reactor around the unstable steady state and the controlled output is defined as:

$$y(t) = \int_0^1 e^{-PeZ} \phi_{11}(z) \bar{x}_1 dz \tag{4.29}$$

and the actuator distribution function is taken to be $b(z) = 1$ (uniform wall temperature in space). A distributed measurement sensor which provides a weighted average value of the temperature throughout the reactor is assumed to be available; the measured output is of the form:

$$q(t) = \int_0^1 e^{-PeZ} \phi_{11}(z) \bar{x}_1 dz. \tag{4.30}$$

Remark 4.1 It is important to note the role of the recycle loop in the reactor system. Specifically, the use of the recycle loop allows reducing the maximum temperature inside the reactor (because a smaller amount of fresh feed of species A is introduced into the reactor), while maintaining the desired rate of production of the species B (because the reactor residence time increases). The disadvantage of adding a recycle loop is that it introduces a feedback mechanism into the process which renders the open-loop steady state unstable, thereby implying the need to operate the system under feedback control.

4.4 Galerkin's Method

We will now review the application of Galerkin's method to the system of Eq. 4.8 to derive an approximate finite-dimensional system. Let $\mathcal{H}_s$, $\mathcal{H}_f$ be modal subspaces of $\mathcal{A}$, defined as $\mathcal{H}_s = span\{\phi_1, \phi_2, \ldots, \phi_m\}$ and $\mathcal{H}_f = span\{\phi_{m+1}, \phi_{m+2}, \ldots, \}$ (the existence of $\mathcal{H}_s$, $\mathcal{H}_f$ follows from assumption 1). Defining the orthogonal projection operators P_s and P_f such that $x_s = P_s x$, $x_f = P_f x$, the state x of the system of Eq. 4.8 can be decomposed as:

$$x = x_s + x_f = P_s x + P_f x. \tag{4.31}$$

Applying P_s and P_f to the system of Eq. 4.8 and using the above decomposition for x, the system of Eq. 4.8 can be equivalently written in the following form:

$$\begin{aligned} \frac{dx_s}{dt} &= \mathcal{A}_s x_s + \mathcal{B}_s u + f_s(x_s, x_f) \\ \frac{\partial x_f}{\partial t} &= \mathcal{A}_f x_f + \mathcal{B}_f u + f_f(x_s, x_f) \\ y &= \mathcal{C} x_s + \mathcal{C} x_f, \quad q = \mathcal{Q} x_s + \mathcal{Q} x_f \\ x_s(0) &= P_s x(0) = P_s x_0, \quad x_f(0) = P_f x(0) = P_f x_0 \end{aligned} \tag{4.32}$$

where $\mathcal{A}_s = P_s \mathcal{A}$, $\mathcal{B}_s = P_s \mathcal{B}$, $f_s = P_s f$, $\mathcal{A}_f = P_f \mathcal{A}$, $\mathcal{B}_f = P_f \mathcal{B}$ and $f_f = P_f f$ and the notation $\partial x_f / \partial t$ is used to denote that the state x_f belongs in an infinite-dimensional space. In the above system, $\mathcal{A}_s$ is a diagonal matrix of dimension $m \times m$ of the form $\mathcal{A}_s = diag\{\lambda_j\}$, $f_s(x_s, x_f)$ and $f_f(x_s, x_f)$ are Lipschitz vector functions, and $\mathcal{A}_f$ is an unbounded differential operator that is exponentially stable and generates a strongly continuous exponentially stable semigroup (this follows from part 3 of assumption 1 and the selection of $\mathcal{H}_s, \mathcal{H}_f$). Neglecting the fast modes, the following finite-dimensional system is derived:

$$\begin{aligned} \frac{dx_s}{dt} &= \mathcal{A}_s x_s + \mathcal{B}_s u + f_s(x_s, 0) \\ y_s &= \mathcal{C} x_s, \quad q_s = \mathcal{Q} x_s \end{aligned} \tag{4.33}$$

where the subscript s in y_s and q_s denotes that these outputs are associated with the slow system. The above system can be directly used for controller design employing standard control methods for ODEs [12, 115, 35].

4.5 Accuracy of ODE System Obtained from Galerkin's Method

In this section, we use singular perturbation methods to establish that if the finite-dimensional system of Eq. 4.33 is exponentially stable, then the system of Eq. 4.32 is also exponentially stable and the discrepancy between the solution of the x_s-subsystem of the system of Eq. 4.32 and the solution of the system of Eq. 4.33, is proportional to the spectral separation of the slow and fast eigenvalues.

Using $\epsilon = \frac{|Re\{\lambda_1\}|}{|Re\{\lambda_{m+1}\}|}$ and multiplying the x_f-subsystem of Eq. 4.32 by ϵ, we obtain the following system:

$$\begin{aligned} \frac{dx_s}{dt} &= \mathcal{A}_s x_s + \mathcal{B}_s u + f_s(x_s, x_f) \\ \epsilon \frac{\partial x_f}{\partial t} &= \mathcal{A}_{f\epsilon} x_f + \epsilon \mathcal{B}_f u + \epsilon f_f(x_s, x_f) \end{aligned} \tag{4.34}$$

where $\mathcal{A}_{f\epsilon}$ is an unbounded differential operator defined as $\mathcal{A}_{f\epsilon} = \epsilon \mathcal{A}_f$. Since ϵ is a small positive number less than unity (assumption 4.1, part 3), and the operators $\mathcal{A}_s, \mathcal{A}_{f\epsilon}$ generate semigroups with growth rates which are of the same order of magnitude, the system of Eq. 4.34 is in the standard singularly perturbed form, with x_s being the slow states and x_f being the fast states. The purpose of rewriting the system of Eq. 4.32 in singularly perturbed form is to extract information about its solutions from simplified systems, for sufficiently small ϵ. In particular, it is possible to derive two separate models that describe the evolution of the states of the system of Eq. 4.34 in different time scales.

Introducing the fast time scale $\tau = \frac{t}{\epsilon}$ and setting $\epsilon = 0$, we obtain the following infinite-dimensional fast subsystem from the system of Eq. 4.34:

$$\frac{\partial x_f}{\partial \tau} = \mathcal{A}_{f\epsilon} x_f. \tag{4.35}$$

From the fact that $Re\ \{\lambda_{m+1}\} < 0$ and the definition of ϵ, we have that the above system is globally exponentially stable. Setting $\epsilon = 0$ in the system of Eq. 4.34 and using the fact that the inverse operator $\mathcal{A}_{f\epsilon}{}^{-1}$ exists and is also bounded (it follows from the fact that zero is in the resolvent of $\mathcal{A}_{f\epsilon}$), we have that:

$$x_f = 0 \tag{4.36}$$

and thus the finite-dimensional slow system takes the form:

$$\frac{dx_s}{dt} = \mathcal{A}_s x_s + \mathcal{B}_s u + f_s(x_s, 0). \tag{4.37}$$

We note that the above system is identical to the one obtained by applying the standard Galerkin's method to the system of Eq. 4.8, keeping the first m ODEs and completely neglecting the x_f-subsystem. Assumption 4.2 that follows states a stability requirement on the system of Eq. 4.37.

Assumption 4.2 *The finite-dimensional system of Eq. 4.37 with $u(t) \equiv 0$ is exponentially stable that is, there exist positive real numbers (K_2, a_2, a_4), with $a_4 > K_2 \geq 1$ such that for all $x_s \in \mathcal{H}_s$ that satisfy $|x_s| \leq a_4$, the following bound holds:*

$$|x_s(t)| \leq K_2 e^{-a_2 t} |x_s(0)|, \quad \forall\, t \geq 0 \tag{4.38}$$

where the notation $|x_s|$ denotes that x_s belongs in a finite-dimensional Hilbert space.

Proposition 4.1 establishes that the solutions of the open-loop systems of Eqs. 4.37–4.35, after a short finite-time interval required for the trajectories of the system of Eq. 4.34 to approach the quasi steady state of Eq. 4.36, consist an $O(\epsilon)$ approximation of the solutions of the open-loop system of Eq. 4.34. The proof of the proposition is given in Appendix C.

Proposition 4.1 *Consider the system of Eq. 4.34 with $u(t) \equiv 0$ and suppose that assumptions 4.1 and 4.2 hold. Then, there exist positive real numbers μ_1, μ_2, ϵ^* such that if $|x_s(0)| \leq \mu_1$, $||x_f(0)||_2 \leq \mu_2$ and $\epsilon \in (0, \epsilon^*]$, then the solution $x_s(t), x_f(t)$ of the system of Eq. 4.34 satisfies for all $t \in [0, \infty)$:*

$$\begin{aligned} x_s(t) &= \bar{x}_s(t) + O(\epsilon) \\ x_f(t) &= \bar{x}_f\left(\frac{t}{\epsilon}\right) + O(\epsilon) \end{aligned} \tag{4.39}$$

where $\bar{x}_s(t), \bar{x}_f(t)$ are the solutions of the slow and fast subsystems of Eqs. 4.37–4.35 with $u(t) \equiv 0$, respectively.

Remark 4.2 An estimate of ϵ^* can be obtained, in principle, from the proof of the proposition. However, such an estimate is typically conservative, and thus, it is useful to check its appropriateness through computer simulations.

Remark 4.3 The counterpart of the result of proposition 4.1 in finite-dimensional spaces is well known (Tikhonov's theorem, [136]), while a similar result has also been established for linear infinite-dimensional systems [15]. The main technical difference in establishing this result between linear and quasi-linear infinite-dimensional systems is that, for quasi-linear systems the proof is based on Lyapunov arguments, while for linear systems the proof is obtained using combination of estimates of the states, obtained from the application of variations of constants formula [15]. This is a consequence of the fact that for quasi-linear systems it is not possible to derive a coordinate change that transforms the system of Eq. 4.34 into a cascaded interconnection where the fast modes are decoupled from the slow modes. Such a coordinate change allows us to derive an exponentially decaying estimate, for sufficiently small ϵ, for the fast state, which is independent of the one of the slow state, and thus to prove the result of the proposition through a direct combination of these estimates.

Remark 4.4 Proposition 4.1 allows characterizing the class of infinite-dimensional systems of the form of Eq. 4.8 whose solutions can be approximated, after a short time interval, up to $O(\epsilon)$ from the solutions of a finite-dimensional system. More specifically, given a parabolic PDE of Eq. 4.1, one has first to compute the eigenvalues of the spatial differential operator of the system and calculate the value of ϵ. Then, the procedure followed in the proof of proposition 4.1 can be used to derive an estimate of the value of ϵ^*. If $\epsilon \leq \epsilon^*$, then a finite dimensional model of the form of Eq. 4.34 yields solutions that are $O(\epsilon)$ close to the ones of the system of Eq. 4.8 for almost all times.

Remark 4.5 We note that it is possible, using standard results from center manifold theory for infinite-dimensional systems of the form of Eq. 4.8 [33], to show that if the system of Eq. 4.37 is asymptotically stable, then the system of Eq. 4.8 is also asymptotically stable and the discrepancy between the solution of the system of Eq. 4.37 and the x_s-subsystem of the closed-loop full-order system is asymptotically (as $t \to \infty$) proportional to ϵ. Although this result is important because it allows establishing asymptotic stability of the closed-loop infinite-dimensional system by performing a stability analysis on a low-order finite-dimensional system, it does not provide any information about the discrepancy between the solutions of these two systems for finite t.

4.6 Construction of ODE Systems of Desired Accuracy via AIMs

In this section, we propose an approach originating from the theory of inertial manifolds for the construction of ODE systems of dimension m, which yield solutions that are arbitrarily close (closer than $O(\epsilon)$) to the ones of the infinite-dimensional system of Eq. 4.8, for almost all times. An inertial manifold $\mathcal{M}$ for the system of Eq. 4.8 is a subset of $\mathcal{H}$, which satisfies the following properties [133]: (i) $\mathcal{M}$ is a finite-dimensional Lipschitz manifold, (ii) $\mathcal{M}$ is a graph of a Lipschitz function $\Sigma(x_s, u, \epsilon)$ mapping $\mathcal{H}_s \times \mathrm{IR}^l \times (0, \epsilon^*]$ into $\mathcal{H}_f$ and for every solution $x_s(t), x_f(t)$ of Eq. 4.34 with $x_f(0) = \Sigma(x_s(0), u, \epsilon)$, then

$$x_f(t) = \Sigma(x_s(t), u, \epsilon), \quad \forall t \geq 0 \tag{4.40}$$

and (iii) $\mathcal{M}$ attracts every trajectory exponentially. The evolution of the state x_f on $\mathcal{M}$ is given by Eq. 4.40, while the evolution of the state x_s is governed by the following finite-dimensional inertial form:

$$\frac{dx_s}{dt} = \mathcal{A}_s x_s + \mathcal{B}_s u + f_s(x_s, \Sigma(x_s, u, \epsilon)). \tag{4.41}$$

Assuming that $u(t)$ is smooth, differentiating Eq. 4.40 and utilizing Eq. 4.34, $\Sigma(x_s, u, \epsilon)$ can be computed as the solution of the following partial differential equation:

$$\epsilon \frac{\partial \Sigma}{\partial x_s} [\mathcal{A}_s x_s + \mathcal{B}_s u + f_s(x_s, \Sigma)] + \epsilon \frac{\partial \Sigma}{\partial u} \dot{u} = \mathcal{A}_{f\epsilon} \Sigma + \epsilon \mathcal{B}_f u + \epsilon f_f(x_s, \Sigma) \tag{4.42}$$

which Σ has to satisfy for all $x_s \in \mathcal{H}_s$, $u \in \mathrm{IR}^l$, $\epsilon \in (0, \epsilon^*]$. However, even for parabolic PDEs for which it is known that $\mathcal{M}$ exists, the derivation of an explicit analytic form of $\Sigma(x_s, u, \epsilon)$ is an extremely difficult (if not impossible) task.

Motivated by this, we will now propose a procedure, motivated by singular perturbations [95], to compute approximations of $\Sigma(x_s, u, \epsilon)$ (approximate inertial manifolds) of desired accuracy. To this end, consider an expansion of $\Sigma(x_s, u, \epsilon)$ and u in a power series in ϵ:

$$\begin{aligned} u &= u_0 + \epsilon u_1 + \epsilon^2 u_2 + \cdots + \epsilon^k u_k + O(\epsilon^{k+1}) \\ \Sigma(x_s, u, \epsilon) &= \Sigma_0(x_s, u) + \epsilon \Sigma_1(x_s, u) + \epsilon^2 \Sigma_2(x_s, u) \\ &\quad + \cdots + \epsilon^k \Sigma_k(x_s, u) + O(\epsilon^{k+1}) \end{aligned} \tag{4.43}$$

where u_k, Σ_k are smooth functions. Substituting the expressions of Eq. 4.43 into Eq. 4.42, and equating terms of the same power in ϵ, one can obtain approximations of $\Sigma(x_s, u, \epsilon)$ up to a desired order. Substituting the expansion for $\Sigma(x_s, u, \epsilon)$ and u up to order k into Eq. 4.41, the following approximation of the inertial form is obtained:

$$\frac{dx_s}{dt} = \mathcal{A}_s x_s + \mathcal{B}_s(u_0 + \epsilon u_1 + \epsilon^2 u_2 + \cdots + \epsilon^k u_k) \\ + f_s(x_s, \Sigma_0(x_s, u) + \epsilon \Sigma_1(x_s, u) + \epsilon^2 \Sigma_2(x_s, u) + \cdots + \epsilon^k \Sigma_k(x_s, u)). \tag{4.44}$$

In order to characterize the discrepancy between the solution of the open-loop finite-dimensional system of Eq. 4.44 and the solution of the x_s-subsystem of the open-loop infinite-dimensional system of Eq. 4.34, we will impose a stability requirement on the system of Eq. 4.44.

Assumption 4.3 *The finite-dimensional system of Eq. 4.44 with $u(t) \equiv 0$ is exponentially stable that is, there exist positive real numbers $(\bar{K}_2, \bar{a}_2, \bar{a}_4)$, with $\bar{a}_4 > \bar{K}_2 \geq 1$ such that for all $x_s \in \mathcal{H}_s$ that satisfy $|x_s| \leq \bar{a}_4$, the following bound holds:*

$$|x_s(t)| \leq \bar{K}_2 e^{-\bar{a}_2 t} |x_s(0)|, \quad \forall\, t \geq 0. \tag{4.45}$$

Proposition 4.2 establishes that the discrepancy between the solutions obtained from the open-loop system of Eq. 4.44 with the expansion of $\Sigma(x_s, u, \epsilon)$ of Eq. 4.43, and the solutions of the infinite-dimensional open-loop system of Eq. 4.34 is of $O(\epsilon^{k+1})$, for almost all times. The proof can be found in Appendix C.

Proposition 4.2 *Consider the system of Eq. 4.34 with $u(t) \equiv 0$ and suppose that assumptions 4.1 and 4.3 hold. Then, there exist positive real numbers $\bar{\mu}_1, \bar{\mu}_2, \bar{\epsilon}^*$ such that if $|x_s(0)| \leq \bar{\mu}_1$, $||x_f(0)||_2 \leq \bar{\mu}_2$ and $\epsilon \in (0, \bar{\epsilon}^*]$, then the solution $x_s(t), x_f(t)$ of the system of Eq. 4.34 satisfies for all $t \in [t_b, \infty)$:*

$$\begin{aligned} x_s(t) &= \tilde{x}_s(t) + O(\epsilon^{k+1}) \\ x_f(t) &= \tilde{x}_f(t) + O(\epsilon^{k+1}) \end{aligned} \tag{4.46}$$

where t_b is the time required for $x_f(t)$ to approach $\tilde{x}_f(t)$, $\tilde{x}_s(t)$ is the solution of Eq. 4.44 with $u(t) \equiv 0$, and $\tilde{x}_f(t) = \epsilon \Sigma_1(\tilde{x}_s, 0) + \epsilon^2 \Sigma_2(\tilde{x}_s, 0) + \cdots + \epsilon^k \Sigma_k(\tilde{x}_s, 0)$.

Remark 4.6 The result of proposition 4.2 provides the means for characterizing the discrepancy between the solution of the open-loop infinite-dimensional system of Eq. 4.8, $x(t)$, (and thus, the solution of the parabolic PDE system of Eq. 4.1 with $u(t) \equiv 0$) and the solution $\tilde{x}(t) = \tilde{x}_s(t) + \tilde{x}_f(t) = \tilde{x}_s(t) + \epsilon \Sigma_1(\tilde{x}_s(t), 0) + \cdots + \epsilon^k \Sigma_k(\tilde{x}_s(t), 0)$ which is obtained from the $O(\epsilon^k)$ approximation of the open-loop inertial form (i.e., Eq. 4.44 with $u(t) \equiv 0$). In particular, substituting Eq. 4.46 into the equation $x(t) = x_s(t) + x_f(t)$, we have that $x(t) = \tilde{x}_s(t) + \tilde{x}_f(t) + O(\epsilon^{k+1})$ for $t \geq t_b$. Utilizing the definition of order of magnitude, we finally obtain the following characterization for the discrepancy between $x(t)$ and $\tilde{x}(t)$: $||x(t) - \tilde{x}(t)||_2 \leq k_1 \epsilon^{k+1}$ for $t \geq t_b$, where k_1 is a positive real number.

Remark 4.7 Following the proposed approximation procedure, we can show that the $O(\epsilon)$ approximation of $\Sigma(x_s, 0, \epsilon)$ is $\Sigma_0(x_s, 0) = 0$ and the corresponding approximate inertial form is identical to the system of Eq. 4.37 (obtained via Galerkin's method) with $u(t) \equiv 0$. This system does not utilize any information about the structure of the fast subsystem, thus yielding solutions which are only $O(\epsilon)$ close to the solutions of the open-loop system of Eq. 4.8 (proposition 4.1). On the other hand, the $O(\epsilon^2)$ approximation of $\Sigma(x_s, 0, \epsilon)$ can be shown to be of the form:

$$\Sigma(x_s, 0, \epsilon) = \Sigma_0(x_s, 0) + \epsilon\Sigma_1(x_s, 0) = \epsilon(\mathcal{A}_{f\epsilon})^{-1}[-f_f(x_s, 0)]. \tag{4.47}$$

The corresponding open-loop approximate inertial form does utilize information about the structure of the fast subsystem, and thus, allows us to obtain solutions that are $O(\epsilon^2)$ close to the solutions of the open-loop system of Eq. 4.8 (proposition 4.2).

Remark 4.8 The standard approach followed in the literature for the construction of AIMs for systems of the form of Eq. 4.32 with $u(t) \equiv 0$ (see, for example, [25]) is to directly set $\partial x_f/\partial t \equiv 0$, solve the resulting algebraic equations for x_f, and substitute the solution for x_f to the x_s−subsystem of Eq. 4.32 to derive the following ODE system:

$$\frac{dx_s}{dt} = \mathcal{A}_s x_s + f_s(x_s, (\mathcal{A}_f)^{-1}[-f_f(x_s)]). \tag{4.48}$$

It is straightforward to show that the slow system of Eq. 4.48 is identical to the one obtained by using the $O(\epsilon^2)$ approximation for $\Sigma(x_s, 0, \epsilon)$ for the construction of the approximate inertial form. From these arguments, it follows that the singular perturbation formulation that we introduced for the construction of approximate inertial manifolds includes the standard approach as special case.

Remark 4.9 The expansion of u in a power series in ϵ is motivated by our intention to modify the synthesis of the feedback controller appropriately, such that the output of the $O(\epsilon^{k+1})$ approximation of the closed-loop inertial form will be arbitrarily close to the output of the closed-loop PDE system for almost all times (see also remark 4.10).

4.7 Nonlinear Output Feedback Control

4.7.1 A general result

In this subsection, we use the result of proposition 4.2 to establish that a nonlinear finite-dimensional output feedback controller that guarantees stability and enforces output tracking in the ODE system of Eq. 4.44, exponentially stabilizes the closed-loop PDE system and ensures that the discrepancy between the output of the closed-loop ODE system and the output of the closed-loop PDE system is of $O(\epsilon^{k+1})$, provided that ϵ is sufficiently small.

The finite-dimensional output feedback controller which achieves the desired objectives for the system of Eq. 4.44 is constructed through a standard combination of a state feedback controller with a state observer. In particular, we consider a state feedback control law of the general form:

$$\begin{aligned} u &= u_0 + \epsilon u_1 + \cdots + \epsilon^k u_k \\ &= p_0(x_s) + Q_0(x_s)v + \epsilon[p_1(x_s) + Q_1(x_s)v] \\ &\quad + \cdots + \epsilon^k[p_k(x_s) + Q_k(x_s)v] \end{aligned} \tag{4.49}$$

where $p_0(x_s), \ldots, p_k(x_s)$ are smooth vector functions, $Q_0(x_s), \ldots, Q_k(x_s)$ are smooth matrices, and $v \in \mathbb{R}^l$ is the constant reference input vector (see remark 4.10 for a procedure for the synthesis of the control law, i.e., the explicit computation of $[p_0(x_s), \ldots, p_k(x_s), Q_0(x_s), \ldots, Q_k(x_s)]$). The following m-dimensional state observer is also considered for the implementation of the state feedback law of Eq. 4.49:

$$\begin{aligned} \frac{d\eta}{dt} &= \mathcal{A}_s\eta + \mathcal{B}_s(p_0(\eta) + Q_0(\eta)v + \epsilon[p_1(\eta) + Q_1(\eta)v] + \cdots \\ &\quad + \epsilon^k[p_k(\eta) + Q_k(\eta)v]) + f_s(\eta, \epsilon\Sigma_1(\eta, u) + \epsilon^2\Sigma_2(\eta, u) \\ &\quad + \cdots + \epsilon^k\Sigma_k(\eta, u)) + L(q - [\mathcal{Q}\eta + \mathcal{Q}\{\epsilon\Sigma_1(\eta, u) \\ &\quad + \epsilon^2\Sigma_2(\eta, u) + \cdots + \epsilon^k\Sigma_k(\eta, u)\}]) \end{aligned} \tag{4.50}$$

where $\eta \in \mathcal{H}_s$ denotes the observer state vector, and L is a matrix chosen so that the eigenvalues of the matrix

$$\begin{aligned} C_L = \mathcal{A}_s + \frac{\partial f_s}{\partial \eta}_{(\eta=\eta_s, u=u_s)} - L\Big[\mathcal{Q} + \mathcal{Q}\Big\{\frac{\partial}{\partial \eta}(\epsilon\Sigma_1(\eta, u(\eta)) \\ + \epsilon^2\Sigma_2(\eta, u(\eta)) + \cdots + \epsilon^k\Sigma_k(\eta, u(\eta))_{(\eta=\eta_s, u=u_s)}\Big\}\Big] \end{aligned}$$

lie in the open left half of the complex plane, where η_s, u_s denote the steady-state values around which the linearization of the system of Eq. 4.50 takes place, and

$$\frac{\partial f_s}{\partial \eta}, \quad \frac{\partial}{\partial \eta}(\epsilon\Sigma_1(\eta, u(\eta)) + \epsilon^2\Sigma_2(\eta, u(\eta)) + \cdots + \epsilon^k\Sigma_k(\eta, u(\eta)))$$

are Jacobian matrices of appropriate dimensions. The finite-dimensional output feedback controller resulting from the combination of the state feedback controller of Eq. 4.49 with the state observer of Eq. 4.50 takes the form:

$$\begin{aligned} \frac{d\eta}{dt} &= \mathcal{A}_s\eta + \mathcal{B}_s\,(p_0(\eta) + Q_0(\eta)v + \epsilon[p_1(\eta) + Q_1(\eta)v] \\ &\quad + \cdots + \epsilon^k[p_k(\eta) + Q_k(\eta)v])) + f_s(\eta, \epsilon\Sigma_1(\eta, u) + \epsilon^2\Sigma_2(\eta, u) \end{aligned}$$

$$
\begin{aligned}
&+\cdots+\epsilon^k\Sigma_k(\eta,u))+L(q-[\mathcal{Q}\eta+\mathcal{Q}\{\Sigma_0(\eta,u)\\
&+\epsilon\Sigma_1(\eta,u)+\epsilon^2\Sigma_2(\eta,u)+\cdots+\epsilon^k\Sigma_k(\eta,u)\}])\\
u=&\ p_0(\eta)+Q_0(\eta)v+\epsilon[p_1(\eta)+Q_1(\eta)v]+\cdots+\epsilon^k[p_k(\eta)+Q_k(\eta)v].
\end{aligned}
\tag{4.51}
$$

We note that the static component of the above controller does not use feedback of the fast state vector x_f in order to avoid destabilization of the fast modes of the closed-loop system. Assumption 4.4 states the desired control objectives under the controller of Eq. 4.51.

Assumption 4.4 *The finite-dimensional output feedback controller of the form of Eq. 4.51 exponentially stabilizes the $O(\epsilon^{k+1})$ approximation of the closed-loop inertial form and ensures that its outputs $y_s^i(t)$, $i=1,\ldots,l$, are the solutions of a known l-dimensional ODE system of the form $\phi(y_s^{i(r_i)}, y_s^{i(r_i-1)}, \ldots, y_s^i, v)=0$, where ϕ is a vector function and r_i is an integer.*

Theorem 4.1 provides a precise characterization of the stability and closed-loop transient performance enforced by the controller of Eq. 4.51 in the closed-loop PDE system (the proof can be found in Appendix C).

Theorem 4.1 *Consider the parabolic infinite dimensional system of Eq. 4.8, for which assumptions 4.1 and 4.4 hold. Then, there exist positive real numbers $\tilde{\mu}_1, \tilde{\mu}_2, \tilde{\epsilon}^*$ such that if $|x_s(0)|\leq\tilde{\mu}_1$, $||x_f(0)||_2\leq\tilde{\mu}_2$ and $\epsilon\in(0,\tilde{\epsilon}^*]$, then the controller of Eq. 4.51:*

(a) guarantees exponential stability of the closed-loop system, and

(b) ensures that the outputs of the closed-loop system satisfy for all $t\in[t_b,\infty)$:

$$
y^i(t)=y_s^i(t)+O(\epsilon^{k+1}),\quad i=1,\ldots,l \tag{4.52}
$$

where $y_s^i(t)$ is the i-th output of the $O(\epsilon^{k+1})$ approximation of the closed-loop inertial form.

Remark 4.10 The construction of the state feedback law of Eq. 4.49, to ensure that the control objectives stated in assumption 4.4 are enforced in the $O(\epsilon^{k+1})$ approximation of the closed-loop inertial form, can be performed following a sequential procedure. Specifically, the component $u_0=p_0(x_s)+Q_0(x_s)v$ can be initially synthesized on the basis of the $O(\epsilon)$ approximation of the inertial form (Eq. 4.37); then the component $u_1=p_1(x_s)+Q_1(x_s)v$ can be synthesized on the basis of the $O(\epsilon^2)$ approximation of the inertial form. In general, at the k-th step, the component $u_k=p_k(x_s)+Q_k(x_s)v$ can be synthesized on the basis of the $O(\epsilon^{k+1})$ approximation of the inertial form (Eq. 4.44). The synthesis of $[p_\nu(x_s), Q_\nu(x_s)]$, $\nu=0,\ldots,k$, can be performed, at each step, utilizing standard geometric control methods for nonlinear ODEs (see subsection 4.7.2 below).

Remark 4.11 The implementation of the controller of Eq. 4.51 requires the explicit computation of the vector function $\Sigma_k(\eta, u)$. However, $\Sigma_k(\eta, u)$ has an infinite-dimensional range and therefore cannot be implemented in practice. Instead, a finite-dimensional approximation of $\Sigma_k(\eta, u)$, say $\Sigma_{kt}(\eta, u)$, can be derived by keeping the first $\bar{m}$ elements of $\Sigma_k(\eta, u)$ and neglecting the remaining infinite ones. Clearly, as $\bar{m} \to \infty$, $\Sigma_{kt}(\eta, u)$ approaches $\Sigma_k(\eta, u)$. This implies that by picking $\bar{m}$ to be sufficiently large, the controller of Eq. 4.51 with $\Sigma_{kt}(\eta, u)$ instead of $\Sigma_k(\eta, u)$ guarantees stability and enforces the requirement of Eq. 4.52 in the closed-loop infinite-dimensional system.

4.7.2 Controller synthesis

In this subsection, we synthesize a finite-dimensional output feedback controller for the system of Eq. 4.8 on the basis of the system of Eq. 4.44, using geometric control methods. To this end, we will initially review the concepts of relative order and characteristic matrix that will be used in the subsequent subsection to synthesize the controller. Referring to the system of Eq. 4.37, we set, in order to simplify the notation, $\mathcal{A}_s x_s + f_s(x_s, 0) = f_0(x_s)$, $\mathcal{B}_s = g_0$, $\mathcal{C} x_s = h_0(x_s)$. The relative order of the output y_s^i with respect to the vector of manipulated inputs u is defined as the smallest integer r_i for which

$$\left[L_{g_0^1} L_{f_0}^{r_i-1} h_0^i(x_s) \ \cdots \ L_{g_0^l} L_{f_0}^{r_i-1} h_0^i(x_s)\right] \not\equiv [0 \ \cdots \ 0] \tag{4.53}$$

or $r_i = \infty$ if such an integer does not exist. Furthermore, the matrix:

$$\mathcal{C}_0(x_s) = \begin{bmatrix} L_{g_0^1} L_{f_0}^{r_1-1} h_0^1(x_s) & \cdots & L_{g_0^l} L_{f_0}^{r_1-1} h_0^1(x_s) \\ L_{g_0^1} L_{f_0}^{r_2-1} h_0^2(x_s) & \cdots & L_{g_0^l} L_{f_0}^{r_2-1} h_0^2(x_s) \\ \vdots & & \\ L_{g_0^1} L_{f_0}^{r_l-1} h_0^l(x_s) & \cdots & L_{g_0^l} L_{f_0}^{r_l-1} h_0^l(x_s) \end{bmatrix} \tag{4.54}$$

is the characteristic matrix of the system. For simplicity, we will assume that $det(\mathcal{C}_0(x_s)) \neq 0$. This assumption is made to simplify the presentation of the controller synthesis results and can be relaxed if instead of static output feedback we use dynamic output feedback (the reader may refer to [86] for details).

Theorem 4.2 below provides the synthesis formula of the output feedback controller and conditions that guarantee closed-loop stability in the case of considering an $O(\epsilon^2)$ approximation of the exact slow system for the synthesis of the controller. The derivation of synthesis formulas for higher-order approximations of the output feedback controller is notationally complicated, although conceptually straightforward, and thus will be omitted for reasons of brevity. The proof of theorem 4.2 follows from a direct application of the result of theorem 4.1 and is omitted for brevity.

Theorem 4.2 *Consider the parabolic PDE system of Eq. 4.8, for which assumptions 4.1 and 4.2 hold. Consider also the $O(\epsilon^2)$ approximation of the inertial form and assume that its characteristic matrix $C_1(x_s, \epsilon)$ is invertible $\forall\ x_s \in \mathcal{H}_s,\ \epsilon \in [0, \epsilon^*]$. Suppose also that the following conditions hold:*

i) The roots of the equation:

$$det(B(s)) = 0 \tag{4.55}$$

where $B(s)$ is a $l \times l$ matrix, whose (i,j)-th element is of the form $\sum_{k=0}^{r_i} \beta_{jk}^i s^k$, lie in the open left half of the complex plane, and

ii) The unforced ($v \equiv 0$) zero dynamics of the $O(\epsilon^2)$ approximation of the inertial form are locally exponentially stable.

Then, there exist constants μ_1, μ_2, ϵ^* such that if $|x_s(0)| \leq \mu_1$, $\|x_f(0)\|_2 \leq \mu_2$ and $\epsilon \in (0, \epsilon^*]$. Then if $\eta(0) = x_s(0)$, the dynamic output feedback controller:

$$\begin{aligned}
\frac{d\eta}{dt} &= \mathcal{A}_s\eta + \mathcal{B}_s u(\eta) + f_s(\eta, \epsilon(\mathcal{A}_{f\epsilon})^{-1}[-\mathcal{B}_f u_0(\eta) - f_f(\eta, 0)]) \\
&\quad + L(q - [\mathcal{Q}\eta + \epsilon(\mathcal{A}_{f\epsilon})^{-1}[-\mathcal{B}_f u_0(\eta) - f_f(\eta, 0)]]) \\
u &= u_0(\eta) + \epsilon u_1(\eta) := \left\{[\beta_{1r_1} \cdots \beta_{lr_l}] C_0(\eta)\right\}^{-1} \left\{ v - \sum_{i=1}^{l}\sum_{k=0}^{r_i} \beta_{ik} L_{f_0}^k h_0^i(\eta) \right\} \\
&\quad + \epsilon \left\{[\beta_{1r_1} \cdots \beta_{lr_l}] C_1(\eta, \epsilon)\right\}^{-1} \left\{ v - \sum_{i=1}^{l}\sum_{k=0}^{r_i} \beta_{ik} L_{f_1}^k h_1^i(\eta, \epsilon) \right\}
\end{aligned} \tag{4.56}$$

(a) guarantees exponential stability of the closed-loop system, and

(b) ensures that the outputs of the closed-loop system satisfy for all $t \in [t_b, \infty)$:

$$y^i(t) = y_s^i(t) + O(\epsilon^2), \quad i = 1, \ldots, l \tag{4.57}$$

where t_b is the time required for the off-manifold fast transients to decay to zero exponentially, and $y_s^i(t)$ is the solution of:

$$\sum_{i=1}^{l}\sum_{k=0}^{r_i} \beta_{ik} \frac{d^k y_s^i}{dt^k} = v, \quad i = 1, \ldots, l. \tag{4.58}$$

Remark 4.12 Note that in the presence of small initialization errors of the observer states (i.e., $\eta(0) \neq x_s(0)$), uncertainty in the model parameters and external disturbances, although a slight deterioration of the performance may occur, (i.e., the requirement of Eq. 4.57 will not be exactly imposed in the closed-loop system), the output feedback controller of theorem 4.2 will enforce exponential stability and asymptotic output tracking in the closed-loop system.

Remark 4.13 Whenever the eigenfunctions ϕ_j of the operator $\mathcal{A}$ cannot be calculated analytically, one can still use Galerkin's method to perform model reduction by using the "empirical eigenfunctions" of the PDE system as basis functions in $\mathcal{H}_s$ and $\mathcal{H}_f$ (such "empirical eigenfunctions" can be extracted from numerical simulations or experimental data using Karhunen–Loève expansion; see Appendix F for a description of this method and [124, 125, 66, 35, 80, 107, 19] for more details). Whenever empirical eigenfunctions are used as basis functions within Galerkin's model reduction framework, the interconnection of Eq. 4.32 takes the form

$$
\begin{aligned}
\frac{dx_s}{dt} &= \mathcal{A}_s x_s + \mathcal{A}_{sf} x_f + \mathcal{B}_s u + f_s(x_s, x_f) \\
\frac{\partial x_f}{\partial t} &= \mathcal{A}_{fs} x_s + \mathcal{A}_f x_f + \mathcal{B}_f u + f_f(x_s, x_f) \\
y &= \mathcal{C} x_s + \mathcal{C} x_f, \quad q = \mathcal{Q} x_s + \mathcal{Q} x_f
\end{aligned}
\tag{4.59}
$$

where $\mathcal{A}_{sf}$, $\mathcal{A}_{fs}$ are bounded operators, and the terms $\mathcal{A}_{sf} x_f$, $\mathcal{A}_{fs} x_s$ represent the modeling errors resulting from the use of empirical eigenfunctions in the model reduction instead of the exact eigenfunctions of $\mathcal{A}$. The problem of synthesizing low-dimensional output feedback controllers for systems of Eq. 4.59 can be addressed by employing the proposed method, provided that the modeling errors $\mathcal{A}_{fs} x_f$, $\mathcal{A}_{sf} x_s$ are sufficiently small.

Remark 4.14 We note that the validity of the approach followed here to synthesize nonlinear controllers relies on the large separation of slow and fast modes of the spatial differential operator of the parabolic PDE system. Clearly, this approach is not applicable to hyperbolic PDE systems (i.e., convection-reaction processes) where the eigenmodes cluster along vertical or nearly vertical asymptotes in the complex plane. An alternative approach to control of nonlinear parabolic PDE systems, which leads to distributed control laws, involves the use of generalized invariants [105].

4.8 Applications

In this section, we illustrate, through computer simulations, the application of the developed method for control of nonlinear parabolic PDE systems to the catalytic rod and the non-isothermal tubular reactor.

4.8.1 Catalytic rod

In this simulation, we study the stabilization of the rod temperature at the unstable spatially uniform steady state, $\bar{x}(z,t) = 0$. For this system, we consider the first eigenvalue as the dominant one (and thus, $\epsilon = \frac{\lambda_1}{\lambda_2} = 0.25$) and use Galerkin's method to derive a scalar ODE that is used for controller design. The controller of Eq. 4.56 is implemented in the simulations with $\beta_0 = 2.0$, $\beta_1 = 1.0$ and $L = 4$.

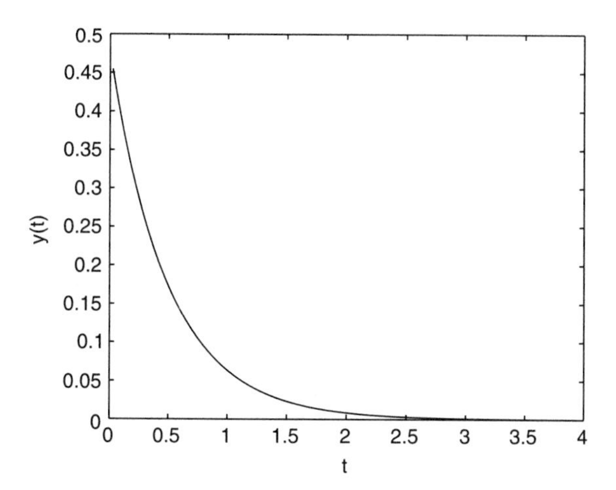

FIGURE 4.5. Closed-loop output profile for nonlinear controller (catalytic rod).

Figures 4.5 and 4.6 show the closed-loop output and manipulated input profiles, respectively, and Figure 4.7 shows the evolution of the closed-loop rod temperature profile for the nonlinear controller starting from a nonzero initial condition. Clearly, the nonlinear controller regulates the temperature profile at $\bar{x}(z,t) = 0$, achieving the control objective. We also implement a linear controller obtained from the Taylor linearization of the nonlinear controller around the operating steady state $\bar{x}(z,t) = 0$. Figures 4.8 and 4.9 show the closed-loop output and manipulated input profiles, respectively, and Figure 4.10 shows the evolution of the closed-loop rod temperature

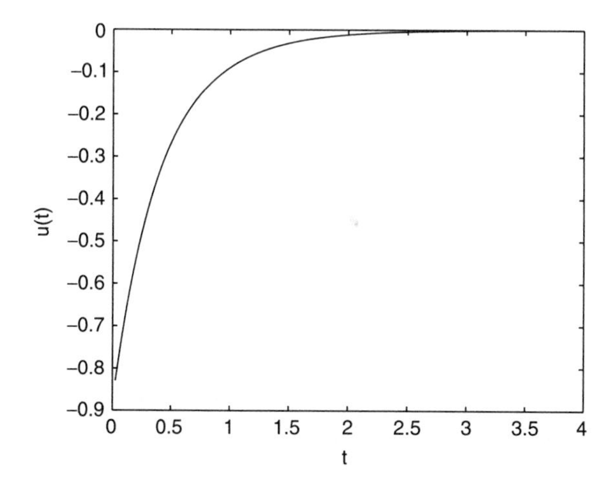

FIGURE 4.6. Manipulated input profile for nonlinear controller (catalytic rod).

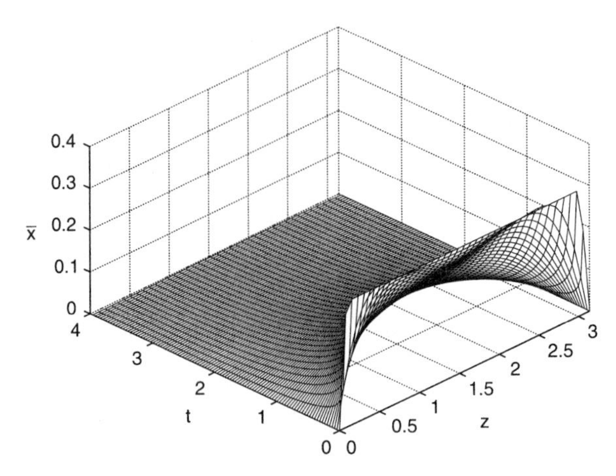

FIGURE 4.7. Profile of evolution of rod temperature for nonlinear controller.

profile for the linear controller starting from the same initial condition as in the previous simulation run. It is clear that the linear controller fails to regulate the temperature profile at the desired steady state, $\bar{x}(z,t) = 0$; and the process moves to another stable steady state characterized by a maximum in the temperature profile (*hot-spot*) in the middle of the rod.

4.8.2 Nonisothermal tubular reactor with recycle

In this simulation, we evaluate the output tracking capabilities of nonlinear controllers synthesized on the basis of the $O(\epsilon)$ and $O(\epsilon^2)$ approximations of the exact slow system. In all the simulations runs, the process is

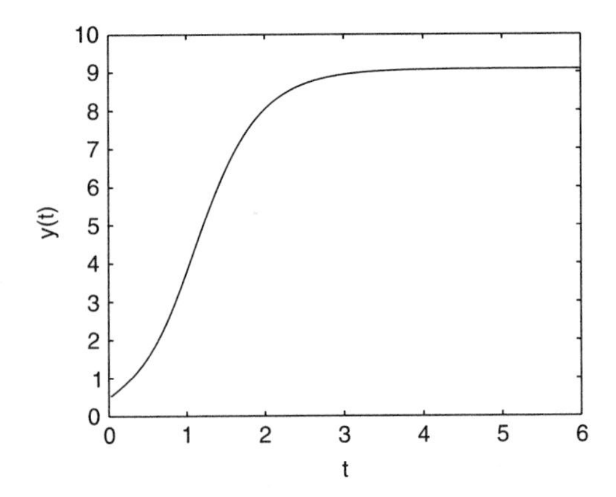

FIGURE 4.8. Closed-loop output profile for linear controller (catalytic rod).

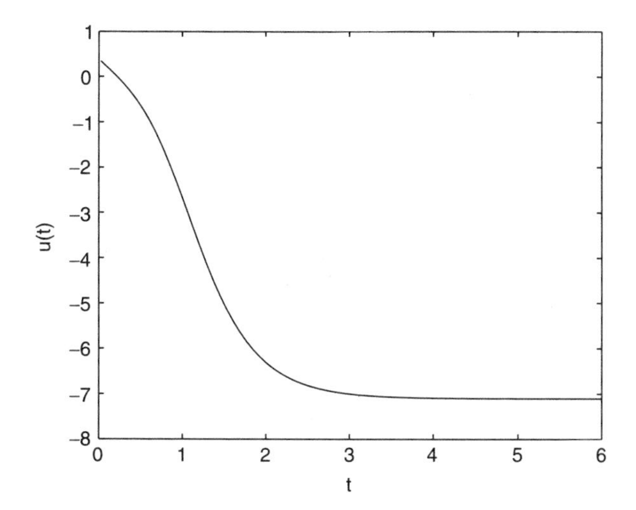

FIGURE 4.9. Manipulated input profile for linear controller (catalytic rod).

initially ($t = 0$) assumed to be at the unstable steady state, and the reference input value is set at $v = 0.12$. The first five modes for each PDE were considered as the dominant ones, which implies that $\epsilon = 0.06$. We use the model reduction procedure of section 4.6 to derive $O(\epsilon)$ and $O(\epsilon^2)$ approximations of the exact slow system. The $O(\epsilon^2)$ approximation of $\Sigma(x_s, u, \epsilon)$ is constructed by retaining the first 6 of the fast modes for each PDE, and discarding the remaining infinite ones (this is because the use of more

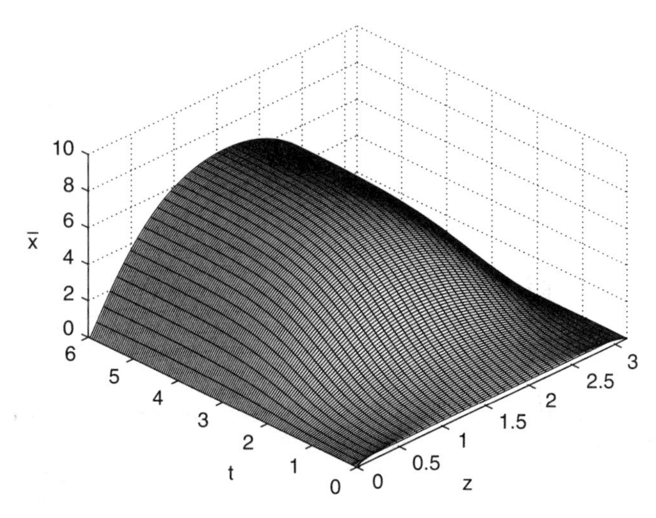

FIGURE 4.10. Profile of evolution of rod temperature for linear controller.

than 6 fast modes provides negligible improvement in the accuracy of the $O(\epsilon^2)$ approximation of the slow system). The relative orders of the nonlinear systems corresponding to the $O(\epsilon)$ and $O(\epsilon^2)$ approximations of the exact slow system are 1, and the zero dynamics of these nonlinear systems are exponentially stable. The nonlinear finite-dimensional controller of Eq. 4.56 is implemented in the simulations with $\beta_1 = 0.5$, $\beta_0 = 1.0$, and $L = [100.0\ 66.2\ 39.4\ 24.0\ 15.6\ 183.5\ 94.8\ 48.0\ 26.9\ 16.7]^T$. The process is initially assumed to be at a spatially nonuniform steady state.

Figure 4.11 shows the output and manipulated input profiles. It is clear that the controller synthesized on the basis of the system which uses an $O(\epsilon^2)$ approximation for $\Sigma(x_s, u, \epsilon)$ provides an excellent performance

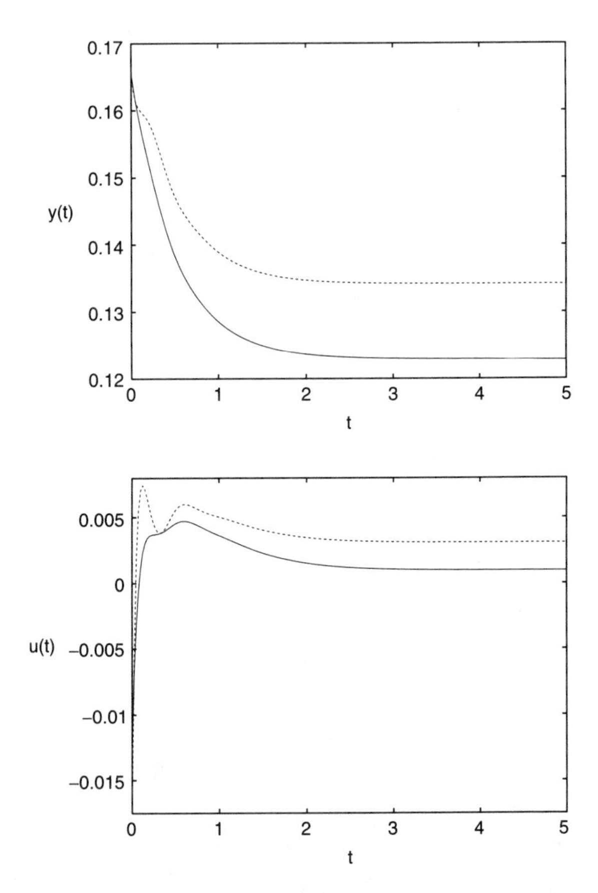

FIGURE 4.11. Comparison of output (top figure) and manipulated input (bottom figure) profiles of the closed-loop system. The dotted lines correspond to a controller based on the slow ODE system with an $O(\epsilon)$ approximation for Σ, while the solid lines correspond to a controller based on the slow ODE system with an $O(\epsilon^2)$ approximation for Σ (tubular reactor with recycle).

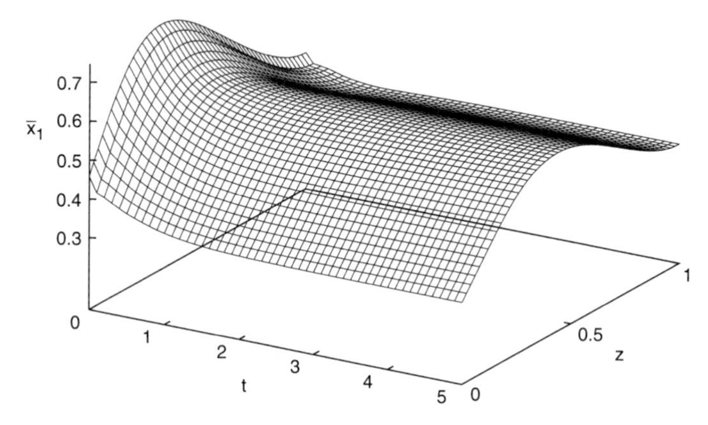

FIGURE 4.12. Profile of evolution of closed-loop reactor temperature ($O(\epsilon^2)$ approximation for Σ).

driving the output (solid line) very close to the new set point (note that as expected, $lim_{t\rightarrow\infty}|y - v| = O(\epsilon^2)$). On the other hand, the controller of the form of Eq. 4.56 with $\epsilon = 0$ drives the output (dotted line) to a neighborhood of the set point (note that $lim_{t\rightarrow\infty}|y - v| = O(\epsilon)$) leading to significant offset (compare with set-point value). Figure 4.12 displays the evolution of the dimensionless temperature of the reactor for the case of using an $O(\epsilon^2)$ approximation for $\Sigma(x_s, u, \epsilon)$. The controller achieves excellent performance, regulating the temperature of the reactor at a spatially nonuniform steady state.

4.9 Conclusions

In this chapter, we presented a method for the synthesis of nonlinear low-dimensional output feedback controllers for nonlinear parabolic PDE systems, for which the eigenspectrum of the spatial differential operator can be partitioned into a finite-dimensional slow one and an infinite-dimensional stable fast one. Combination of Galerkin's method with a novel procedure for the construction of AIMs was used for the derivation of ODE systems of dimension equal to the number of slow modes that yield solutions which are close, up to a desired accuracy, to the ones of the PDE system, for almost all times. These ODE systems were used as the basis for the synthesis of output feedback controllers that guarantee stability and enforce the output of the closed-loop system to follow, up to a desired accuracy, a prespecified response for almost all times. The method was successfully employed to control the temperature profiles of a catalytic rod and a nonisothermal tubular reactor with recycle around unstable steady states.

Chapter 5

Robust Control of Parabolic PDE Systems

5.1 Introduction

This chapter focuses on the control problem for quasi-linear parabolic PDEs with time-varying uncertain variables, for which the eigenspectrum of the spatial differential operator can be partitioned into a finite-dimensional (possibly unstable) slow one and an infinite-dimensional stable fast complement. The objective is to develop a general and practical methodology for the synthesis of nonlinear robust state and output feedback controllers that guarantee boundedness of the state and output tracking with arbitrary degree of asymptotic attenuation of the effect of the uncertain variables on the output of the closed-loop system.

This chapter is structured as follows. Initially, the class of parabolic PDE systems with uncertainty is introduced, and Galerkin's method is used to derive an ODE system of dimension equal to the number of slow modes. This ODE system is subsequently used to synthesize robust state feedback controllers via Lyapunov's direct method. Singular perturbation methods are employed to establish that the degree of asymptotic attenuation of the effect of uncertain variables on the output, enforced by these controllers, is proportional to the degree of separation of the fast and slow modes of the spatial differential operator. For processes in which such a degree of uncertainty attenuation is not sufficient, a sequential procedure, based on the concept of approximate inertial manifold, is developed for the synthesis of robust controllers that achieve arbitrary degree of asymptotic uncertainty attenuation in the closed-loop parabolic PDE system. Then, under the assumption that the number of measurements is equal to the number of slow modes, we propose a procedure for obtaining estimates for the states of the approximate ODE model from the measurements. We show that the use of these estimates in the robust state feedback controller leads to a robust output feedback controller, which enforces the desired properties in the closed-loop system, provided that the separation between the slow and fast eigenvalues is sufficiently large. The developed robust output feedback controllers are successfully applied to a catalytic rod with uncertainty. The results of this chapter were first presented in [36, 37].

5.2 Preliminaries

5.2.1 Parabolic PDE systems with uncertainty

We consider quasi-linear parabolic PDE systems with uncertain variables, with a state-space description of the form:

$$
\begin{aligned}
\frac{\partial \bar{x}}{\partial t} &= A\frac{\partial \bar{x}}{\partial z} + B\frac{\partial^2 \bar{x}}{\partial z^2} + wb(z)u + f(\bar{x}) + W(\bar{x}, r(z)\theta(t)) \\
y^i &= \int_{\alpha}^{\beta} c^i(z)k\bar{x}\, dz, \quad i = 1, \ldots, l \\
q^{\kappa} &= \int_{\alpha}^{\beta} s^{\kappa}(z)\omega\bar{x}\, dz, \quad \kappa = 1, \ldots, p
\end{aligned}
\tag{5.1}
$$

subject to the boundary conditions:

$$
\begin{aligned}
C_1\bar{x}(\alpha, t) + D_1\frac{\partial \bar{x}}{\partial z}(\alpha, t) &= R_1 \\
C_2\bar{x}(\beta, t) + D_2\frac{\partial \bar{x}}{\partial z}(\beta, t) &= R_2
\end{aligned}
\tag{5.2}
$$

and the initial condition:

$$
\bar{x}(z, 0) = \bar{x}_0(z) \tag{5.3}
$$

where $W(\bar{x}, r(z)\theta(t))$ is a nonlinear vector function, $\theta(t) = [\theta_1(t)\theta_2(t)\cdots\theta_q(t)] \in \mathbb{R}^q$ denotes the vector of uncertain variables, which may include uncertain process parameters or exogenous disturbances, and $r(z) = [r_1(z)\ \cdots\ r_q(z)]$, where $r_k(z)$ is a known smooth function of z that specifies the position of action of the uncertain variable θ_k on $[\alpha, \beta]$. The rest of the notation was given in subsection 4.2.1.

Systems of the form of Eq. 5.1 describe the majority of diffusion-convection-reaction processes with uncertain variables for which the diffusion coefficient and the thermal conductivity are independent of temperature and concentrations, and thus, the corresponding spatial differential operators are linear. They are characterized by linear appearances of the manipulated input and the controlled and measured output. These features are common in most practical applications where the wall temperature is usually chosen to be the manipulated input (see the packed-bed reactor example of subsection 4.3.2), and the controlled and measured outputs are usually some of the state variables of the process (e.g., concentrations and temperature, respectively). The nonlinear and additive appearance of the vector of uncertain variables in Eq. 5.1 allows accounting for most uncertain process parameters and exogenous disturbances including heat of reactions, pre-exponential constants, activation energies, and temperature and concentration of lateral inlet streams.

Similar to the presentation of subsection 4.2.2, we formulate the parabolic PDE system of Eq. 5.1 as an infinite dimensional system in the Hilbert space $\mathcal{H}([\alpha,\beta],\mathbb{R}^n)$. Defining the state function x on $\mathcal{H}([\alpha,\beta],\mathbb{R}^n)$ as:

$$x(t)=\bar{x}(z,t), \quad t>0, \quad z\in[\alpha,\beta], \tag{5.4}$$

the differential operator:

$$\mathcal{A}x = A\frac{\partial \bar{x}}{\partial z}+B\frac{\partial^2 \bar{x}}{\partial z^2},$$
$$x\in D(\mathcal{A})=\Big\{x\in\mathcal{H}([\alpha,\beta];\mathbb{R}^n);\quad C_1\bar{x}(\alpha,t)+D_1\frac{\partial \bar{x}}{\partial z}(\alpha,t)=R_1, \tag{5.5}$$
$$C_2\bar{x}(\beta,t)+D_2\frac{\partial \bar{x}}{\partial z}(\beta,t)=R_2\Big\}$$

the manipulated input, controlled, and measured output operators:

$$\mathcal{B}u=wbu, \quad \mathcal{C}x=(c,kx), \quad \mathcal{Q}x=(s,\omega x) \tag{5.6}$$

and the nonlinear terms $f(x(t))=f(\bar{x}(z,t))$ and $\mathcal{W}(x,\theta)=W(\bar{x},r\theta)$, the system of Eqs. 5.1–5.3 takes the form:

$$\begin{aligned}\dot{x}&=\mathcal{A}x+\mathcal{B}u+f(x)+\mathcal{W}(x,\theta), \quad x(0)=x_0\\ y&=\mathcal{C}x, \quad q=\mathcal{Q}x\end{aligned} \tag{5.7}$$

where $x_0=\bar{x}_0(z)$. We assume that the nonlinear terms $f(x),\mathcal{W}(x,\theta)$ are locally Lipschitz with respect to their arguments and satisfy $f(0)=0$, $\mathcal{W}(0,0)=0$.

We also assume that $\sigma(\mathcal{A})$ satisfies the properties of assumption 4.1, which implies that $\mathcal{A}$ generates a strongly continuous semigroup of bounded linear operators $U(t)$ which implies that the generalized solution of the system of Eq. 5.7 is given by:

$$x(t)=U(t)x_0+\int_0^t U(t-\tau)(\mathcal{B}u(\tau)+f(x(\tau))+\mathcal{W}(x(\tau),\theta(\tau)))\,d\tau. \tag{5.8}$$

$U(t)$ satisfies the following growth property:

$$\|U(t)\|_2\le K_1e^{a_1t}, \quad \forall\, t\ge 0 \tag{5.9}$$

where K_1, a_1 are positive real numbers, with $K_1\ge 1$ and $a_1\ge Re[\lambda_1]$.

The following example illustrates modeling of a diffusion-reaction process in the form of Eq. 5.7.

5.2.2 Illustrative example: catalytic rod with uncertainty

We consider the catalytic rod of subsection 4.3.1 and assume that both sides are insulated and the heat of reaction is unknown and time varying. In this case, the mathematical model which describes the spatiotemporal evolution of the rod temperature consists of the following quasi-linear parabolic PDE:

$$\frac{\partial \bar{x}}{\partial t} = \frac{\partial^2 \bar{x}}{\partial z^2} + \beta_T e^{-\frac{\gamma}{1+\bar{x}}} + \beta_U(b(z)u(t) - \bar{x}) - \beta_{T,n} e^{-\gamma} \tag{5.10}$$

subject to the nonflux boundary conditions:

$$\frac{\partial \bar{x}}{\partial z}(0,t) = 0, \quad \frac{\partial \bar{x}}{\partial z}(\pi,t) = 0 \tag{5.11}$$

and the initial condition:

$$\bar{x}(z,0) = \bar{x}_0(z) \tag{5.12}$$

where $\bar{x}$ denotes the dimensionless temperature of the rod, β_T denotes a dimensionless heat of reaction, $\beta_{T,n}$ denotes a *nominal* dimensionless heat of reaction, γ denotes a dimensionless activation energy, β_U denotes a dimensionless heat transfer coefficient, and u denotes the temperature of the cooling medium.

The control objective is the regulation of the temperature profile in the rod through manipulation of the temperature of the cooling medium in the presence of time-varying uncertainty in the heat of the reaction β_T. For the control of the process, we assume that there is available one control actuator, with distribution function $b(z) = 1$, and one point measurement sensor placed at the center of the rod. Therefore, the uncertain variable, measured, and controlled outputs are defined as:

$$\theta(t) = \beta_T - \beta_{T,n}, \quad y(t) = \int_0^1 \bar{x}(z,t)\,dz, \quad q(t) = \int_0^1 \delta(z-0.5)\bar{x}(z,t)\,dz. \tag{5.13}$$

where $\delta(\cdot)$ is the standard Dirac function. For this example, the Hilbert space is $\mathcal{H}([0,1],\mathbb{R})$, the state function on $\mathcal{H}([0,1],\mathbb{R})$ is defined as $x(t) = \bar{x}(z,t)$, and the operator $\mathcal{A}$ takes the form:

$$\mathcal{A}x = \frac{\partial^2 \bar{x}}{\partial z^2},$$
$$x \in D(\mathcal{A}) = \left\{x \in \mathcal{H}([0,1];\mathbb{R});\ \frac{\partial \bar{x}}{\partial z}(0,t) = 0; \quad z = 1,\ \frac{\partial \bar{x}}{\partial z}(1,t) = 0\right\}. \tag{5.14}$$

Furthermore, the manipulated input, controlled output, and measured output operators can be defined as:

$$\mathcal{B}u = u, \quad \mathcal{C}x = (1,x), \quad \mathcal{S}x = (\delta(z-0.5),x) \tag{5.15}$$

and the nonlinear terms $f(x)$, $\mathcal{W}(x,\theta)$ as:

$$f(x) = \beta_{T,n} e^{-\frac{\gamma}{1+\bar{x}}} - \beta_{T,n} e^{-\gamma}, \quad \mathcal{W}(x,\theta) = e^{-\frac{\gamma}{1+\bar{x}}}\theta(t) \tag{5.16}$$

to write the PDE system of Eqs. 5.10–5.12 together with the appropriate specifications for manipulated input, and uncertain variable, controlled, and measured output in the form of Eq. 5.7. The eigenvalue problem for $\mathcal{A}$

can be solved analytically and its solution is of the form:

$$\begin{aligned} \lambda_j &= -j^2\pi^2, \quad j = 0, \ldots, \infty, \\ \phi_0 = 1, \quad \phi_j &= \sqrt{2}\ cos(j\pi z), \quad j = 1, \ldots, \infty. \end{aligned} \tag{5.17}$$

In the remainder of this chapter, we will first proceed with the presentation of the formulation and solution of the robust state feedback control problem, and then continue with the robust output feedback control problem.

5.3 Robust State Feedback Control of Parabolic PDE Systems

5.3.1 Problem formulation—Finite-dimensional approximation

The objective of this section is to synthesize nonlinear robust state feedback controllers for the infinite-dimensional system of Eq. 5.7 on the basis of appropriate finite-dimensional systems, which enforce the following properties in the closed-loop system: (a) boundedness of the state, (b) output tracking for changes in the reference input, and (c) asymptotic attenuation of the effect of uncertain variables on the output.

To develop a solution to the above problem, we will initially transform the system of Eq. 5.7 into an equivalent set of infinite ordinary differential equations. Letting $\mathcal{H}_s$, $\mathcal{H}_f$ be two subspaces of $\mathcal{H}$, defined as $\mathcal{H}_s = span\{\phi_1, \phi_2, \ldots, \phi_m\}$ and $\mathcal{H}_f = span\{\phi_{m+1}, \phi_{m+2}, \ldots,\}$, and defining the orthogonal projection operators P_s and P_f such that $x_s = P_s x$, $x_f = P_f x$, the state x of the system of Eq. 5.7 can be decomposed to:

$$x = x_s + x_f = P_s x + P_f x. \tag{5.18}$$

Applying P_s and P_f to the system of Eq. 5.7 and using the above decomposition for x, the system of Eq. 5.7 can be equivalently written in the following form:

$$\begin{aligned} \frac{dx_s}{dt} &= \mathcal{A}_s x_s + \mathcal{B}_s u + f_s(x_s, x_f) + \mathcal{W}_s(x_s, x_f, \theta) \\ \frac{\partial x_f}{\partial t} &= \mathcal{A}_f x_f + \mathcal{B}_f u + f_f(x_s, x_f) + \mathcal{W}_f(x_s, x_f, \theta) \\ y &= \mathcal{C} x_s + \mathcal{C} x_f \\ x_s(0) = P_s x(0) &= P_s x_0, \quad x_f(0) = P_f x(0) = P_f x_0 \end{aligned} \tag{5.19}$$

where $\mathcal{A}_s = P_s\mathcal{A}$, $\mathcal{B}_s = P_s\mathcal{B}$, $f_s = P_s f$, $\mathcal{W}_s = P_s\mathcal{W}$, $\mathcal{A}_f = P_f\mathcal{A}$, $\mathcal{B}_f = P_f\mathcal{B}$, $f_f = P_f f$, $\mathcal{W}_f = P_f\mathcal{W}$. In the above system, $\mathcal{A}_s$ is a diagonal matrix of dimension $m \times m$ of the form $\mathcal{A}_s = diag\{\lambda_j\}$, $f_s(x_s, x_f)$, $f_f(x_s, x_f)$, $\mathcal{W}_s(x_s, x_f, \theta)$, $\mathcal{W}_f(x_s, x_f, \theta)$ are Lipschitz vector functions, and $\mathcal{A}_f$ is an unbounded differential operator that generates a strongly continuous exponentially stable semigroup (following from part 3 of assumption 4.1 and the selection of $\mathcal{H}_s$ and $\mathcal{H}_f$). In the remainder of this subsection, we will assume, in order to simplify the presentation of our results, that the performance specification functions $c^i(z)$ are chosen such that $\mathcal{C}x_f \equiv 0$ (a selection for $c^i(z)$ which ensures that $\mathcal{C}x_f \equiv 0$ is $c^i(z) = \phi_j(z)$ for $j = 1, \ldots, m$).

In order to derive precise conditions that guarantee that the proposed controllers enforce the desired properties in the infinite-dimensional closed-loop system and precisely characterize the degree of asymptotic attenuation of the uncertain variables on the output, we will formulate the system of Eq. 5.19 within the framework of singular perturbations. Such a formulation is motivated by the fact that the system of Eq. 5.19 exhibits two-time-scale behavior (which is a consequence of part 3 of assumption 4.1). Using that $\epsilon = \frac{|Re\ \{\lambda_1\}|}{|Re\ \{\lambda_{m+1}\}|}$, the system of Eq. 5.19 can be written in the following form:

$$\begin{aligned}
\frac{dx_s}{dt} &= \mathcal{A}_s x_s + \mathcal{B}_s u + f_s(x_s, x_f) + \mathcal{W}_s(x_s, x_f, \theta) \\
\epsilon \frac{\partial x_f}{\partial t} &= \mathcal{A}_{f\epsilon} x_f + \epsilon \mathcal{B}_f u + \epsilon f_f(x_s, x_f) + \epsilon \mathcal{W}_f(x_s, x_f, \theta) \\
y &= \mathcal{C} x_s
\end{aligned} \tag{5.20}$$

where $\mathcal{A}_{f\epsilon}$ is an unbounded differential operator defined as $\mathcal{A}_{f\epsilon} = \epsilon \mathcal{A}_f$. Introducing the fast time scale $\tau = \frac{t}{\epsilon}$ and setting $\epsilon = 0$, we obtain the following infinite-dimensional fast subsystem from the system of Eq. 5.20:

$$\frac{\partial x_f}{\partial \tau} = \mathcal{A}_{f\epsilon} x_f. \tag{5.21}$$

From the fact that $Re\{\lambda_{m+1}\} < 0$ and the definition of ϵ, we have that the above system is globally exponentially stable. Setting $\epsilon = 0$ in the system of Eq. 5.20, we have that $x_f = 0$, and thus, the finite-dimensional slow system takes the form:

$$\begin{aligned}
\frac{dx_s}{dt} &= \mathcal{A}_s x_s + f_s(x_s, 0) + \mathcal{B}_s u + \mathcal{W}_s(x_s, 0, \theta) \\
&=: F_0(x_s) + \sum_{i=1}^{l} \mathcal{B}_0^i u_0^i + \mathcal{W}_0(x_s, 0, \theta) \\
y_s^i &= \mathcal{C}^i x_s =: h_0^i(x_s)
\end{aligned} \tag{5.22}$$

where the subscript s in y_s^i denotes that this output is associated with an ODE system and the subscript 0 in $(F_0, \mathcal{B}_0^i, u_0^i, \mathcal{W}_0, h_0^i)$ denotes that they are elements of the $O(\epsilon)$ approximation of the x_s-subsystem. In the above system, a new notation was introduced to facilitate the presentation of the robust controller synthesis results in the next subsections. We finally note that the above system is identical to the one obtained by applying the standard Galerkin's method to the system of Eq. 5.7, keeping the first m ODEs and completely neglecting the x_f-subsystem.

5.3.2 Robust state feedback controller synthesis

In this subsection, we synthesize robust controllers for the system of Eq. 5.1 on the basis of the finite-dimensional system of Eq. 5.22, using Lyapunov's direct method and precisely characterize the ultimate uncertainty attenuation level. Motivated by the requirement of output tracking with attenuation of the effect of the uncertainty on the output and the fact that the system of Eq. 5.22 includes only uncertain variables that appear in an additive fashion, we consider the synthesis of robust control laws of the form:

$$u_0 = p_0(x_s) + Q_0(x_s)\bar{v} + r_0(x_s, t) \tag{5.23}$$

where $p_0(x_s)$, $r_0(x_s, t)$ are vector functions, $Q_0(x_s)$ is a matrix, and $\bar{v}$ is a vector of the form $\bar{v} = \mathcal{V}(v_i, v_i^{(1)}, \ldots, v_i^{(r_i)})$, where $\mathcal{V}(v_i, v_i^{(1)}, \ldots, v_i^{(r_i)})$ is a smooth vector function, $v_i^{(k)}$ is the k-th time derivative of the external reference input v_i (which is assumed to be a smooth function of time), and r_i is a positive integer. The control law of Eq. 5.23 comprises the component $p_0(x_s) + Q_0(x_s)\bar{v}$, which is responsible for the output tracking and stabilization of the closed-loop slow system, and the component $r_0(x_s, t)$, which is responsible for the asymptotic attenuation of the effect of the uncertain variables on the outputs of the closed-loop slow system.

In order to derive an explicit formula of the control law of Eq. 5.23, we will impose the following three assumptions on the system of Eq. 5.22. We initially assume that there exists a coordinate transformation that renders the system of Eq. 5.22 partially linear and that the time derivatives of the output y^i up to order $r_i - 1$ are independent of the vector of uncertain variables θ. Assumption 5.1 that follows states precisely this requirement:

Assumption 5.1 *Referring to the system of Eq. 5.22, there exist a set of integers $(r_1, r_2, \ldots, r_l)$ and a coordinate transformation $(\zeta, \eta) = T(x_s, \theta)$ such that the representation of the system, in the coordinates (ζ, η), takes the form:*

$$
\begin{aligned}
\dot{\zeta}_1^{(1)} &= \zeta_2^{(1)} \\
&\vdots \\
\dot{\zeta}_{r_1-1}^{(1)} &= \zeta_{r_1}^{(1)} \\
\dot{\zeta}_{r_1}^{(1)} &= L_{F_0}^{r_1} h_0^1(T^{-1}(\zeta, \eta, \theta)) + \sum_{i=1}^{l} L_{B_0^i} L_{F_0}^{r_1-1} h_0^1(T^{-1}(\zeta, \eta, \theta)) u_0^i \\
&\quad + L_{W_0} L_{F_0}^{r_1-1} h_0^1(T^{-1}(\zeta, \eta, \theta)) \\
&\vdots \\
\dot{\zeta}_1^{(l)} &= \zeta_2^{(l)}
\end{aligned}
$$

$$\begin{aligned}
&\vdots \\
\dot{\zeta}^{(l)}_{r_l-1} &= \zeta^{(l)}_{r_l} \\
\dot{\zeta}^{(l)}_{r_l} &= L^{r_l}_{F_0} h^l_0(T^{-1}(\zeta,\eta,\theta)) + \sum_{i=1}^{l} L_{\mathcal{B}^i_0} L^{r_l-1}_{F_0} h^l_0(T^{-1}(\zeta,\eta,\theta)) u^i_0 \\
&\quad + L_{\mathcal{W}_0} L^{r_l-1}_{F_0} h^l_0(T^{-1}(\zeta,\eta,\theta)) \\
\dot{\eta}_1 &= \Psi_1(\zeta,\eta,\theta,\dot{\theta}) \\
&\vdots \\
\dot{\eta}_{m-\sum_i r_i} &= \Psi_{m-\sum_i r_i}(\zeta,\eta,\theta,\dot{\theta}) \\
y^i_s &= \zeta^{(i)}_1, \quad i = 1,\ldots,l
\end{aligned} \tag{5.24}$$

where $x_s = T^{-1}(\zeta,\eta,\theta)$, $\zeta = [\zeta^{(1)} \cdots \zeta^{(l)}]^T$, $\eta = [\eta_1 \cdots \eta_{m-\sum_i r_i}]^T$.

We note that the above assumption is always satisfied for systems for which $r_i = 1$, for all $i = 1,\ldots,l$. In most practical applications, this requirement can be easily achieved by selecting the form of the actuator distribution functions $b^i(z)$ to be different than the form of the eigenfunctions ϕ_j for $j = 2,\ldots,\infty$. Finally, we note that we do not require $\theta(t)$ to enter the system of Eq. 5.20 in the same differential equations as u, which is the case in many robust controller design methods (e.g., [48]).

Referring to the system of Eq. 5.24, we assume, similar to the presentation in chapter 4, that the matrix:

$$C_0(x_s) = \begin{bmatrix} L_{\mathcal{B}^1_0} L^{r_1-1}_{F_0} h^1_0(x_s) & \cdots & L_{\mathcal{B}^l_0} L^{r_1-1}_{F_0} h^1_0(x_s) \\ \vdots & \cdots & \vdots \\ L_{\mathcal{B}^1_0} L^{r_l-1}_{F_0} h^l_0(x_s) & \cdots & L_{\mathcal{B}^l_0} L^{r_l-1}_{F_0} h^l_0(x_s) \end{bmatrix} \tag{5.25}$$

is nonsingular uniformly in $x_s \in \mathcal{H}_s$.

Assumption 5.2 states that the system describing the unforced (i.e., $\eta(t) = \theta(t) = \dot{\theta}(t) \equiv 0$) inverse dynamics of the system of Eq. 5.24 is locally exponentially stable (in other words the system of Eq. 5.24 is assumed to be minimum phase). This assumption is standard in most nonlinear control methods for ODE systems (e.g., [96]) and is satisfied by many practical applications. This assumption is needed to establish that the state of the closed-loop slow system is locally bounded.

Assumption 5.2 *The dynamical system:*

$$\begin{aligned}
\dot{\eta}_1 &= \Psi_1(\zeta,0,0,0) \\
&\vdots \\
\dot{\eta}_{m-\sum_i r_i} &= \Psi_{m-\sum_i r_i}(\zeta,0,0,0)
\end{aligned} \tag{5.26}$$

is locally exponentially stable.

Assumption 5.3 that follows, requires the existence of a nonlinear time-varying bounding function that captures the size of the uncertain terms in the system of Eq. 5.24. Such a bounding function is typically obtained from physical considerations, preliminary simulations, or experimental data. The requirement of existence of a bounding function is standard in most Lyapunov-based robust control methods (see, for example, [48, 46, 41]).

Assumption 5.3 *There exists a known function $c_0(x_s, t)$ such that the following condition holds:*

$$\left|\left[L_{\mathcal{W}_0} L_{F_0}^{r_1-1} h_0^1(T^{-1}(\zeta,\eta,\theta)) \cdots L_{\mathcal{W}_0} L_{F_0}^{r_l-1} h_0^l(T^{-1}(\eta,\zeta,\theta))\right]^T\right| \leq c_0(x_s, t) \tag{5.27}$$

for all $x_s \in \mathcal{H}_s$, $\theta \in \mathbb{R}^q$, $t \geq 0$.

Whenever assumptions 5.1, 5.2, and 5.3 are satisfied, it is possible to synthesize a robust controller on the basis of the system of Eq. 5.22 using Lyapunov's direct method (the reader may refer to [46, 41] for details on robust controller design for nonlinear ODE systems). Theorem 5.1 provides an explicit formula of the robust controller, conditions that ensure boundedness of the state, and a precise characterization of the ultimate uncertainty attenuation level. The proof of the theorem is given in Appendix D. To simplify the statement of the theorem, we set $\bar{v}_i = [v_i \; v_i^{(1)} \; \cdots \; v_i^{(r_i)}]^T$ and $\bar{v} = [\bar{v}_1^T \; \bar{v}_2^T \; \cdots \; \bar{v}_m^T]^T$.

Theorem 5.1 *Consider the parabolic infinite dimensional system of Eq. 5.7 for which assumption 4.1 holds, and the finite-dimensional system of Eq. 5.22, for which assumptions 5.1, 5.2, and 5.3 hold, under the robust controller:*

$$\begin{aligned} u_0 = [\mathcal{C}_0(x_s)]^{-1} \Bigg\{ & \sum_{i=1}^{l}\sum_{k=1}^{r_i} \frac{\beta_{ik}}{\beta_{ir_i}} \left(v_i^{(k)} - L_{F_0}^k h_0^i(x_s)\right) \\ & + \sum_{i=1}^{l}\sum_{k=1}^{r_i} \frac{\beta_{ik}}{\beta_{ir_i}} \left(v_i^{(k-1)} - L_{F_0}^{k-1} h_0^i(x_s)\right) - \chi[c_0(x_s,t)] \\ & \times \frac{\displaystyle\sum_{i=1}^{l}\sum_{k=1}^{r_i} \frac{\beta_{ik}}{\beta_{ir_i}} \left(L_{F_0}^{k-1} h_0^i(x_s) - v_i^{(k-1)}\right)}{\left|\displaystyle\sum_{i=1}^{l}\sum_{k=1}^{r_i} \frac{\beta_{ik}}{\beta_{ir_i}} \left(L_{F_0}^{k-1} h_0^i(x_s) - v_i^{(k-1)}\right)\right| + \phi} \Bigg\} \end{aligned} \tag{5.28}$$

where $\frac{\beta_{ik}}{\beta_{ir_i}} = [\frac{\beta_{ik}^1}{\beta_{ir_i}^1} \; \cdots \; \frac{\beta_{ik}^l}{\beta_{ir_i}^l}]^T$ are column vectors of parameters chosen so that the roots of the equation det $(B(s)) = 0$, where $B(s)$ is an $l \times l$ matrix, whose (i, j)-th element is of the form $\sum_{k=1}^{r_i} \frac{\beta_{jk}^i}{\beta_{jr_i}^i} s^{k-1}$, lie in the open left half of the complex plane, and χ, ϕ are adjustable parameters with $\chi > 1$

and $\phi > 0$. Then, there exist positive real numbers (δ, ϕ^*) such that for each $\phi \leq \phi^*$, there exists $\epsilon^*(\phi)$ such that if $\phi \leq \phi^*$, $\epsilon \leq \epsilon^*(\phi)$ and $\max\{|x_s(0)|, \|x_f(0)\|_2, \|\theta\|, \|\dot{\theta}\|, \|\bar{v}\|\} \leq \delta$,

(a) the state of the infinite-dimensional closed-loop system is bounded, and

(b) the outputs of the infinite-dimensional closed-loop system satisfy:

$$\limsup_{t \to \infty} |y^i - v_i| \leq d_0, \quad i = 1, \dots, l \tag{5.29}$$

where $d_0 = O(\phi + \epsilon)$ is a positive real number.

Remark 5.1 Theorem 5.1 establishes that the ultimate uncertainty attenuation level in the case of synthesizing a robust controller on the basis of the system of Eq. 5.22 is $d_0 = O(\phi + \epsilon)$. This result is intuitively expected because the controller of Eq. 5.28 ensures that the outputs of the $O(\epsilon)$ approximation of the closed-loop parabolic PDE system satisfy $\limsup_{t \to \infty} |y_s^i - v_i| \leq d_\phi = O(\phi)$, $i = 1, \dots, l$ (the fact that $d_\phi = O(\phi)$ is rigorously established in the proof of the theorem; see Eqs. D.5–D.13).

Remark 5.2 Regarding the structure of the controller of Eq. 5.28, we remark that the nonlinear term,

$$\left(-\chi[c_0(x_s, t)] \frac{\displaystyle\sum_{i=1}^{l} \sum_{k=1}^{r_i} \frac{\beta_{ik}}{\beta_{ir_i}} \left(L_{F_0}^{k-1} h_0^i(x_s) - v_i^{(k-1)} \right)}{\left| \displaystyle\sum_{i=1}^{l} \sum_{k=1}^{r_i} \frac{\beta_{ik}}{\beta_{ir_i}} \left(L_{F_0}^{k-1} h_0^i(x_s) - v_i^{(k-1)} \right) \right| + \phi} \right)$$

could have been replaced by a sufficiently large positive constant (high gain). Although this modification would lead to a simplification in the practical implementation of the controller and the analysis of the properties of the closed-loop slow subsystem, we select to use the nonlinear term because the use of a large positive constant results in a controller that computes very large control action, when the tracking error is far from being zero. The controller that uses the nonlinear term avoids this problem and does not compute unnecessarily large control action (see the manipulated input profiles in example of section 5.5). Large control actions could lead to serious stability and performance degradation problems in the presence of constraints on the manipulated inputs and should be avoided.

Remark 5.3 Regarding the practical application of theorem 5.1, one has to initially verify assumption 4.1 (separation of eigenvalues of $\mathcal{A}$ into slow and fast ones), and then verify assumptions 5.1, 5.2, and 5.3 on the basis of the system of Eq. 5.22. Then, the synthesis formula of Eq. 5.28 can be directly used to derive the explicit form of the controller (see section 5.5 for an application of this procedure to a catalytic rod example). Moreover, since the value of ϵ in a diffusion-reaction process is typically fixed, say ϵ_p, there is a limit on how small the ultimate bound d can be chosen. For example, one can initially compute, through simulations, a ϕ^* from the desired (δ, d) and, in turn, the value ϵ^* for $\phi \leq \phi^*$. If this ϵ^* is less than ϵ_p, then d may need to be readjusted (increased) so that $\epsilon^* \geq \epsilon_p$. Of course, if ϵ_p is too large, there may be no value of d that works, in which case one will have to employ the methodology of the next section for robust controller design with improved uncertainty attenuation.

5.3.3 Improving uncertainty attenuation using approximate inertial manifolds

The controller of Eq. 5.28, synthesized on the basis of the slow system of Eq. 5.22, enforces an ultimate uncertainty attenuation level $d_0 = O(\phi + \epsilon)$. Even though this degree of attenuation may be sufficient for several practical applications where the degree of separation of slow and fast modes is large (i.e., ϵ is very small), it may not be sufficient for diffusion-convection-reaction processes for which ϵ is close to one. The objective of this subsection is to propose a procedure for the synthesis of robust controllers that achieve *arbitrary* degree of asymptotic attenuation of the effect of the uncertain variables on the outputs. The proposed procedure is based on the construction of higher-order (higher than $O(\epsilon)$) m-dimensional approximations of the x_s-subsystem of Eq. 5.19 (which will be used for controller design) by utilizing a geometric framework based on the concept of inertial manifold for systems of the form of Eq. 5.7. The inertial manifold is an appropriate tool for improving uncertainty attenuation because if the trajectories of the infinite dimensional system of Eq. 5.7 are on the manifold, then this system is exactly described by an m-dimensional slow system. We note that the concept of inertial manifold used in this work is a direct generalization of the one introduced by Temam for quasi-linear parabolic PDE systems without time-varying inputs (see, for example, [133]), to systems with time-varying inputs. A concept of inertial manifold for Navier–Stokes equations with time-dependent inputs, similar to the one used here, has been introduced in [89].

An inertial manifold $\mathcal{M}$ for the system of Eq. 5.7 is defined as a subset of $\mathcal{H}$, which satisfies the following properties: (*i*) $\mathcal{M}$ is a finite-dimensional Lipschitz manifold, (*ii*) $\mathcal{M}$ is a graph of a Lipschitz function $\Sigma\,(x_s, u, \theta, \epsilon)$ mapping $\mathcal{H}_s \times \mathrm{I\!R}^l \times \mathrm{I\!R}^q \times (0, \epsilon^*]$ into $\mathcal{H}_f$ and for every solution $x_s(t)$, $x_f(t)$ of Eq. 5.20 with $x_f(0) = \Sigma\,(x_s(0), u(0), \theta(0), \epsilon)$, then

$$x_f(t) = \Sigma\,(x_s(t), u(t), \theta(t), \epsilon), \quad \forall\; t \geq 0 \tag{5.30}$$

and (*iii*) $\mathcal{M}$ attracts every trajectory exponentially. Owing to the second and third properties of the IM, the dynamics of the system of Eq. 5.7 restricted to $\mathcal{M}$ are exactly described by the following m-dimensional system (called inertial form):

$$\begin{aligned}\frac{dx_s}{dt} &= \mathcal{A}_s x_s + \mathcal{B}_s u + f_s(x_s, \Sigma(x_s, u, \theta, \epsilon)) + \mathcal{W}_s(x_s, \Sigma(x_s, u, \theta, \epsilon), \theta) \\ y &= \mathcal{C} x_s\end{aligned} \tag{5.31}$$

where $\Sigma(x_s, u, \theta, \epsilon)$ is the solution of the following partial differential equation (which was obtained by differentiating Eq. 5.30 and utilizing Eq. 5.20):

$$\begin{aligned}&\epsilon \frac{\partial \Sigma}{\partial x_s}[\mathcal{A}_s x_s + \mathcal{B}_s u + f_s(x_s, \Sigma) + \mathcal{W}_s(x_s, \Sigma)] + \epsilon \frac{\partial \Sigma}{\partial u}\dot{u} + \epsilon \frac{\partial \Sigma}{\partial \theta}\dot{\theta} \\ &= \mathcal{A}_{f\epsilon}\Sigma + \epsilon \mathcal{B}_f u + \epsilon f_f(x_s, \Sigma) + \epsilon \mathcal{W}_f(x_s, \Sigma, \theta)\end{aligned} \tag{5.32}$$

Since the inertial form of Eq. 5.31 exactly describes the long-term dynamics of the system of Eq. 5.7, it follows that a robust feedback controller synthesized on the basis of the inertial form, using the methodology described in the previous subsection, will ensure that the outputs of the closed-loop parabolic PDE system satisfy $\limsup_{t\to\infty} |y^i - v_i| \leq d = O(\phi)$ (i.e., the ultimate uncertainty attenuation level in the closed-loop system is independent of ϵ), provided that ϵ is sufficiently small. However, because of the complexity present in computing the exact form of $\Sigma(x_s, u, \theta, \epsilon)$ from Eq. 5.32, it is impossible to directly utilize the inertial form of Eq. 5.31 for controller synthesis. In order to overcome the problems associated with the computation of $\Sigma(x_s, u, \theta, \epsilon)$, we will use a procedure which involves expansion of $\Sigma(x_s, u, \theta, \epsilon)$ and u in a power series in ϵ, to compute approximations of $\Sigma(x_s, u, \theta, \epsilon)$ (approximate inertial manifolds) and approximations of the inertial form, of desired accuracy.

Specifically, we consider an expansion of $\Sigma(x_s, u, \theta, \epsilon)$ and u in a power series in ϵ:

$$\begin{aligned}u &= u_0 + \epsilon u_1 + \epsilon^2 u_2 + \cdots + \epsilon^k u_k + O(\epsilon^{k+1}) \\ \Sigma(x_s, u, \theta, \epsilon) &= \Sigma_0(x_s, u, \theta) + \epsilon \Sigma_1(x_s, u, \theta) + \epsilon^2 \Sigma_2(x_s, u, \theta) \\ &\quad + \cdots + \epsilon^k \Sigma_k(x_s, u, \theta) + O(\epsilon^{k+1})\end{aligned} \tag{5.33}$$

where u_k, Σ_k are smooth functions. Substituting the expressions of Eq. 5.33 into Eq. 5.32, and equating terms of the same power in ϵ, one can obtain approximations of $\Sigma(x_s, u, \theta, \epsilon)$ up to a desired order. By substituting the expansion for $\Sigma(x_s, u, \theta, \epsilon)$ and u up to order k into Eq. 5.31, we obtain the following approximation of the inertial form:

$$\begin{aligned}\frac{dx_s}{dt} &= \mathcal{A}_s x_s + \mathcal{B}_s\left(u_0 + \epsilon u_1 + \epsilon^2 u_2 + \cdots + \epsilon^k u_k\right) + f_s(x_s, \Sigma_0(x_s, u, \theta) \\ &\quad + \epsilon \Sigma_1(x_s, u, \theta) + \epsilon^2 \Sigma_2(x_s, u, \theta) + \cdots + \epsilon^k \Sigma_k(x_s, u, \theta)) \\ &\quad + \mathcal{W}_s(x_s, \Sigma_0(x_s, u, \theta) + \epsilon \Sigma_1(x_s, u, \theta) + \epsilon^2 \Sigma_2(x_s, u, \theta) \\ &\quad + \cdots + \epsilon^k \Sigma_k(x_s, u, \theta), \theta) \\ y &= \mathcal{C} x_s.\end{aligned} \tag{5.34}$$

The above approximation procedure is motivated and validated from the fact that the inertial form of Eq. 5.31 reduces to the system of Eq. 5.22 as $\epsilon \to 0$, which ensures that the inertial form is well posed with respect to ϵ. The expansion of u in a power series in ϵ in Eq. 5.33 is motivated by our intention to appropriately modify the synthesis of the controller such that the outputs of the $O(\epsilon^{k+1})$ approximation of the closed-loop inertial form satisfy $\limsup_{t\to\infty} |y_s^i - v_i| \leq d$, where $d = d_\phi$. The construction of the robust control law of Eq. 5.33 to achieve this objective can be performed following a sequential procedure. Specifically, the component $u_0 = p_0(x_s) + Q_0(x_s)v + r_0(x_s, t)$ can be initially synthesized on the basis of the $O(\epsilon)$ approximation of the inertial form (Eq. 5.22); then the component $u_1 = p_1(x_s) + Q_1(x_s)v + r_1(x_s, t)$ can be synthesized on the basis of the $O(\epsilon^2)$ approximation of the inertial form. In general, at the k-th step, the component $u_k = p_k(x_s) + Q_k(x_s)v + r_k(x_s, t)$ can be synthesized on the basis of the $O(\epsilon^k)$ approximation of the inertial form (Eq. 5.34). The synthesis of $[p_\nu(x_s),\, Q_\nu(x_s),\, r_\nu(x_s, t)]$, $\nu = 0, \ldots, k$, can be performed, at each step, utilizing the methodology presented in the previous subsection for the synthesis of the component $u_0 = p_0(x_s) + Q_0(x_s)v + r_0(x_s, t)$.

Theorem 5.2 that follows provides conditions that ensure boundedness of the state and a precise characterization of the ultimate uncertainty attenuation level in the case of using robust control law of the form of Eq. 5.33 synthesized following the above procedure (the proof of the theorem is similar to the one of theorem 5.1 and thus, will be omitted for brevity).

Theorem 5.2 *Consider the parabolic infinite dimensional system of Eq. 5.7 for which assumption 4.1 holds, under the robust control law:*

$$\begin{aligned} u = &\; p_0(x_s) + Q_0(x_s)v + r_0(x_s, t) + \epsilon(p_1(x_s) + Q_1(x_s)v + r_1(x_s, t)) \\ &+ \cdots + \epsilon^k(p_k(x_s) + Q_k(x_s)v + r_k(x_s, t)) \end{aligned} \tag{5.35}$$

where $[p_\nu(x_s),\, Q_\nu(x_s),\, r_\nu(x_s, t)]$, $\nu = 0, \ldots, k$ are synthesized, under the assumptions 5.1, 5.2, and 5.3, following the aforementioned procedure. Then, there exist positive real numbers (δ, ϕ^*) such that for each $\phi \leq \phi^*$, there exists $\epsilon^*(\phi)$, such that if $\phi \leq \phi^*$, $\epsilon \leq \epsilon^*(\phi)$ and $\max\{|x_s(0)|, \|x_f(0)\|_2, \|\theta\|, \|\dot{\theta}\|, \|\bar{v}\|\} \leq \delta$,

(a) the state of the infinite dimensional closed-loop system is bounded, and

(b) the outputs of the infinite dimensional closed-loop system satisfy:

$$\limsup_{t\to\infty} |y^i - v_i| \leq d_k, \quad i = 1, \ldots, l \tag{5.36}$$

where $d_k = O(\phi + \epsilon^{k+1})$ is a positive real number.

Remark 5.4 Since $\epsilon < 1$ and thus, $\epsilon^2 \ll \epsilon$, an uncertainty attenuation level $d_1 = O(\phi + \epsilon^2)$ should be satisfactory for most practical applications. In such a case, an application of the proposed approximation procedure yields that the $O(\epsilon^2)$ approximation of $\Sigma(x_s, u, \theta, \epsilon)$ is of the form:

$$\Sigma(x_s, u, \theta, \epsilon) = (\mathcal{A}_{f\epsilon})^{-1}[-\epsilon f_f(x_s) - \epsilon \mathcal{B}_f u_0 - \epsilon \mathcal{W}_f(x_s, \theta)] \tag{5.37}$$

and the corresponding $O(\epsilon^2)$ approximation of the inertial form is given by:

$$\begin{aligned}\frac{dx_s}{dt} &= \mathcal{A}_s x_s + \mathcal{B}_s u_0 + \epsilon \mathcal{B}_s u_1 + f_s(x_s, \mathcal{A}_{f\epsilon})^{-1}[-\epsilon f_f(x_s) - \epsilon \mathcal{B}_f u_0 \\ &\quad - \epsilon \mathcal{W}_f(x_s, \theta)]) + \mathcal{W}_s(x_s, \mathcal{A}_{f\epsilon})^{-1}[-\epsilon f_f(x_s) - \epsilon \mathcal{B}_f u_0 \\ &\quad - \epsilon \mathcal{W}_f(x_s, \theta)], \theta) \\ &=: F_1(x_s, \epsilon) + \sum_{i=1}^{l} \mathcal{B}_1^i(\epsilon) u_1^i + \mathcal{W}_1(x_s, \theta, \epsilon) \\ y_s^i &= \mathcal{C}^i x_s =: h_1^i(x_s).\end{aligned} \tag{5.38}$$

The necessary robust controller ensuring that $\limsup_{t\to\infty} |y^i - v_i| \le d_1 = O(\phi + \epsilon^2)$, $i = 1, \ldots, l$ then takes the form:

$$\begin{aligned} u = [C_0(x_s)]^{-1} \Bigg\{ & \sum_{i=1}^{l}\sum_{k=1}^{r_i} \frac{\beta_{ik}}{\beta_{ir_i}} \left(v_i^{(k)} - L_{F_0}^k h_0^i(x_s)\right) \\ & + \sum_{i=1}^{l}\sum_{k=1}^{r_i} \frac{\beta_{ik}}{\beta_{ir_i}} \left(v_i^{(k-1)} - L_{F_0}^{k-1} h_0^i(x_s)\right) \\ & - \chi[c_0(x_s, t)] \frac{\sum_{i=1}^{l}\sum_{k=1}^{r_i} \frac{\beta_{ik}}{\beta_{ir_i}} \left(L_{F_0}^{k-1} h_0^i(x_s) - v_i^{(k-1)}\right)}{\left|\sum_{i=1}^{l}\sum_{k=1}^{r_i} \frac{\beta_{ik}}{\beta_{ir_i}} \left(L_{F_0}^{k-1} h_0^i(x_s) - v_i^{(k-1)}\right)\right| + \phi} \Bigg\} \\ & + \epsilon \Bigg\{ [C_1(x_s, \epsilon)]^{-1} \Bigg\{ \sum_{i=1}^{l}\sum_{k=1}^{r_i} \frac{\beta_{ik}}{\beta_{ir_i}} \left(v_i^{(k)} - L_{F_1}^k h_1^i(x_s)\right) \\ & + \sum_{i=1}^{l}\sum_{k=1}^{r_i} \frac{\beta_{ik}}{\beta_{ir_i}} \left(v_i^{(k-1)} - L_{F_1}^{k-1} h_1^i(x_s)\right) \\ & - \chi[c_1(x_s, t)] \frac{\sum_{i=1}^{l}\sum_{k=1}^{r_i} \frac{\beta_{ik}}{\beta_{ir_i}} \left(L_{F_1}^{k-1} h_1^i(x_s) - v_i^{(k-1)}\right)}{\left|\sum_{i=1}^{l}\sum_{k=1}^{r_i} \frac{\beta_{ik}}{\beta_{ir_i}} \left(L_{F_1}^{k-1} h_1^i(x_s) - v_i^{(k-1)}\right)\right| + \phi} \Bigg\}\Bigg\} \end{aligned} \tag{5.39}$$

where the explicit form of $C_1(x_s, \epsilon)$, $c_1(x_s, t)$ is omitted for brevity.

Remark 5.5 Following the proposed approximation procedure, we can be show that $\Sigma_0(x_s, u, \theta) = 0$ and the corresponding approximate inertial form is identical to the system of Eq. 5.22 (obtained via Galerkin's method) with $u(t) \equiv 0$. This system does not utilize any information about the structure of the fast subsystem, thus yielding an ultimate degree of attenuation for the closed-loop parabolic PDE system $d_0 = O(\phi + \epsilon)$ (theorem 5.1). On the other hand, the $O(\epsilon^2)$ approximation of $\Sigma(x_s, u, \epsilon, \theta)$ is of the form of Eq. 5.37 and the corresponding open-loop approximate inertial form does utilize information about the structure of the fast subsystem, and thus allows us to obtain an ultimate degree of attenuation for the closed-loop parabolic PDE system $d_1 = O(\phi + \epsilon^2)$ (theorem 5.2).

Remark 5.6 Owing to the infinite-dimensional range of the vector function $\Sigma_k(x_s, u, \theta)$, the implementation of the controller of Eq. 5.35 requires computing a finite-dimensional approximation of $\Sigma_k(x_s, u, \theta)$, say $\Sigma_{kt}(x_s, u, \theta)$, by keeping the first $\bar{m}$ elements of $\Sigma_k(x_s, u, \theta)$ and neglecting the remaining infinite ones. Since as $\bar{m} \to \infty$, $\Sigma_{kt}(x_s, u, \theta)$ approaches $\Sigma_k(x_s, u, \theta)$, the controller of Eq. 5.39 with $\Sigma_{kt}(x_s, u, \theta)$ guarantees stability and enforces the requirement of Eq. 5.36 in the closed-loop infinite-dimensional system, provided that $\bar{m}$ is sufficiently large.

Remark 5.7 The robust controllers of Eqs. 5.28–5.35 possess a robustness property with respect to fast and asymptotically stable unmodeled dynamics (i.e., the controllers enforce boundedness, output tracking, and uncertainty attenuation in the closed-loop system, despite the presence of additional dynamics in the process model, as long as they are stable and sufficiently fast). This property of the controllers can be rigorously established by analyzing the closed-loop system with the unmodeled dynamics using singular perturbations (see also Chapter 3 for a similar analysis in the case of robust control of hyperbolic PDE systems). This robustness property of the controllers is of particular importance for many practical applications where unmodeled dynamics often occur due to actuator and sensor dynamics, fast process dynamics, and so forth (see, for example, the application in section 5.5).

5.4 Robust Output Feedback Controller Synthesis

The robust nonlinear controllers of theorem 5.1 and 5.2 were derived under the assumption that measurements of the state variables, $x(z, t)$, are available at all positions and times. However, there are many practical applications where these measurements are not available (for example, concentrations of certain species in a chemical reactor may not be measured on-line). Motivated by this practical problem, we address in this section the synthesis of robust output feedback controllers that use measurements of the process outputs, q, to enforce the following properties in the closed-loop system: (a) boundedness of the state, (b) output tracking for changes in the reference input, and (c) asymptotic attenuation of the effect of uncertain variables on the output.

Specifically, we consider the synthesis of robust static output feedback control laws of the form:

$$u_0 = p_0(q) + Q_0(q)\bar{v} + r_0(q,t) \tag{5.40}$$

where $p_0(q)$, $r_0(q,t)$ are vector functions, $Q_0(q)$ is a matrix, q is the vector of measured outputs, and $\bar{v}$ is a vector function of the external reference inputs and their time derivatives. A discussion that explains the choice of static versus dynamic output feedback is given in remark 5.9.

The synthesis of the controllers of Eq. 5.40 will be achieved by combining the state feedback controller synthesis result of theorem 5.1 with a procedure for obtaining estimates for the states of the approximate ODE model of Eq. 5.22 from the measurements. To this end, we need to impose the following requirement on the number of measured outputs in order to obtain estimates of the states x_s of the system of Eq. 5.22 from the measurements q^κ, $\kappa = 1, \ldots, p$.

Assumption 5.4 $p = m$ *(i.e., the number of measurements is equal to the number of slow modes), and the inverse of the operator* $\mathcal{Q}$ *exists so that* $\hat{x}_s = \mathcal{Q}^{-1}q$*, where* $\hat{x}_s$ *is an estimate of* x_s*.*

We note that the requirement that the inverse of the operator $\mathcal{Q}$ exists can be achieved by appropriate choice of the location of the measurement sensors (i.e., functions $q^\kappa(z)$).

Theorem 5.3 that follows provides an explicit formula for the robust controller, conditions that ensure boundedness of the state, and a precise characterization of the ultimate uncertainty attenuation level. The proof is given in Appendix D.

Theorem 5.3 *Consider the parabolic infinite-dimensional system of Eq. 5.7, for which assumption 4.1 holds, and the finite-dimensional system of Eq. 5.22, for which assumptions 5.1, 5.2, 5.3, and 5.4 hold, under the robust output feedback controller:*

$$\begin{aligned}
u_0 &= a_0(x_s, x_f, \bar{v}, t) \\
&:= [C_0(\hat{x}_s)]^{-1} \left\{ \sum_{i=1}^{l}\sum_{k=1}^{r_i} \frac{\beta_{ik}}{\beta_{ir_i}} \left(v_i^{(k)} - L_{F_0}^{k} h_0^i(\hat{x}_s) \right) \right. \\
&\quad + \sum_{i=1}^{l}\sum_{k=1}^{r_i} \frac{\beta_{ik}}{\beta_{ir_i}} \left(v_i^{(k-1)} - L_{F_0}^{k-1} h_0^i(\hat{x}_s) \right)
\end{aligned}$$

$$- \chi[c_0(\hat{x}_s,t)] \frac{\sum_{i=1}^{l}\sum_{k=1}^{r_i} \frac{\beta_{ik}}{\beta_{ir_i}} \left(L_{F_0}^{k-1} h_0^i(\hat{x}_s) - v_i^{(k-1)}\right)}{\left|\sum_{i=1}^{l}\sum_{k=1}^{r_i} \frac{\beta_{ik}}{\beta_{ir_i}} \left(L_{F_0}^{k-1} h_0^i(\hat{x}_s) - v_i^{(k-1)}\right)\right| + \phi} \Bigg\} \quad (5.41)$$

where $\hat{x}_s = \mathcal{Q}^{-1}q$, $\frac{\beta_{ik}}{\beta_{ir_i}} = [\frac{\beta^1_{ik}}{\beta^1_{ir_i}} \cdots \frac{\beta^l_{ik}}{\beta^l_{ir_i}}]^T$ *are column vectors of parameters chosen so that the roots of the equation* $\det(B(s)) = 0$, *where* $B(s)$ *is an* $l \times l$ *matrix, whose* (i,j)*-th element is of the form* $\sum_{k=1}^{r_i} \frac{\beta^i_{jk}}{\beta^i_{jr_i}} s^{k-1}$, *lie in the open left half of the complex plane, and* χ, ϕ *are adjustable parameters with* $\chi > 1$ *and* $\phi > 0$. *Then, there exist positive real numbers* (δ, ϕ^*) *such that for each* $\phi \le \phi^*$, *there exists* $\epsilon^*(\phi)$ *such that if* $\phi \le \phi^*$, $\epsilon \le \epsilon^*(\phi)$ *and* $\max\{|x_s(0)|, \|x_f(0)\|_2, \|\theta\|, \|\dot{\theta}\|, \|\tilde{v}\|\} \le \delta$,

(a) the state of the infinite-dimensional closed-loop system is bounded, and

(b) the outputs of the infinite-dimensional closed-loop system satisfy:

$$\limsup_{t\to\infty} |y^i - v_i| \le d_0, \quad i = 1,\ldots,l \quad (5.42)$$

where $d_0 = O(\phi + \epsilon)$ *is a positive real number.*

Remark 5.8 We note that the controller of Eq. 5.41 uses static feedback of the measured outputs q^κ, $\kappa = 1,\ldots,p$, and thus, it feeds back both x_s and x_f (this is in contrast to the robust state feedback controllers of theorems 5.1 and 5.2 which use only feedback of the slow state x_s). However, even though the use of x_f feedback could lead to destabilization of the stable fast subsystem, the large separation of the slow and fast modes of the spatial differential operator (i.e., the assumption that ϵ is sufficiently small) and the fact that the controller does not include terms of the form $O(\frac{1}{\epsilon})$ do not allow such a destabilization to occur.

Remark 5.9 Even though static output feedback is more sensitive to measurement noise than dynamic output feedback, we prefer to use static feedback of q in the controller of Eq. 5.41 because the presence of the unknown variables does not allow the design of a robust state observer to obtain estimates of the slow-state variables without imposing very restrictive conditions on the way $\theta(t)$ enters the finite-dimensional system of Eq. 5.22.

5.5 Application to a Catalytic Rod with Uncertainty

We consider the catalytic rod example studied in subsection 4.3.1 and assume that the heat of reaction is unknown and time varying (uncertain variable). The spatiotemporal evolution of the dimensionless rod temperature is described by the following parabolic PDE:

$$\frac{\partial \bar{x}}{\partial t} = \frac{\partial^2 \bar{x}}{\partial z^2} + \beta_T e^{-\frac{\gamma}{1+\bar{x}}} + \beta_U(b(z)u(t) - \bar{x}) - \beta_{T,n} e^{-\gamma} \tag{5.43}$$

subject to the Dirichlet boundary conditions:

$$\bar{x}(0,t) = 0, \quad \bar{x}(\pi,t) = 0 \tag{5.44}$$

and the initial condition:

$$\bar{x}(z,0) = \bar{x}_0(z) \tag{5.45}$$

where $\bar{x}$ denotes the dimensionless temperature of the rod, β_T denotes a dimensionless heat of reaction, $\beta_{T,n}$ denotes a *nominal* dimensionless heat of reaction, γ denotes a dimensionless activation energy, β_U denotes a dimensionless heat transfer coefficient, and u denotes the manipulated input (temperature of the cooling medium). The following typical values are given to the process parameters:

$$\beta_{T,n} = 50.0, \quad \beta_U = 2.0, \quad \gamma = 4.0. \tag{5.46}$$

Introducing the Hilbert space $\mathcal{H}$ of square integrable functions that satisfy the boundary conditions of Eq. 5.44 and defining $x \in \mathcal{H}$ as:

$$x(t) = \bar{x}(z,t), \quad \forall\, z \in [0,\pi], \tag{5.47}$$

the system of Eqs. 5.43, 5.44, and 5.45 can be written in the form of Eq. 5.7, where the spatial differential operator takes the form:

$$\begin{aligned} \mathcal{A}x &= \frac{\partial^2 \bar{x}}{\partial z^2}, \\ x \in D(\mathcal{A}) &= \left\{x \in \mathcal{H}([0,\pi];\mathbb{R});\ \bar{x}(0,t) = 0, \quad \bar{x}(\pi,t) = 0\right\} \end{aligned} \tag{5.48}$$

The eigenvalue problem for $\mathcal{A}$ can be solved analytically and its solution is of the form:

$$\lambda_j = -j^2, \quad \phi_j(z) = \sqrt{\frac{2}{\pi}}\ sin(jz), \quad j = 1,\ldots,\infty. \tag{5.49}$$

Even though the eigenvalues of $\mathcal{A}$ are all stable, the spatially uniform operating steady state $\bar{x}(z,t) = 0$ of the system of Eq. 5.43 is unstable (i.e., the linearization of the system of Eq. 5.43 around $\bar{x}(z,t) = 0$ possesses one stable eigenvalue owing to the exothermicity of the reaction). Due to the stability of the spatially uniform operating steady-state $\bar{x}(z,t) = 0$, we rmulate the control problem as the one of stabilizing the rod tempera- ıre profile at $\bar{x}(z,t) = 0$ in the presence of time-varying uncertainty in he dimensionless heat of the reaction β_T, (i.e., $\beta_T = \beta_{T,n} + \theta(t)$, where $\theta(t) = \beta_{T,n}\ sin(0.524\ t)$). We note that this selection for $\theta(t)$ satisfies the requirements of theorem 5.3 that $\theta(t), \dot{\theta}(t)$ should be sufficiently small (see closed-loop simulations below), while it leads to a very poor open-loop behavior for $\bar{x}(z,t)$ (see Figure 5.1). Since the maximum open-loop

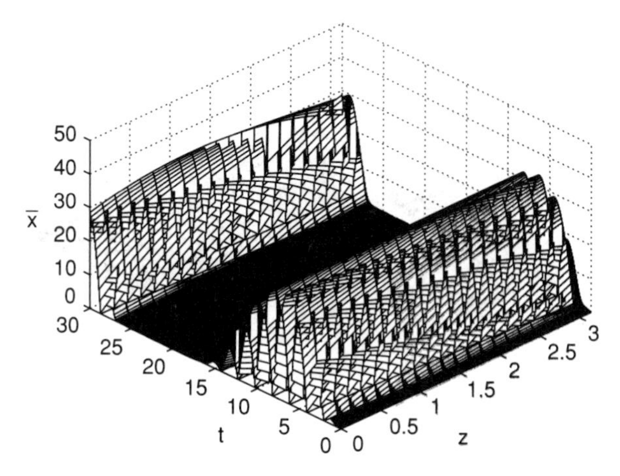

FIGURE 5.1. Open-loop profile of $\bar{x}$ with $\theta(t) = \beta_{T,n}\, sin(0.524\, t)$.

temperature occurs in the middle of the rod, the controlled output is defined as:

$$y(t) = \int_0^{\pi} \sqrt{\frac{2}{\pi}} sin(z) \bar{x}(z,t) dz \tag{5.50}$$

and the actuator distribution function is taken to be $b(z) = \sqrt{\frac{2}{\pi}}\, sin(z)$, in order to apply maximum cooling towards the middle of the rod. One measurement of x at $z = \frac{\pi}{4}$ is assumed to be available.

For the system of Eq. 5.43, we consider the first eigenvalue as the dominant one ($\epsilon = 0.25$) and use Galerkin's method to derive a scalar ODE that is used for the synthesis of a nonlinear robust output feedback controller through an application of the formula of Eq. 5.41. This controller is implemented in the simulations with $\chi = 1.2$, $c_0(\hat{x}_s, t) = \beta_{T,n} \int_0^{\pi} \phi_1(z) e^{-\frac{\gamma}{1+\hat{x}_s}}\, dz$ and $\phi = 0.01$ to achieve an uncertainty attenuation level $d_0 = 0.1$ (note that $0.1 = O(\epsilon + \phi) = O(0.25 + 0.01)$).

Figure 5.2 shows the evolution of the closed-loop rod temperature profile under the nonlinear robust output feedback controller, while Figure 5.3 shows the corresponding manipulated input profile. Clearly, the proposed controller regulates the temperature profile at $\bar{x}(z,t) = 0$, attenuating the effect of the uncertain variable (note that the requirement $\limsup_{t\to\infty} |y| \leq 0.1$ is enforced in the closed-loop system; Figure 5.4). For the sake of comparison, we also implement on the process the same controller as before without the term, which compensates for the effect of the uncertainty (i.e., $c_0(\hat{x}_s, t) \equiv 0$). Figure 5.5 shows the evolution of the closed-loop rod temperature profile. It is clear that this controller cannot regulate the temperature profile at the desired steady state, $\bar{x}(z,t) = 0$ because it does not compensate for the effect of the uncertainty.

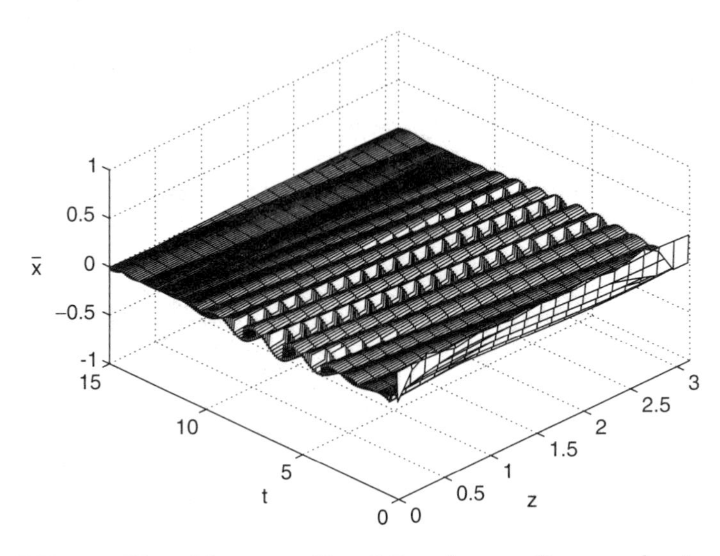

FIGURE 5.2. Closed-loop profile of $\bar{x}$ under nonlinear robust output feedback control—distributed control actuation.

Finally, in order to show that the proposed control method can be readily applied to the case of point-control actuation, we consider the same control problem as above but with $b(z) = \delta(z - \frac{\pi}{2})$ (i.e., a point-control actuator influencing the rod at $z = \frac{\pi}{2}$ is used to stabilize the system at $\bar{x}(z,t) = 0$ in the presence of the uncertain variable). A nonlinear robust output feedback controller is synthesized on the basis of a scalar ODE model obtained from application of Galerkin's method to the system of Eq. 5.43 and implemented with $\chi = 1.2$, $c_0(\hat{x}_s, t) = \beta_{T,n} \int_0^{\pi} \phi_1(z) e^{-\frac{\gamma}{1+\hat{x}_s}}\, dz$ and $\phi = 0.01$ to achieve an

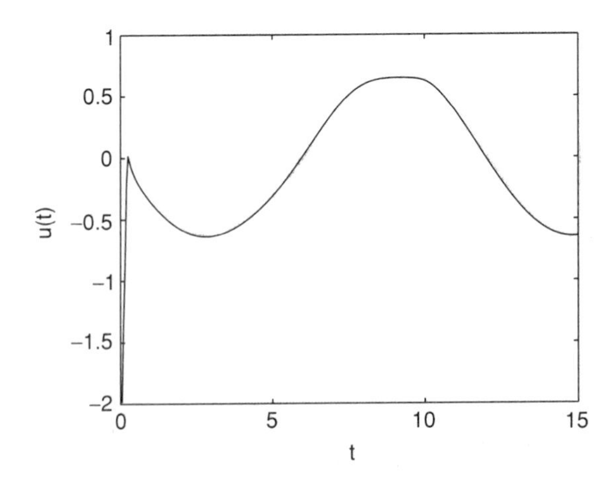

FIGURE 5.3. Manipulated input profile for nonlinear robust output feedback controller—distributed control actuation.

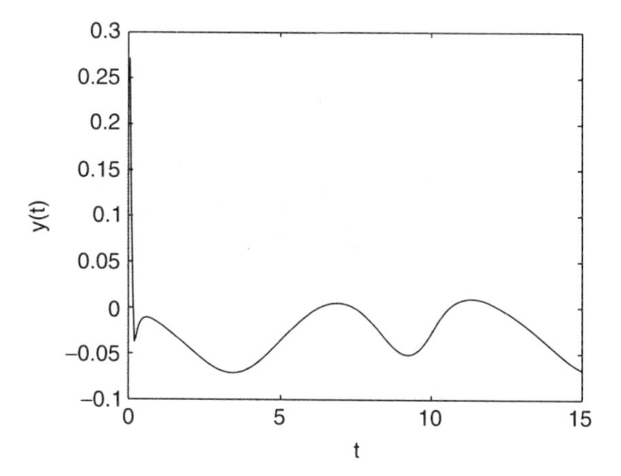

FIGURE 5.4. Closed-loop output profile under nonlinear robust output feedback control—distributed control actuation.

uncertainty attenuation level $d_0 = 0.4$ (note that $0.4 = O(0.25 + 0.01)$). Figure 5.6 shows the evolution of the closed-loop rod temperature profile, and Figure 5.7 shows the corresponding manipulated input profile for this case. The stabilization of the system at $\bar{x}(z,t) = 0$ with uncertainty attenuation is achieved (the requirement $\limsup_{t\to\infty} |y| \leq 0.4$ is satisfied; Figure 5.8). Note that as expected, in the case of point-control actuation $u(t)$ influences the states of the closed-loop system that are not used in the controller design model (spill-over effect), and thus, $\bar{x}(z,t)$ exhibits more oscillatory behavior (compare Figure 5.6 with Figure 5.2). From the results of the simulation study, it is evident that the proposed methodology is a

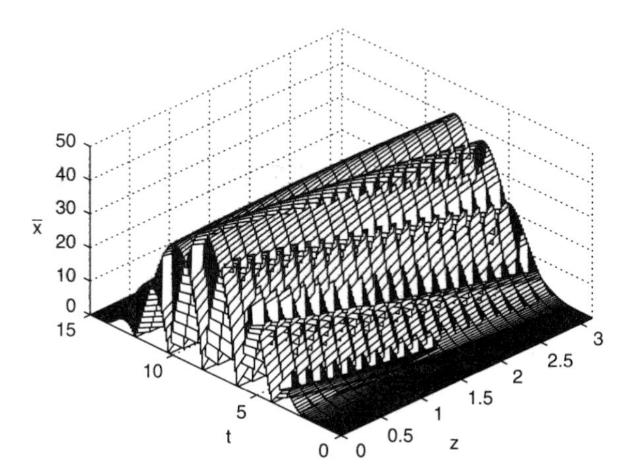

FIGURE 5.5. Closed-loop profile of $\bar{x}$ under nonlinear output feedback control (no uncertainty compensation)—distributed control actuation.

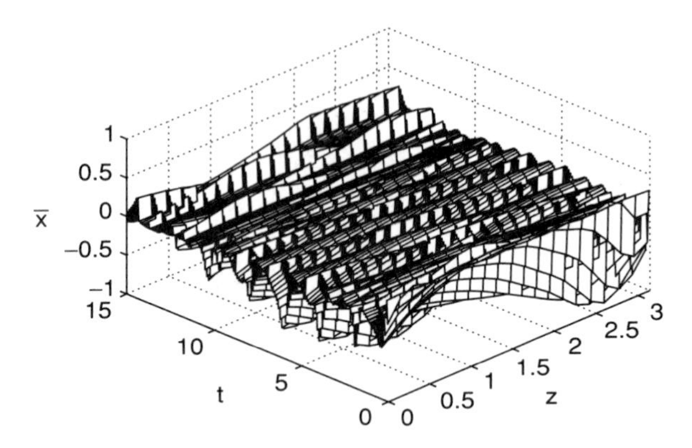

FIGURE 5.6. Closed-loop profile of $\bar{x}$ under nonlinear robust output feedback control—point control actuation.

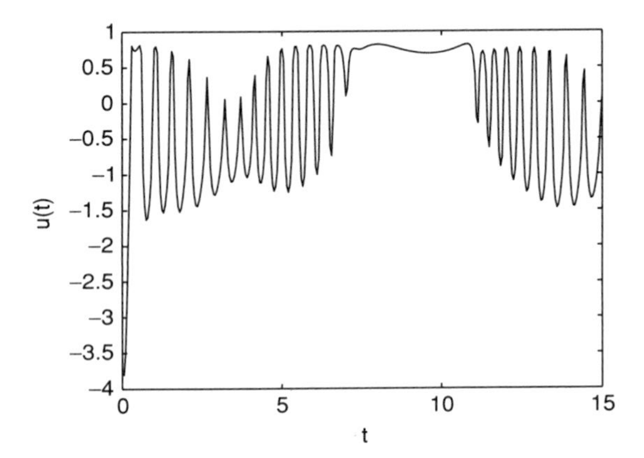

FIGURE 5.7. Manipulated input profile for nonlinear robust output feedback controller—point control actuation.

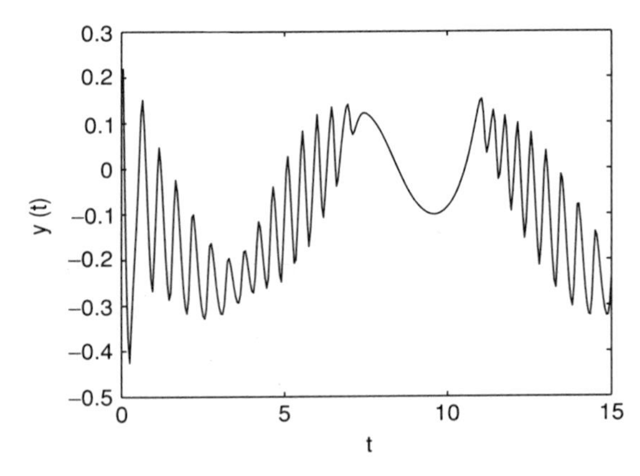

FIGURE 5.8. Closed-loop output profile under nonlinear robust output feedback control—point control actuation.

useful tool for the synthesis of robust controllers that attenuate the effect of uncertain variables for quasi-linear parabolic PDE systems.

5.6 Conclusions

In this chapter, we presented an approach for the synthesis of robust state and output feedback controllers for quasi-linear parabolic PDEs with time-varying uncertain variables, for which the eigenspectrum of the spatial differential operator can be partitioned into a finite-dimensional (possibly unstable) slow one and an infinite-dimensional stable fast complement. Initially, nonlinear Galerkin's method was used to derive an ODE system of dimension equal to the number of slow modes, which was subsequently used to synthesize robust state feedback controllers via Lyapunov's direct method. The controllers enforce asymptotic attenuation of the effect of uncertain variables on the output, provided that the degree of separation of the fast and slow modes of the spatial differential operator is sufficiently large. Then, under the assumption that the number of measurements is equal to the number of slow modes, we proposed a procedure for obtaining estimates for the states of the approximate ODE model from the measurements. We showed that the use of these estimates in the robust state feedback controller leads to a robust output feedback controller, which enforces the desired properties in the closed-loop system, provided that the separation between the slow and fast eigenvalues is sufficiently large. The developed robust output feedback controllers were successfully applied to a catalytic rod with uncertainty.

Chapter 6

Nonlinear and Robust Control of Parabolic PDE Systems with Time-Dependent Spatial Domains

6.1 Introduction

There is a large number of industrial control problems which involve highly nonlinear transport-reaction processes with moving boundaries such as crystal growth, metal casting, gas-solid reaction systems, and coatings. In these processes, nonlinear behavior typically arises from complex reaction mechanisms and their Arrhenius dependence on temperature, while motion of boundaries is usually a result of phase change (like melting or solidification), chemical reaction, and/or mass and heat transfer. The mathematical models of transport-reaction processes with moving boundaries are usually obtained from the dynamic conservation equations and consist of nonlinear parabolic PDEs with time-dependent spatial domains.

Few results are available on control and estimation of parabolic PDE systems with time-dependent spatial domains. In this area, important contributions include the synthesis of linear optimal controllers (e.g., [143, 144]), stochastic controllers [76] estimators [110], their applications to crystal growth [145] casting processes [76] and flow through porous media [103].

This chapter focuses on a broad class of quasi-linear parabolic PDE systems with time-dependent spatial domains whose dynamics can be partitioned into slow and fast ones. Such systems arise naturally in the modeling of diffusion-reaction processes with moving boundaries. The objective is to develop general methods for the synthesis of nonlinear and robust time-varying output feedback controllers that can be readily implemented in practice.

The chapter is structured as follows: initially, we focus on the synthesis of nonlinear controllers for quasi-linear parabolic PDE systems with time-dependent spatial domains. The class of PDE systems is first presented and formulated as an evolution equation in an appropriate Hilbert space. Then, a nonlinear model reduction scheme, similar to the one introduced in Chapter 4, which is based on combination of Galerkin's method with the concept of approximate inertial manifold, is employed for the derivation of ODE systems that yield solutions which are close, up to a desired accuracy, to the ones of the PDE system, for almost all times. Then, these ODE systems are used as the basis for the explicit construction of nonlinear output feedback controllers via geometric control methods. The nonlinear model reduction and control methods are applied to a catalytic rod with time-dependent spatial domain and are shown to outperform nonlinear control methods that do not account for the variation of the spatial domain.

Subsequently, we address the problem of synthesizing robust time-varying output feedback controllers for quasi-linear parabolic PDE systems with time-dependent spatial domains and uncertain variables. Under similar assumptions to the ones employed in Chapter 5, we synthesize controllers, using Lyapunov techniques, that enforce stability, output tracking, and uncertainty attenuation in the closed-loop system. The robust controllers are successfully tested on a catalytic rod with time-dependent spatial domain and uncertainty. The results of this chapter were first presented in [6, 7].

6.2 Preliminaries

6.2.1 Parabolic PDE systems with time-dependent spatial domains

We consider quasi-linear parabolic PDE systems with time-dependent spatial domains with the following state-space description:

$$\begin{aligned}\frac{\partial \bar{x}}{\partial t} &= A\frac{\partial \bar{x}}{\partial z} + B\frac{\partial^2 \bar{x}}{\partial z^2} + wb(z,t)u + f(t,\bar{x}) \\ y^i &= \int_0^{l(t)} c^i(z,t)k\bar{x}\,dz, \quad i = 1,\dots,l \\ q^\kappa &= \int_0^{l(t)} s^\kappa(z,t)\omega\bar{x}\,dz, \quad \kappa = 1,\dots,p\end{aligned} \tag{6.1}$$

subject to the boundary conditions:

$$\begin{aligned}C_1\bar{x}(0,t) + D_1\frac{\partial \bar{x}}{\partial z}(0,t) &= R_1 \\ C_2\bar{x}(l(t),t) + D_2\frac{\partial \bar{x}}{\partial z}(l(t),t) &= R_2\end{aligned} \tag{6.2}$$

and the initial condition:

$$\bar{x}(z,0) = \bar{x}_0(z) \tag{6.3}$$

where the rate of change of the length of the domain, $l(t)$, is governed by the following ordinary differential equation:

$$\frac{dl}{dt} = \mathcal{G}\left(t, l, \int_0^{l(t)} \bar{a}\left(z,t,l,\bar{x},\frac{\partial \bar{x}}{\partial z}\right)dz\right) \tag{6.4}$$

where $\bar{x}(z,t) = [\bar{x}_1(z,t) \; \cdots \; \bar{x}_n(z,t)]^T$ denotes the vector of state variables, $[0, l(t)] \subset \mathbb{R}$ is the domain of definition of the process, $z \in [0, l(t)]$ is the spatial coordinate, $t \in [0, \infty)$ is the time, $u = [u^1 \; u^2 \; \cdots \; u^l]^T \in \mathbb{R}^l$ denotes the vector of manipulated inputs, $y^i \in \mathbb{R}$ denotes the i-th controlled output and $q^\kappa \in \mathbb{R}$ denotes the κ-th measured output. $\frac{\partial \bar{x}}{\partial z}$, $\frac{\partial^2 \bar{x}}{\partial z^2}$ denote the first- and second-order spatial derivatives of $\bar{x}$, $f(t, \bar{x})$, $\mathcal{G}(t, l, \int_0^{l(t)} \bar{a}(z, t, l, \bar{x}, \frac{\partial \bar{x}}{\partial z}) dz)$ are nonlinear vector functions, $\bar{a}(z, t, l, \bar{x}, \frac{\partial \bar{x}}{\partial z})$ is a nonlinear scalar function, k, w, ω are constant vectors, A, B, C_1, D_1, C_2, D_2 are constant matrices, R_1, R_2 are column vectors, and $\bar{x}_0(z)$ is the initial condition. $b(z,t)$ is a known smooth vector function of (z,t) of the form $b(z,t) = [b^1(z,t) \; b^2(z,t) \; \cdots \; b^l(z,t)]$, where $b^i(z,t)$ describes how the control action $u^i(t)$ is distributed in the interval $[0, l(t)]$ (e.g., point/distributed actuation); $c^i(z,t)$ is a known smooth function of (z,t), which is determined by the desired performance specifications in the interval $[0, l(t)]$ (e.g., regulation of the entire temperature profile of a crystal or regulation of the temperature at a specific point); and $s^\kappa(z,t)$ is a known smooth function of (z,t) which is determined by the location and type of the κ-th measurement sensor (e.g., point/distributed sensing). We note that in contrast to the case of parabolic PDE systems defined on a fixed spatial domain considered in Chapters 4 and 5, we allow the actuator, performance specification, and measurement sensor functions to depend explicitly on time (i.e., moving control actuators and objectives, and measurement sensors). The value of using moving control actuators and sensors in certain applications will be illustrated in the example of section 6.5. A schematic of a typical process with moving boundaries is shown in Figure 6.1, in the case of point-control actuators and sensors. In order to simplify the notation of this manuscript, we assume that $l(t)$ is a known and smooth function of time. Our assumptions on the properties of $l(t)$ are precisely stated below:

Assumption 6.1 *$l(t)$ is a known smooth (i.e., $\dot{l}$ exists and is bounded, $\forall\, t \in [0, \infty)$) function of time which satisfies $l(t) \in (0, l_{max}]$, $\forall\, t \in [0, \infty)$, where l_{max} denotes the maximum length of the spatial domain.*

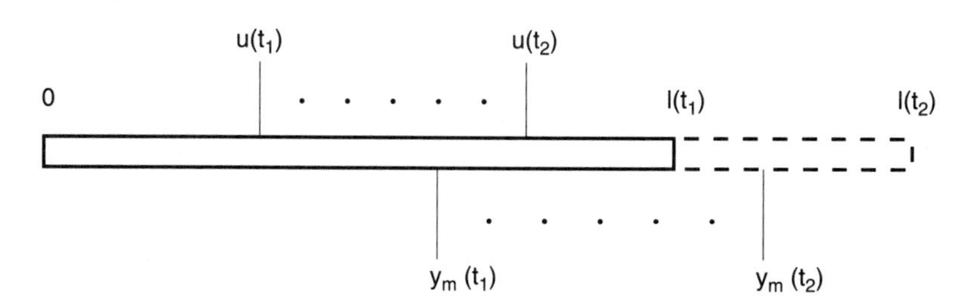

FIGURE 6.1. Schematic of a prototype system with time-dependent spatial domain, moving control actuators and moving measurement sensors.

Remark 6.1 Referring to assumption 6.1, we note that the requirement that the length of the domain is a smooth function of time is needed for the system of Eq. 6.9 to be well posed in $\mathcal{H}(t)$, while the requirement that the length of the domain is always finite is necessary for the dominant dynamics of the parabolic PDE system to be described by a finite number of degrees of freedom (see assumption 6.2 below). We note that both requirements are usually satisfied in practice: a typical example is the Czochralski crystal growth process (see Chapter 7) where a seed crystal is pulled smoothly from the melt within a well-regulated thermal environment with the final crystal size being finite.

6.2.2 Formulation of the parabolic PDE system in Hilbert space

We formulate the system of Eqs. 6.1–6.3 in a Hilbert space $\mathcal{H}(t)$ consisting of n-dimensional vector functions defined on $[0, l(t)]$ that satisfy the boundary conditions of Eq. 6.2, with inner product and norm:

$$
\begin{aligned}
(\omega_1, \omega_2) &= \int_0^{l(t)} (\omega_1(z), \omega_2(z))_{\mathbb{R}^n} \, dz \\
\|\omega_1\|_2 &= (\omega_1, \omega_1)^{\frac{1}{2}}
\end{aligned}
\tag{6.5}
$$

where ω_1, ω_2 are two elements of $\mathcal{H}(t)$ and the notation $(\cdot, \cdot)_{\mathbb{R}^n}$ denotes the standard inner product in $\mathbb{R}^n$. To this end, we define the state function x on $\mathcal{H}(t)$ as

$$
x(t) = \bar{x}(z, t), \quad t > 0, \quad z \in [0, l(t)], \tag{6.6}
$$

the time-varying operator as:

$$
\begin{aligned}
\mathcal{A}(t)x &= A\frac{\partial \bar{x}}{\partial z} + B\frac{\partial^2 \bar{x}}{\partial z^2} + \dot{l}(t)\frac{z}{l(t)}\frac{\partial \bar{x}}{\partial z}, \\
x \in D(\mathcal{A}) &= \Big\{ x \in \mathcal{H}(t) : C_1\bar{x}(0, t) + D_1\frac{\partial \bar{x}}{\partial z}(0, t) = R_1, \\
& \qquad C_2\bar{x}(l(t), t) + D_2\frac{\partial \bar{x}}{\partial z}(l(t), t) = R_2 \Big\}
\end{aligned}
\tag{6.7}
$$

and the input, controlled output, and measurement operators as:

$$
\mathcal{B}(t)u = wb(t)u, \quad \mathcal{C}(t)x = (c(t), kx), \quad \mathcal{Q}(t)x = (s(t), \omega x) \tag{6.8}
$$

where $c(t) = [c^1(t) \; c^2(t) \; \cdots \; c^l(t)]^T$ and $s(t) = [s^1(t) \; s^2(t) \; \cdots \; s^p(t)]^T$, and $c^i(t) \in \mathcal{H}(t)$, $s^\kappa(t) \in \mathcal{H}(t)$. The system of Eqs. 6.1–6.3 can then be written as:

$$
\begin{aligned}
\dot{x} &= \mathcal{A}(t)x + \mathcal{B}(t)u + f(t, x), \quad x(0) = x_0 \\
y &= \mathcal{C}(t)x, \quad q = \mathcal{Q}(t)x
\end{aligned}
\tag{6.9}
$$

where $f(t, x(t)) = f(t, \bar{x}(z, t))$ and $x_0 = \bar{x}_0(z)$. We assume that the nonlinear term $f(t, x)$ satisfies $f(t, 0) = 0$ and is also locally Lipschitz continuous uniformly in t, i.e., there exist positive real numbers a_0, K_0 such that for any $x_1, x_2 \in \mathcal{H}(t)$ that satisfy $\max\{\|x_1\|_2, \|x_2\|_2\} \leq a_0$, we have that:

$$\|f(t, x_1) - f(t, x_2)\|_2 \leq K_0 \|x_1 - x_2\|_2, \quad \forall\, t \in [0, \infty). \tag{6.10}$$

Remark 6.2 In the formulation of the PDE system of Eqs. 6.1–6.3 in $\mathcal{H}(t)$, the time-varying term $\dot{l}(t)\frac{z}{l(t)}\frac{\partial \bar{x}}{\partial z}$ in the expression of $\mathcal{A}(t)$ (Eq. 6.7) accounts for convective transport owing to the motion of the domain. This term was not present in the expression of the differential operator in the case of parabolic PDE systems with fixed spatial domains (where $\dot{l}(t) \equiv 0$; see Chapter 4), and makes $\mathcal{A}(t)$ an explicit function of time.

6.2.3 Singular perturbation formulation

In this subsection, we state precisely our assumption that the dynamics of the infinite-dimensional system of Eq. 6.9 can be partitioned into slow (finite dimensional) and fast (infinite dimensional) ones. We again note that this assumption is usually satisfied by most diffusion-reaction processes. Assumption 6.2 that follows states this requirement.

Assumption 6.2 *Let $\{\phi_j(t)\}$, $j = 1, \ldots, \infty$, be an orthogonal and countable, $\forall\, t \in [0, \infty)$, basis of $\mathcal{H}(t)$. Let also $\mathcal{H}_s(t) = span\{\phi_1(t), \phi_2(t), \ldots, \phi_m(t)\}$ and $\mathcal{H}_f(t) = span\{\phi_{m+1}(t), \phi_{m+2}(t), \ldots,\}$ be two modal subspaces of $\mathcal{H}(t)$ such that $\mathcal{H}_s(t) \oplus \mathcal{H}_f(t) = \mathcal{H}(t)$, and define the orthogonal (pointwise in time) projection operators $P_s : \mathcal{H}(t) \to \mathcal{H}_s(t)$ and $P_f : \mathcal{H}(t) \to \mathcal{H}_f(t)$ so that the state x of the system of Eq. 6.9 can be written as $x = x_s + x_f = P_s x + P_f x$. Using this decomposition for x, we assume that the system of Eq. 6.9 can be written in the following singularly perturbed form:*

$$\begin{aligned}
\frac{dx_s}{dt} &= \mathcal{A}_s(t)x_s + \mathcal{B}_s(t)u + f_s(t, x_s, x_f), \quad x_s(0) = P_s x_0 \\
\epsilon\frac{\partial x_f}{\partial t} &= \mathcal{A}_{f\epsilon}(t)x_f + \epsilon\mathcal{B}_f(t)u + \epsilon f_f(t, x_s, x_f), \quad x_f(0) = P_f x_0 \\
y &= \mathcal{C}(t)(x_s + x_f), \quad q = \mathcal{Q}(t)(x_s + x_f)
\end{aligned} \tag{6.11}$$

where $\mathcal{A}_s(t) = P_s\mathcal{A}(t)$, $\mathcal{B}_s(t) = P_s\mathcal{B}(t)$, $f_s(t, x_s, x_f) = P_s f(t, x) + P_s\mathcal{A}(t)x_f$, $\mathcal{A}_{f\epsilon}(t) = \epsilon P_f\mathcal{A}(t)$ is an unbounded differential operator, $\mathcal{B}_f(t) = P_f\mathcal{B}(t)$, $f_f(t, x_s, x_f) = P_f f(t, x) + P_f\mathcal{A}(t)x_s$, $\epsilon \ll 1$ and the operators $\mathcal{A}_s(t)$, $\mathcal{A}_{f\epsilon}(t)$ generate semigroups with growth rates that are of the same order of magnitude.

Remark 6.3 The statement "the operators $\mathcal{A}_s(t)$, $\mathcal{A}_{f\epsilon}(t)$ generate semigroups with growth rates that are of the same order of magnitude" in assumption 6.2 means that the solutions x_s and x_f of the systems $\dot{x}_s = \mathcal{A}_s(t)x_s$ and $\dot{x}_f = \mathcal{A}_{f\epsilon}(t)x_f$, respectively, satisfy: $|x_s(t)| \leq K|x_s(0)|e^{k_1 t}$,

$\|x_f(t)\|_2 \leq K\|x_f(0)\|_2 e^{k_2 t}$, where K, k_1, k_2 are real numbers, and $O(k_1) = O(k_2)$. The norm notation $|x_s|$ denotes that x_s belongs in a finite-dimensional Hilbert space. This assumption ensures that the system of Eq. 6.11 is in the standard singularly perturbed form [95].

Remark 6.4 In assumption 6.2, the basis, $\{\phi_j(t)\}$, $j = 1, \ldots, \infty$, of $\mathcal{H}(t)$ can be chosen from standard basis functions sets, or it can be computed by solving an eigenvalue problem of the form $\mathcal{A}(t)\phi_j(t) = \lambda_j(t)\phi_j(t)$ or by applying Karhunen–Loève expansion on an appropriately chosen ensemble of solutions of the system of Eq. 6.9 (see [19, 35] for details on Karhunen–Loève expansion). The terms $P_s\mathcal{A}(t)x_f$ in $f_s(t, x_s, x_f)$ and $P_f\mathcal{A}(t)x_s$ in $f_f(t, x_s, x_f)$ account for the use of alternative basis function sets (e.g., empirical eigenfunctions).

Remark 6.5 Assumption 6.2 can be thought of as a natural generalization of assumption 4.1 to parabolic PDE systems with time-dependent spatial domains. Furthermore, for parabolic PDE systems with fixed spatial domains, ϵ was defined in assumption 4.1 as $\epsilon = \frac{|Re\{\lambda_1\}|}{|Re\{\lambda_{m+1}\}|}$, where λ_1, λ_{m+1} are the eigenvalues of the operator $\mathcal{A}$, and are obtained by solving the eigenvalue problem $\mathcal{A}\phi_j = \lambda_j\phi_j$, where ϕ_j is an eigenfunction. Generalizing this definition to systems with time-dependent spatial domains, one can define $\epsilon = \sup_{t\in[0,\infty)} \frac{|Re\{\lambda_1(t)\}|}{|Re\{\lambda_{m+1}(t)\}|}$, where $\lambda_1(t)$ is the largest eigenvalue of the matrix $\mathcal{A}_s(t)$ and $\lambda_{m+1}(t)$ is the largest eigenvalue of the operator (infinite-range matrix) $P_f\mathcal{A}(t)$. Moreover, the presence of ϵ in the right-hand side of the x_f-subsystem in Eq. 6.11 is also consistent with the development in the case of parabolic PDE systems with fixed spatial domains, where a structurally similar system to the one of Eq. 6.11 was derived by using $\epsilon = \frac{|Re\{\lambda_1\}|}{|Re\{\lambda_{m+1}\}|}$.

6.2.4 Illustrative example: Catalytic rod with moving boundary

We consider the catalytic rod of section 4.3 with moving boundary. The process is described by the following parabolic PDE:

$$\frac{\partial \bar{x}}{\partial t} = \frac{\partial^2 \bar{x}}{\partial z^2} + \beta_T e^{-\frac{\gamma}{1+\bar{x}}} - \beta_T e^{-\gamma} + \beta_u(b(z,t)u(t) - \bar{x}) \tag{6.12}$$

subject to the Dirichlet boundary conditions:

$$\bar{x}(0,t) = 0, \quad \bar{x}(l(t),t) = 0 \tag{6.13}$$

and the initial condition:

$$\bar{x}(z,0) = \bar{x}_0(z) \tag{6.14}$$

where $\bar{x}$ is the state, β_T, γ, β_u are dimensionless process parameters, $b(z,t) = [b^1(z,t) \;\; b^2(z,t)]$ are the actuator distribution functions, and $u = [u^1 \;\; u^2]^T$ is the manipulated input vector. The spatial domain is assumed

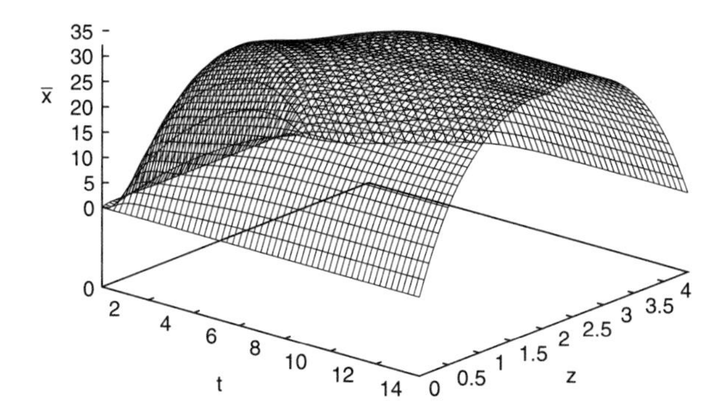

FIGURE 6.2. Evolution of state of open-loop system.

to change according to the relation:

$$l(t) = \pi\left(1.4 - 0.4e^{(-0.02t^{2.7})}\right) \tag{6.15}$$

(it can be easily seen that the above function satisfies the requirements of assumption 6.1) and the following typical values were given to the process parameters:

$$\beta_T = 85.0, \quad \beta_U = 2.0, \quad \gamma = 4.0. \tag{6.16}$$

A 20-th order Galerkin truncation of the system of Eq. 6.12 is used in our simulations (it was verified that further increase in the order of the Galerkin model provides no substantial improvement on the accuracy of the simulation results). It was found that the operating steady state $\bar{x}(z,t) = 0$ is an unstable one (Figure 6.2 shows the evolution of the open-loop state starting from initial conditions close to the steady state $\bar{x}(z,t) = 0$; the system moves to another stable steady state characterized by a maximum at $z = \frac{l(t)}{2}$). Moreover, the linearization of the system of Eq. 6.12 around the steady state $\bar{x}(z,t) = 0$ possesses two positive eigenvalues $\forall\, l(t) \in [\pi, 1.4\pi]$. The control objective is to stabilize the system at the unstable steady state $\bar{x}(z,t) = 0$ by employing a nonlinear output feedback controller that uses a point measurement of the state at $z = 0.7l(t)$ (i.e., moving sensor with $s(z,t) = \delta(z - 0.7l(t))$, where $\delta(\cdot)$ is the standard Dirac function). Since the maximum open-loop value of $\bar{x}(z,t)$ occurs at $z = l(t)/2$ and the first two modes of the process are unstable $\forall\, l(t) \in [\pi, 1.4\pi]$, the controlled outputs are defined as:

$$\begin{aligned} y^1(t) &= \int_0^{l(t)} \sqrt{\frac{2}{l(t)}} sin\left(\frac{\pi}{l(t)}z\right)\bar{x}(z,t)\,dz \\ y^2(t) &= \int_0^{l(t)} \sqrt{\frac{2}{l(t)}} sin\left(2\frac{\pi}{l(t)}z\right)\bar{x}(z,t)\,dz. \end{aligned} \tag{6.17}$$

The actuator distribution functions are taken to be $b^1(z,t) = 1$ (uniform in space, distributed control action) and $b^2(z,t) = \delta(z - \frac{2}{3}l(t))$ (moving point-control actuator).

For the system of Eq. 6.12, the differential operator is of the form:

$$\mathcal{A}(t)x = \frac{\partial^2 \bar{x}}{\partial z^2} + \dot{l}(t)\frac{z}{l(t)}\frac{\partial \bar{x}}{\partial z},$$
$$x \in D(\mathcal{A}) = \{x \in \mathcal{H}([0,l(t)];\mathbb{R}); \quad \bar{x}(0,t) = 0; \quad \bar{x}(l(t),t) = 0\} \tag{6.18}$$

and a countable, $\forall\, t \in [0,\infty)$, orthogonal basis of $\mathcal{H}(t)$ is:

$$\phi_j(t,z) = \sqrt{\frac{2}{l(t)}}\, sin\left(j\frac{\pi}{l(t)}z\right), \qquad j = 1,\ldots,\infty. \tag{6.19}$$

Note that the above set of basis functions satisfies the conditions of assumption 6.2.

6.3 Nonlinear Model Reduction

In this section, we construct nonlinear low-dimensional ODE systems that accurately reproduce the dynamics and solutions of the infinite dimensional system of Eq. 6.11. The construction of the ODE systems is achieved by generalizing the nonlinear model reduction procedure introduced in Chapter 4 for systems with fixed spatial domains, to the class of systems of Eq. 6.11. The nonlinear model reduction procedure is based on combination of standard Galerkin's method with the concept of approximate inertial manifold and provides a characterization of the accuracy of the ODE systems in terms of ϵ.

We begin with the introduction of the concepts of inertial manifold and approximate inertial manifold. Our definition of the concept of inertial manifold is a direct generalization of the concept used in Chapter 4 (see also [133]) for systems with time-invariant differential operators, to systems with time-varying operators (the reader may also refer to [89, 126] for concepts of IMs for infinite-dimensional systems with time-varying forcing inputs). An inertial manifold $\mathcal{M}(t)$ for the system of Eq. 6.11 is a subset of $\mathcal{H}(t)$, which satisfies the following properties:

(i) $\mathcal{M}(t)$ is a finite-dimensional Lipschitz manifold,

(ii) $\mathcal{M}(t)$ is a graph of a Lipschitz function $\Sigma(t, x_s, u, \epsilon)$ mapping $[0,\infty) \times \mathcal{H}_s(t) \times \mathbb{R}^l \times (0,\epsilon^*]$ into $\mathcal{H}_f(t)$, and for every solution $x_s(t)$, $x_f(t)$ of Eq. 6.11 with $x_f(0) = \Sigma(0, x_s(0), u(0), \epsilon)$, then

$$x_f(t) = \Sigma(t, x_s(t), u, \epsilon), \quad \forall\, t \geq 0 \tag{6.20}$$

(iii) $\mathcal{M}(t)$ attracts every trajectory exponentially.

The evolution of the state x_f on $\mathcal{M}(t)$ is given by Eq. 6.20, while the evolution of the state x_s is governed by the following finite-dimensional system (called inertial form):

$$
\begin{aligned}
\frac{dx_s}{dt} &= \mathcal{A}_s(t)x_s + \mathcal{B}_s(t)u + f_s(t, x_s, \Sigma(t, x_s, u, \epsilon)) \\
y &= \mathcal{C}(t)(x_s + \Sigma(t, x_s, u, \epsilon)), \quad q = \mathcal{Q}(t)(x_s + \Sigma(t, x_s, u, \epsilon)).
\end{aligned} \tag{6.21}
$$

Assuming that $u(t)$ is smooth, differentiating Eq. 6.20 and utilizing Eq. 6.11, $\Sigma(t, x_s, u, \epsilon)$ can be computed as the solution of the following partial differential equation:

$$
\begin{aligned}
&\epsilon\frac{\partial \Sigma}{\partial t} + \epsilon\frac{\partial \Sigma}{\partial x_s}[\mathcal{A}_s(t)x_s + \mathcal{B}_s(t)u + f_s(t, x_s, \Sigma)] + \epsilon\frac{\partial \Sigma}{\partial u}\dot{u} \\
&\quad = \mathcal{A}_{f\epsilon}(t)\Sigma + \epsilon\mathcal{B}_f(t)u + \epsilon f_f(t, x_s, \Sigma)
\end{aligned} \tag{6.22}
$$

which $\Sigma(t, x_s, u, \epsilon)$ has to satisfy for all $t \in [0, \infty)$, $x_s \in \mathcal{H}_s(t)$, $u \in \mathbb{R}^l$, $\epsilon \in (0, \epsilon^*]$.

From the complex structure of Eq. 6.22, it is obvious that the computation of the explicit form of $\Sigma(t, x_s, u, \epsilon)$ is impossible in most practical applications. To circumvent this problem, a procedure based on singular perturbations is used to compute approximations of $\Sigma(t, x_s, u, \epsilon)$ (approximate inertial manifolds) and approximations of the inertial form, of desired accuracy. More specifically, the vectors $\Sigma(t, x_s, u, \epsilon)$ and u are expanded in a power series in ϵ:

$$
\begin{aligned}
u &= u_0 + \epsilon u_1 + \epsilon^2 u_2 + \cdots + \epsilon^k u_k + O(\epsilon^{k+1}) \\
\Sigma(t, x_s, u, \epsilon) &= \Sigma_0(t, x_s, u) + \epsilon\Sigma_1(t, x_s, u) + \epsilon^2\Sigma_2(t, x_s, u) \\
&\quad + \cdots + \epsilon^k\Sigma_k(t, x_s, u) + O(\epsilon^{k+1})
\end{aligned} \tag{6.23}
$$

where u_k, Σ_k are smooth vector functions. Substituting the expressions of Eq. 6.23 into Eq. 6.22 and equating terms of the same power in ϵ, one can obtain approximations of $\Sigma(t, x_s, u, \epsilon)$ up to a desired order. By substituting the $O(\epsilon^{k+1})$ approximation of $\Sigma(t, x_s, u, \epsilon)$ and u into Eq. 6.21, we can obtain the following approximation of the inertial form:

$$
\begin{aligned}
\frac{dx_s}{dt} &= \mathcal{A}_s(t)x_s + \mathcal{B}_s(t)(u_0 + \epsilon u_1 + \cdots + \epsilon^k u_k) + f_s(t, x_s, \Sigma_0(t, x_s, u) \\
&\quad + \epsilon\Sigma_1(t, x_s, u) + \cdots + \epsilon^k\Sigma_k(t, x_s, u)) \\
y_s &= \mathcal{C}(t)(x_s + \Sigma_0(t, x_s, u) + \epsilon\Sigma_1(t, x_s, u) + \cdots + \epsilon^k\Sigma_k(t, x_s, u)) \\
q_s &= \mathcal{Q}(t)(x_s + \Sigma_0(t, x_s, u) + \epsilon\Sigma_1(t, x_s, u) + \cdots + \epsilon^k\Sigma_k(t, x_s, u))
\end{aligned} \tag{6.24}
$$

where the subscript s in y_s and q_s are used to denote that these outputs are associated with an approximate slow subsystem. In order to characterize the discrepancy between the solution of the open-loop finite-dimensional system of Eq. 6.24 and the solution of the x_s-subsystem of the open-loop infinite-dimensional system of Eq. 6.11, we need to impose the following stability requirements on the slow and fast dynamics of the system of Eq. 6.11.

Assumption 6.3 states that the system of Eq. 6.24 with $u(t) \equiv 0$ and $\epsilon = 0$ is exponentially stable.

Assumption 6.3 *The finite-dimensional system of Eq. 6.24 with $u(t) \equiv 0$ and $\epsilon = 0$ is exponentially stable in the sense that there exists a smooth Lyapunov function $V : \mathcal{H}_s(t) \to \mathbb{R}_{\geq 0}$ and a set of positive real numbers $(a_1, a_2, a_3, a_4, a_5)$ such that for all $x_s \in \mathcal{H}_s(t)$ that satisfy $|x_s| \leq a_5$ the following conditions hold:*

$$\begin{aligned} a_1|x_s|^2 \leq\ & V(t, x_s) \leq a_2|x_s|^2 \\ \dot{V}(t, x_s) =\ & \frac{\partial V}{\partial t} + \frac{\partial V}{\partial x_s}[\mathcal{A}_s(t)x_s + f_s(t, x_s, 0)] \leq -a_3|x_s|^2 \\ \left|\frac{\partial V}{\partial x_s}\right| \leq\ & a_4|x_s|. \end{aligned} \tag{6.25}$$

To state our stability requirements on the fast dynamics of Eq. 6.11, we write the system of Eq. 6.11 in the fast time scale $\tau = t/\epsilon$ and set $\epsilon = 0$ to derive the following infinite-dimensional fast subsystem:

$$\frac{\partial x_f}{\partial \tau} = \mathcal{A}_{f\epsilon}(t)x_f. \tag{6.26}$$

Assumption 6.4, states that the above system is exponentially stable.

Assumption 6.4 *The fast subsystem of Eq. 6.26 is exponentially stable, in the sense that there exists a Lyapunov functional $W : \mathcal{H}_f(t) \to \mathbb{R}_{\geq 0}$ and a set of positive real numbers $(b_1, b_2, b_3, b_4, b_5, b_6)$ such that for all $x_f \in \mathcal{H}_f(t)$ that satisfy $\|x_f\|_2 \leq b_6$ the following conditions hold:*

$$\begin{aligned} b_1\|x_f\|_2^2 \leq W(t, x_f) \leq b_2\|x_f\|_2^2 \\ \frac{\partial W}{\partial x_f}\mathcal{A}_{f\epsilon}(t)x_f \leq -b_3\|x_f\|_2^2 \\ \left\|\frac{\partial W}{\partial x_f}\right\|_2 \leq b_4\|x_f\|_2 \\ \left\|\frac{\partial W}{\partial t}\right\|_2 \leq b_5\|x_f\|_2^2. \end{aligned} \tag{6.27}$$

We are now in a position to state the main result of this section, which characterizes the discrepancy between the solution obtained from the open-loop system of Eq. 6.24 with the expansion for $\Sigma(t, x_s, u, \epsilon)$ of Eq. 6.23 and the solution of the infinite-dimensional open-loop system of Eq. 6.11, in terms of ϵ. The proof is given in Appendix E.

Proposition 6.1 *Consider the system of Eq. 6.9 with $u(t) \equiv 0$ and suppose that assumptions 6.1, 6.2, 6.3, and 6.4 hold. Then, there exist positive real numbers μ_1, μ_2, ϵ^* such that if $|x_s(0)| \leq \mu_1$, $\|x_f(0)\|_2 \leq \mu_2$ and $\epsilon \in$*

$(0, \epsilon^*]$, *then the solutions* $x_s(t)$, $x_f(t)$ *of the system of Eq. 6.11 satisfy* $\forall\, t \in [t_b, \infty)$:

$$\begin{aligned} x_s(t) &= \tilde{x}_s(t) + O(\epsilon^{k+1}) \\ x_f(t) &= \tilde{x}_f(t) + O(\epsilon^{k+1}) \end{aligned} \tag{6.28}$$

where t_b *is the time required for* $x_f(t)$ *to approach* $\tilde{x}_f(t)$, $\tilde{x}_s(t)$ *is the solution of Eq. 6.24 with* $u(t) \equiv 0$, *and* $\tilde{x}_f(t) = \Sigma_0(t, \tilde{x}_s, 0) + \epsilon\Sigma_1(t, \tilde{x}_s, 0) + \epsilon^2\Sigma_2(t, \tilde{x}_s, 0) + \cdots + \epsilon^k\Sigma_k(t, \tilde{x}_s, 0)$.

Remark 6.6 Utilizing the result of proposition 6.1, one can show that $x(t) = \tilde{x}(t) + O(\epsilon^{k+1})$, $\forall\, t \geq t_b$, where $x(t)$ is the solution of the open-loop infinite-dimensional system of Eq. 6.9 and $\tilde{x}(t) = \tilde{x}_s(t) + \tilde{x}_f(t)$ is the solution obtained from the $O(\epsilon^{k+1})$ approximation of the open-loop inertial form (Eq. 6.24) and the inertial manifold (Eq. 6.23).

Remark 6.7 For $k = 0$, the expansion of Eq. 6.23 yields $\Sigma(t, x_s, u, \epsilon) = \Sigma_0(t, x_s, u) = 0$, and the corresponding approximate inertial form is:

$$\begin{aligned} \frac{dx_s}{dt} &= \mathcal{A}_s(t)x_s + \mathcal{B}_s(t)u_0 + f_s(t, x_s, 0) \\ y &= \mathcal{C}(t)x_s, \quad q = \mathcal{Q}(t)x_s. \end{aligned} \tag{6.29}$$

The above system is identical to the one obtained from a direct application of Galerkin's method to the system of Eq. 6.11 and does not utilize any information about the structure of the fast subsystem, thus yielding solutions that are only $O(\epsilon)$ close to the solutions of the open-loop system of Eq. 6.9 (result of proposition 6.1 with $k = 0$). On the other hand, for $k = 1$, the expansion of Eq. 6.23 yields $\Sigma(t, x_s, u, \epsilon) = \epsilon\Sigma_1(t, x_s, u_0) = -\epsilon(\mathcal{A}_{f\epsilon})^{-1}(t)[\mathcal{B}_f(t)u_0 + f_f(t, x_s, 0)]$ (note that from assumption 6.4 we have that $(\mathcal{A}_{f\epsilon})^{-1}(t)$ exists and is bounded, $\forall\, t \geq 0$), and the corresponding approximate inertial form is:

$$\begin{aligned} \frac{dx_s}{dt} &= \mathcal{A}_s(t)x_s + \mathcal{B}_s(t)(u_0 + \epsilon u_1) + f_s(t, x_s, \epsilon\Sigma_1(t, x_s, u_0)) \\ y &= \mathcal{C}(t)(x_s + \epsilon\Sigma_1(t, x_s, u_0)) \\ q &= \mathcal{Q}(t)(x_s + \epsilon\Sigma_1(t, x_s, u_0)). \end{aligned} \tag{6.30}$$

The above system does utilize information about the structure of the fast subsystem, thereby yielding solutions that are $O(\epsilon^2)$ close to the solutions of the open-loop system of Eq. 6.9.

Remark 6.8 Similar to the case of parabolic PDEs with fixed spatial domains, the expansion of $\Sigma(t, x_s, u, \epsilon)$ in a power series in ϵ (Eq. 6.23) is motivated and validated by the fact that as $\epsilon \to 0$, the inertial form of Eq. 6.21 reduces to the system of Eq. 6.29 obtained from the standard Galerkin's method, which ensures that the inertial form is well posed with respect to ϵ. The expansion of u in a power series in ϵ in Eq. 6.23 is motivated by our intention to appropriately modify the synthesis of the controller such that the outputs of the $O(\epsilon^{k+1})$ approximation of the closed-loop inertial form, y_s^i, $i = 1, \ldots, l$, satisfy $\lim_{t \to \infty} |y_s^i - v_i| = 0$, where v_i is the reference input.

6.4 Nonlinear Output Feedback Control

In this section, we synthesize nonlinear finite-dimensional output feedback controllers that guarantee local exponential stability and force the controlled output of the closed-loop PDE system to follow, up to a desired accuracy, a prespecified response, provided that ϵ is sufficiently small. The output feedback controllers are constructed through combination of state feedback controllers with state observers.

More specifically, we use the system of Eq. 6.24 to synthesize nonlinear state feedback controllers of the following general form:

$$\begin{aligned} u &= u_0 + \epsilon u_1 + \epsilon^2 u_2 + \cdots + \epsilon^k u_k \\ &= p_0(t, x_s) + Q_0(t, x_s)v + \epsilon[p_1(t, x_s) + Q_1(t, x_s)v] \\ &\quad + \cdots + \epsilon^k [p_k(t, x_s) + Q_k(t, x_s)v] \end{aligned} \tag{6.31}$$

where $p_0(t, x_s), \ldots, p_k(t, x_s)$ are smooth vector functions, $Q_0(t, x_s), \ldots, Q_k(t, x_s)$ are smooth matrices, and $v \in \mathbb{R}^l$ is the constant reference input vector. The nonlinear controllers are constructed by following a sequential procedure. Specifically, the component $u_0 = p_0(t, x_s) + Q_0(t, x_s)v$ is initially synthesized on the basis of the $O(\epsilon)$ approximation of the inertial form; then the component $u_1 = p_1(t, x_s) + Q_1(t, x_s)v$ is synthesized on the basis of the $O(\epsilon^2)$ approximation of the inertial form. In general, at the k-th step, the component $u_k = p_k(t, x_s) + Q_k(t, x_s)v$ is synthesized on the basis of the $O(\epsilon^{k+1})$ approximation of the inertial form (Eq. 6.24). The synthesis of $[p_\nu(t, x_s), Q_\nu(t, x_s)]$, $\nu = 0, \ldots, k$, so that a nonlinear controller of the form of Eq. 6.31 guarantees local exponential stability and forces the output of the system of Eq. 6.24 to follow a desired linear response, is performed by utilizing geometric control methods for nonlinear ODEs (see theorem 6.1 below).

Since measurements of $\bar{x}(z, t)$ (and thus, $x_s(t)$) are usually not available in practice, we assume that there exists an L so that the nonlinear dynamical system:

$$\begin{aligned} \frac{d\eta}{dt} &= \mathcal{A}_s(t)\eta + \mathcal{B}_s(t)[u_0 + \epsilon u_1 + \epsilon^2 u_2 + \cdots + \epsilon^k \bar{u}_k] + f_s(t, \eta, \epsilon\Sigma_1(t, \eta, u) \\ &\quad + \epsilon^2\Sigma_2(t, \eta, u) + \cdots + \epsilon^k\Sigma_k(t, \eta, u)) + L[q - \mathcal{Q}(t)(\eta + \epsilon\Sigma_1(t, \eta, u) \\ &\quad + \epsilon^2\Sigma_2(t, \eta, u) + \cdots + \epsilon^k\Sigma_k(t, \eta, u))] \end{aligned} \tag{6.32}$$

where η denotes an m-dimensional state vector, is a local exponential observer for the system of Eq. 6.24 (i.e., the discrepancy $|\eta(t) - x_s(t)|$ tends exponentially to zero).

Theorem 6.1 follows which provides the synthesis formula of the output feedback controller and conditions that guarantee closed-loop stability in the case of considering an $O(\epsilon^2)$ approximation of the exact slow system for the synthesis of the controller. The proof of the theorem is given in Appendix E. To state our result, we need the following definitions. First, the Lie derivative of the scalar function $h_0^i(t, x_s)$ with respect to the vector function $f_0(t, x_s)$ is defined as $L_{f_0} h_0^i(t, x_s) = \partial h_0^i / \partial x_s \, f_0(t, x_s) + \partial h_0^i / \partial t$ (this definition of Lie derivative was introduced in [104] and is different than the standard one used in [86] for the case of time invariant h_{0i}, f_0), $L_{f_0}^k h_0^i(t, x_s)$ denotes the k-th order Lie derivative, and $L_{g_{0i}} L_{f_0}^k h_0^i(t, x_s)$ denotes the mixed Lie derivative. Now, referring to the system of Eq. 6.29, we set $\mathcal{A}_s(t)x_s + f_s(t, x_s, 0) = f_0(t, x_s)$, $\mathcal{B}_s(t) = g_0(t, x_s)$, $\mathcal{C}_i(t)x_s = h_0^i(t, x_s)$ to obtain:

$$
\begin{aligned}
\frac{dx_s}{dt} &= f_0(t, x_s) + g_0(t, x_s)u \\
y_s^i &= h_0^i(t, x_s).
\end{aligned}
\tag{6.33}
$$

For the above system, the relative order of the output y_s^i with respect to the vector of manipulated inputs u is defined as the smallest integer r_i for which:

$$
\left[L_{g_{01}} L_{f_0}^{r_i - 1} h_0^i(t, x_s) \quad \cdots \quad L_{g_{0l}} L_{f_0}^{r_i - 1} h_0^i(t, x_s) \right] \not\equiv [0 \quad \cdots \quad 0] \tag{6.34}
$$

or $r_i = \infty$ if such an integer does not exist. Furthermore, the matrix:

$$
\mathcal{C}_0(t, x_s) = \begin{bmatrix}
L_{g_{01}} L_{f_0}^{r_1 - 1} h_0^1(t, x_s) & \cdots & L_{g_{0l}} L_{f_0}^{r_1 - 1} h_0^1(t, x_s) \\
L_{g_{01}} L_{f_0}^{r_2 - 1} h_0^2(t, x_s) & \cdots & L_{g_{0l}} L_{f_0}^{r_2 - 1} h_0^2(t, x_s) \\
\vdots & & \\
L_{g_{01}} L_{f_0}^{r_l - 1} h_0^l(t, x_s) & \cdots & L_{g_{0l}} L_{f_0}^{r_l - 1} h_0^l(t, x_s)
\end{bmatrix} \tag{6.35}
$$

is the characteristic matrix of the system of Eq. 6.33.

Theorem 6.1 *Consider the parabolic infinite dimensional system of Eq. 6.9, for which assumptions 6.1, 6.2, and 6.4 hold. Consider also the* $O(\epsilon^2)$ *approximation of the inertial form and assume that its characteristic matrix* $\mathcal{C}_1(t, x_s, \epsilon)$ *is invertible* $\forall\, t \in [0, \infty)$, $\forall\, x_s \in \mathcal{H}_s(t), \forall \epsilon \in (0, \epsilon^*]$. *Suppose also that the following conditions hold:*

(i) The roots of the equation:

$$
det(B(s)) = 0 \tag{6.36}
$$

where $B(s)$ *is an* $l \times l$ *matrix whose* (i, j)*-th element is of the form* $\sum_{k=0}^{r_i} \beta_{jk}^i s^k$, *lie in the open left half of the complex plane, where* β_{jk}^i *are adjustable controller parameters.*

(ii) The zero dynamics of the $O(\epsilon^2)$ approximation of the inertial form are locally exponentially stable.

Then, there exist positive real numbers $\tilde{\mu}_1, \tilde{\mu}_2, \tilde{\epsilon}^*$ such that if $|x_s(0)| \leq \tilde{\mu}_1$, $\|x_f(0)\|_2 \leq \tilde{\mu}_2$ and $\epsilon \in (0, \tilde{\epsilon}^*]$, and $\eta(0) = x_s(0)$, the dynamic output feedback controller:

$$\begin{aligned} \frac{d\eta}{dt} &= \mathcal{A}_s(t)\eta + \mathcal{B}_s(t)[u_0(t,\eta) + \epsilon u_1(t,\eta)] + f_s(t,\eta,\epsilon(\mathcal{A}_{f\epsilon})^{-1}(t) \\ &\quad \times [-\mathcal{B}_f(t)u_0(t,\eta) - f_f(t,\eta,0)]) + L(q - (\mathcal{Q}(t)\eta \\ &\quad + \mathcal{Q}(t)\epsilon(\mathcal{A}_{f\epsilon})^{-1}(t)[-\mathcal{B}_f(t)\bar{u}_0(t,\eta) - f_f(t,\eta,0)])) \\ u &= u_0(t,\eta) + \epsilon u_1(t,\eta) := \left\{[\beta_{1r_1} \;\; \cdots \;\; \beta_{lr_l}]C_0(t,\eta)\right\}^{-1} \\ &\quad \times \left\{v - \sum_{i=1}^{l}\sum_{k=0}^{r_i} \beta_{ik} L_{f_0}^k h_0^i(t,\eta)\right\} + \epsilon\{[\beta_{1r_1} \;\; \cdots \;\; \beta_{lr_l}] \\ &\quad \times C_1(t,\eta,\epsilon)\}^{-1} \left\{v - \sum_{i=1}^{l}\sum_{k=0}^{r_i} \beta_{ik} L_{f_1}^k h_1^i(t,\eta,\epsilon)\right\} \end{aligned} \tag{6.37}$$

(a) guarantees local exponential stability of the closed-loop system, and

(b) ensures that the outputs of the closed-loop system satisfy for all $t \in [t_b, \infty)$:

$$y^i(t) = y_s^i(t) + O(\epsilon^2), \quad i = 1, \ldots, l \tag{6.38}$$

where t_b is the time required for the off-manifold fast transients to decay to zero exponentially, and $y_s^i(t)$ is the solution of:

$$\sum_{i=1}^{l}\sum_{k=0}^{r_i} \beta_{ik} \frac{d^k y_s^i}{dt^k} = v. \tag{6.39}$$

Remark 6.9 The implementation of the controller of Eq. 6.37 requires computing a finite-dimensional approximation, say $\Sigma_{1t}(t,\eta,u_0)$, of the infinite-dimensional vector $\Sigma_1(t,\eta,u_0)$. Similar to the case of parabolic PDEs with fixed spatial domains, this can be done by keeping the first $\tilde{m}$ elements of $\Sigma_1(t,\eta,u_0)$ and neglecting the remaining infinite ones. Again, the choice of a sufficiently large $\tilde{m}$ ensures that the controller of Eq. 6.37 with $\Sigma_{1t}(t,\eta,u_0)$ instead of $\Sigma_1(t,\eta,u_0)$ guarantees stability and enforces the requirement of Eq. 6.39 in the closed-loop infinite-dimensional system.

Remark 6.10 Note that in the presence of small initialization errors of the observer states (i.e., $\eta(0) \neq x_s(0)$), uncertainty in the model parameters and external disturbances, although a slight deterioration of the performance may occur (i.e., the requirement of Eq. 6.38 will not be exactly imposed in the closed-loop system), the output feedback controller of theorem 6.1 will continue to enforce exponential stability and asymptotic output tracking in the closed-loop system. Furthermore, the assumption that the characteristic matrices $C_0(t,x_s)$, $C_1(t,x_s,\epsilon)$ are invertible

$\forall\, t \in [0, \infty), \forall\, x_s \in \mathcal{H}_s(t), \forall \epsilon \in (0, \epsilon^*]$ is made in order to simplify the development and can be relaxed by using dynamic state feedback instead of static-state feedback (see [86] for details).

6.5 Application to a Catalytic Rod with Moving Boundary

In this section, we apply the developed nonlinear model reduction and control algorithms to the catalytic rod with moving boundary example presented in subsection 6.2.4. Specifically, we perform several simulation runs to evaluate: (a) the reduction in the order of the controller achieved when the controller is synthesized on the basis of ODE models derived from combination of Galerkin's method with approximate inertial manifolds, and (b) the choice of using moving control actuators and measurement sensors. In all the simulation runs, the process is assumed to be at a nonzero initial condition.

We initially employ standard Galerkin's method to derive an approximate ODE system that is used for the synthesis of a nonlinear output feedback controller. The *lowest-order* model obtained from the standard Galerkin's method, which leads to the synthesis of a controller that stabilizes the open-loop system at $\bar{x}(z,t) = 0$ is 8 (i.e., $m = 8$, $\tilde{m} = 0$), as verified by simulations. Figures 6.3 and 6.4 show the evolution of the state of the closed-loop system and the profile of the manipulated input under an 8-th order nonlinear output feedback controller of the form of Eq. 6.37 (the controller parameters are $\epsilon = 0$, $\beta_{10} = 1.0$, $\beta_{11} = 4.0$, $\beta_{20} = 1.0$, $\beta_{21} = 4.0$, and $L = [0.7\ -25.0\ -9.6\ 0.0\ \cdots\ 0.0]^T$). It is clear that this controller stabilizes the state of the system at $\bar{x}(z,t) = 0$.

We now use the proposed combination of Galerkin's method with approximate inertial manifolds to derive an ODE system that is utilized for

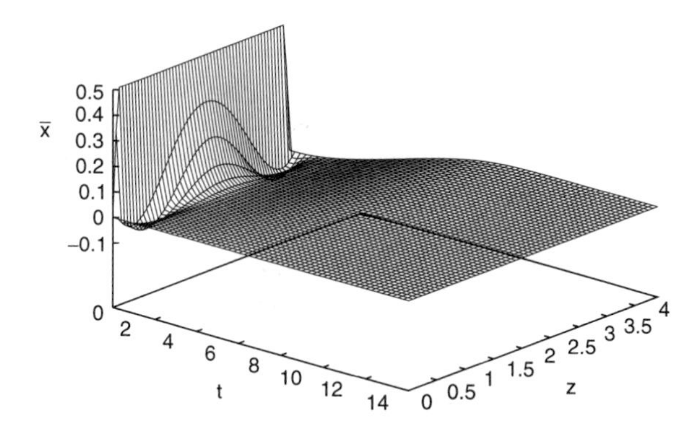

FIGURE 6.3. Evolution of state of closed-loop system under an 8-th order nonlinear time-varying output feedback controller synthesized on the basis of an ODE system obtained from standard Galerkin's method.

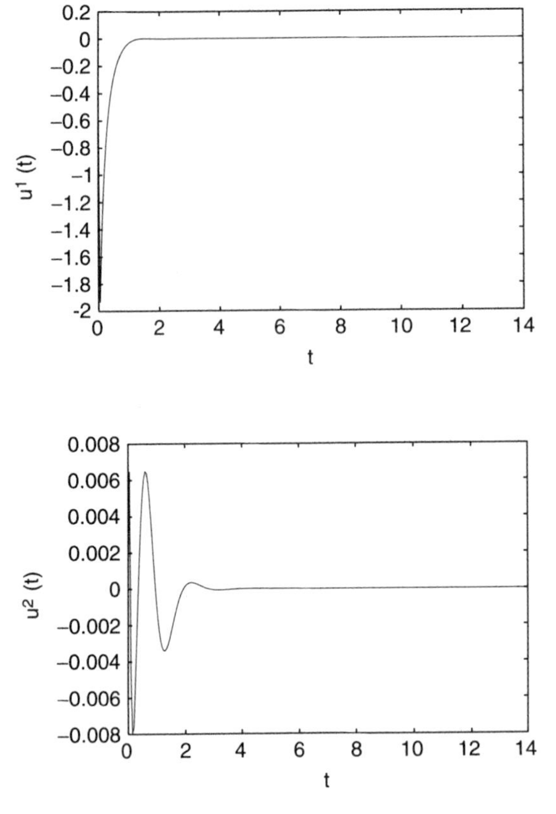

FIGURE 6.4. Manipulated input profiles of 8-th order nonlinear time-varying output feedback controller.

the synthesis of a nonlinear output feedback controller. For the case of using an $O(\epsilon^2)$ approximation of the AIM, the *lowest-order* ODE model, which leads to the synthesis of a controller that stabilizes the open-loop system at $\bar{x}(z,t) = 0$, is one of order 5 which uses a 6-th order approximation for x_f (i.e., $m = 5$ and $\tilde{m} = 6$). Figures 6.5 and 6.6 show the evolution of the state of the closed-loop system and the profile of the manipulated input under a 5-th order controller of the form of Eq. 6.37 (the controller parameters are $\epsilon = 0.0278$, $\beta_{10} = 1.0$, $\beta_{11} = 4.0$, $\beta_{20} = 1.0$, $\beta_{21} = 4.0$, and $L = [0.7 \; -25.0 \; -9.6 \; 0.0 \; \cdots \; 0.0]^T$). The controller clearly regulates the system at $\bar{x}(z,t) = 0$.

We also implement a nonlinear output feedback controller synthesized on the basis of an 8-th order Galerkin truncation of the system of Eq. 6.12 with $l(t) = \pi$ (i.e., the domain is assumed to be fixed in the design of the controller). Figure 6.7 shows the evolution of the state of the closed-loop system, and Figure 6.8 shows the corresponding manipulated input profiles.

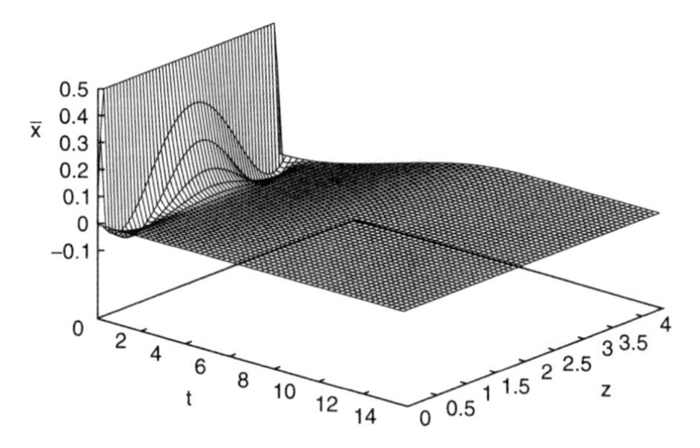

FIGURE 6.5. Evolution of state of closed-loop system under a 5-th order nonlinear time-varying output feedback controller synthesized on the basis of an ODE system obtained from combination of Galerkin's method with approximate inertial manifolds.

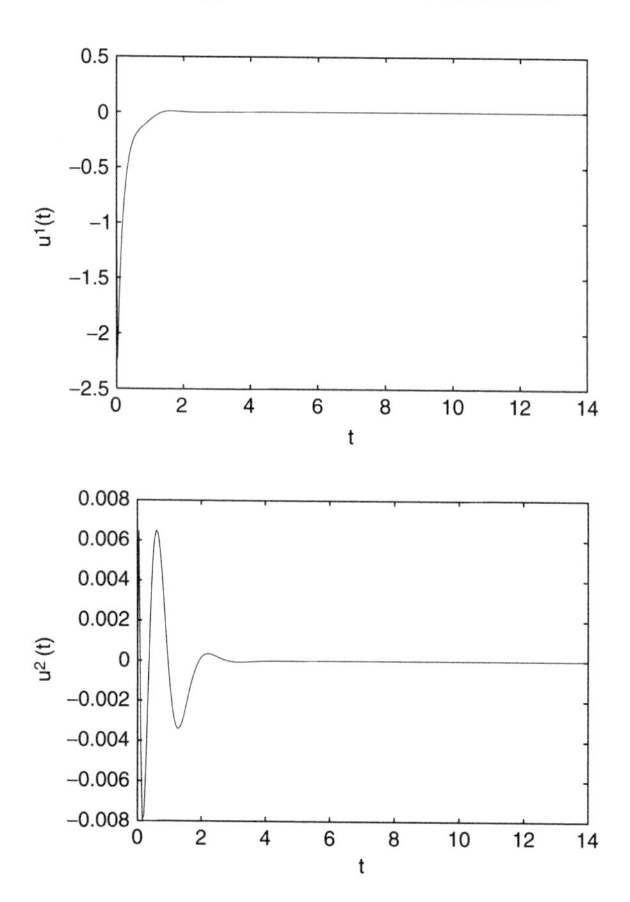

FIGURE 6.6. Manipulated input profiles of 5-th order nonlinear time-varying output feedback controller.

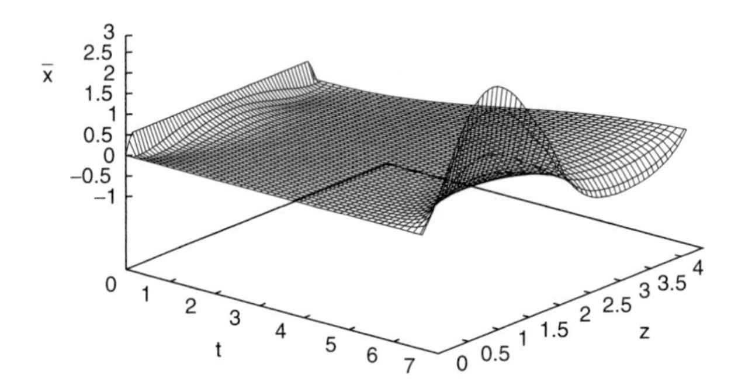

FIGURE 6.7. **Evolution of state of closed-loop system under an 8-th order nonlinear output feedback controller, synthesized under the assumption of fixed spatial domain (i.e., $l(t) = \pi, \forall t \geq 0$).**

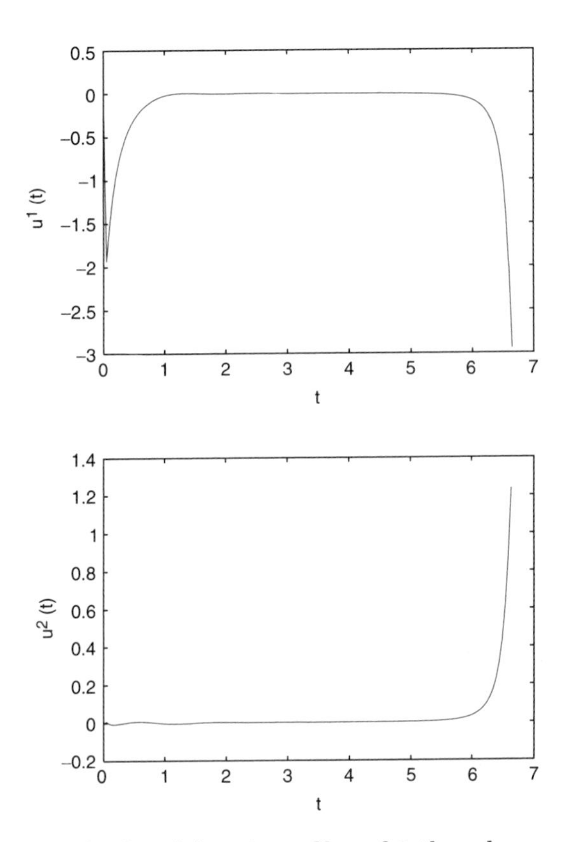

FIGURE 6.8. **Manipulated input profiles of 8-th order nonlinear output feedback controller synthesized under the assumption of fixed spatial domain (i.e., $l(t) = \pi, \forall t \geq 0$).**

Clearly, this controller leads to closed-loop instability because the PDE system becomes "uncontrollable" (i.e., the position of the point-control actuator approaches the location of the zero of the second eigenfunction at $z = 0.5l(5.29)$ which makes the stabilization of the second unstable mode impossible, thereby leading to closed-loop instability). We finally note that similar closed-loop instabilities are observed when: (a) the nonlinear output feedback controller is synthesized on the basis of an 8-th order Galerkin truncation of the system of Eq. 6.12 with $l(t)$ equal to 1.2π and 1.4π, and (b) the nonlinear output feedback controller is synthesized on the basis of a 5-th order model obtained from combination of Galerkin's method with AIMs ($m = 5, \tilde{m} = 6$), for $l(t) = \pi,\ 1.2\pi,\ 1.4\pi$.

From the results of the simulation study, it is evident that the combination of Galerkin's method and approximate inertial manifolds to derive ODE models used for controller synthesis, leads to a significant reduction in the order of the stabilizing controller, while the use of moving control actuators and sensors allows controlling processes with moving boundaries, which are difficult to control with actuators and sensors placed at fixed locations.

6.6 Robust Control of Parabolic PDEs with Time-Dependent Spatial Domains

In the remainder of this chapter, we present a method for the synthesis of robust nonlinear time-varying output feedback controllers for parabolic PDEs with time-dependent spatial domains and uncertain variables. In the same spirit of the robust control results of Chapters 3 and 5, the controllers enforce boundedness of the states, output tracking, and attenuation of the effect of the uncertain variables on the controlled output, provided that the separation between the fast and slow modes of the PDE system is sufficiently large.

6.6.1 Preliminaries

We consider quasi-linear uncertain parabolic PDE systems with time-dependent spatial domain of the form:

$$\begin{aligned}
\frac{\partial \bar{x}}{\partial t} &= A\frac{\partial \bar{x}}{\partial z} + B\frac{\partial^2 \bar{x}}{\partial z^2} + wb(z,t)u + f(t,\bar{x}) + W(t,\bar{x},r(z)\theta(t)) \\
y^i &= \int_0^{l(t)} c^i(z,t)k\bar{x}\,dz, \quad i = 1,\ldots,l \\
q^\kappa &= \int_0^{l(t)} s^\kappa(z,t)\omega\bar{x}\,dz, \quad \kappa = 1,\ldots,q
\end{aligned} \tag{6.40}$$

subject to the boundary conditions:

$$
\begin{aligned}
C_1\bar{x}(0,t) + D_1\frac{\partial \bar{x}}{\partial z}(0,t) &= R_1 \\
C_2\bar{x}(l(t),t) + D_2\frac{\partial \bar{x}}{\partial z}(l(t),t) &= R_2
\end{aligned}
\tag{6.41}
$$

and the initial condition:

$$
\bar{x}(z,0) = \bar{x}_0(z) \tag{6.42}
$$

where $W(t, \bar{x}, r(z)\theta(t))$ is a nonlinear vector function, $\theta = [\theta_1 \ \cdots \ \theta_q] \in \mathbb{R}^q$ denotes the vector of uncertain variables, which may include uncertain process parameters or exogenous disturbances, and $r(z) = [r_1(z) \ \cdots \ r_q(z)]$, where $r_k(z)$ is a known smooth function of z that specifies the position of action of the uncertain variable θ_k on $[\alpha, \beta]$.

We assume that $l(t)$ satisfies the properties of assumption 6.1 and consider the Hilbert space $\mathcal{H}(t)$ (see subsection 6.2.2 for a precise definition of $\mathcal{H}(t)$) in which we define the state x:

$$
x(t) = \bar{x}(z,t), \quad t > 0, \quad z \in [0, l(t)], \tag{6.43}
$$

the time-varying operator:

$$
\begin{aligned}
\mathcal{A}(t)x &= A\frac{\partial \bar{x}}{\partial z} + B\frac{\partial^2 \bar{x}}{\partial z^2} - \dot{l}(t)\frac{z}{l(t)}\frac{\partial \bar{x}}{\partial z}, \\
x \in D(\mathcal{A}(t)) &= \Big\{ x \in \mathcal{H}(t) : C_1\bar{x}(0,t) + D_1\frac{\partial \bar{x}}{\partial z}(0,t) = R_1, \\
&\qquad C_2\bar{x}(l(t),t) + D_2\frac{\partial \bar{x}}{\partial z}(l(t),t) = R_2 \Big\}
\end{aligned}
\tag{6.44}
$$

and the input, controlled output, and measurement operators:

$$
\mathcal{B}(t)u = wb(t)u, \quad \mathcal{C}(t)x = (c(t), kx), \quad \mathcal{Q}(t)x = (s(t), \omega x). \tag{6.45}
$$

Then, in $\mathcal{H}(t)$, the system of Eqs. 6.40–6.42 takes the form:

$$
\begin{aligned}
\dot{x} &= \mathcal{A}(t)x + \mathcal{B}(t)u + f(t,x) + \mathcal{W}(t,x,\theta), \quad x(0) = x_0 \\
y &= \mathcal{C}(t)x, \quad q = \mathcal{Q}(t)x
\end{aligned}
\tag{6.46}
$$

where $f(t, x(t)) = f(t, \bar{x}(z,t)), \mathcal{W}(t, x(t), \theta) = W(t, \bar{x}, r\theta)$, and $x_0 = \bar{x}_0(z)$.

Finally, we state our assumption that the dynamics of the infinite-dimensional system of Eq. 6.46 can be partitioned into slow (finite-dimensional) and fast (infinite-dimensional) ones.

Assumption 6.5 *Let $\{\phi_j(t)\}$, $j = 1, \ldots, \infty$, be an orthogonal and countable, $\forall\, t \in [0, \infty)$, basis of $\mathcal{H}(t)$. Let also $\mathcal{H}_s(t) = span\{\phi_1(t), \phi_2(t), \ldots, \phi_m(t)\}$ and $\mathcal{H}_f(t) = span\{\phi_{m+1}(t), \phi_{m+2}(t), \ldots\}$ be two subspaces of $\mathcal{H}(t)$ such that $\mathcal{H}_s(t) \oplus \mathcal{H}_f(t) = \mathcal{H}(t)$, and define the orthogonal (pointwise in time) projection operators $P_s : \mathcal{H}(t) \to \mathcal{H}_s(t)$ and $P_f : \mathcal{H}(t) \to \mathcal{H}_f(t)$ so that the state x of the system of Eq. 6.46 can be written as $x = x_s + x_f =$*

$P_s x + P_f x$. Using this decomposition for x, we assume that the system of Eq. 6.46 can be written in the following singularly perturbed form:

$$
\begin{aligned}
\frac{dx_s}{dt} &= \mathcal{A}_s(t)x_s + \mathcal{B}_s(t)u + f_s(t,x_s,x_f) + \mathcal{W}_s(t,x_s,x_f,\theta),\; x_s(0) = P_s x_0 \\
\epsilon\frac{\partial x_f}{\partial \tau} &= \mathcal{A}_{f\epsilon}(t)x_f + \epsilon\mathcal{B}_f(t)u + \epsilon f_f(t,x_s,x_f) + \mathcal{W}_f(t,x_s,x_f,\theta), \\
x_f(0) &= P_f x_0 \\
& y = \mathcal{C}(t)(x_s + x_f), \quad q = \mathcal{Q}(t)(x_s + x_f)
\end{aligned}
\tag{6.47}
$$

where $\mathcal{A}_s(t) = P_s\mathcal{A}(t)$, $\mathcal{B}_s(t) = P_s\mathcal{B}(t)$, $f_s(t,x_s,x_f) = P_s f(t,x) + P_s\mathcal{A}(t)x_f$, $\mathcal{W}_s(t,x_s,x_f,\theta) = P_s\mathcal{W}(t,x,\theta)$ $\mathcal{A}_{f\epsilon}(t) = \epsilon P_f\mathcal{A}(t)$ is an unbounded differential operator, $\mathcal{B}_f(t) = P_f\mathcal{B}(t)$, $f_f(t,x_s,x_f) = P_f f(t,x) + P_f\mathcal{A}(t)x_s$, $\mathcal{W}_f(t,x_s,x_f,\theta) = P_f\mathcal{W}(t,x,\theta)$, $\epsilon \ll 1$, and the operators $\mathcal{A}_s(t)$, $\mathcal{A}_{f\epsilon}(t)$ generate semigroups with growth rates which are of the same order of magnitude.

6.6.2 Robust nonlinear output feedback controller synthesis

We consider the synthesis of robust time-varying output feedback control laws of the form:

$$u_0 = p_0(t,q) + Q_0(t,q)\bar{v} + r_0(t,q) \tag{6.48}$$

where $p_0(t,q)$, $r_0(t,q)$ are vector functions, $Q_0(t,q)$ is a matrix, and $\bar{v}$ is a vector of the form $\bar{v} = \mathcal{V}(v_i, v_i^{(1)}, \ldots, v_i^{(r_i)})$, where $\mathcal{V}(v_i, v_i^{(1)}, \ldots, v_i^{(r_i)})$ is a smooth vector function, $v_i^{(k)}$ is the k-th time derivative of the external reference input v_i (which is assumed to be a smooth function of time), and r_i is a positive integer.

We initially state our stability requirements on the fast dynamics of Eq. 6.47. To this end, we write the system of Eq. 6.47 in the fast time scale $\tau = t/\epsilon$ and set $\epsilon = 0$, to derive the following infinite-dimensional fast subsystem:

$$\frac{\partial x_f}{\partial \tau} = \mathcal{A}_{f\epsilon}(t)x_f. \tag{6.49}$$

Assumption 6.6 *The fast subsystem of Eq. 6.49 is exponentially stable.*

Setting $\epsilon = 0$ in the system of Eq. 6.47, we end up with the following m-dimensional slow system:

$$
\begin{aligned}
\frac{dx_s}{dt} &= \mathcal{A}_s(t)x_s + f_s(t,x_s,0) + \mathcal{B}_s(t)u + \mathcal{W}_s(t,x_s,0,\theta) \\
&=: F_0(t,x_s) + \sum_{i=1}^{l}\mathcal{B}_0^i(t)u_0^i + \mathcal{W}_0(t,x_s,0,\theta) \\
& y^i = \mathcal{C}^i(t)x_s =: h_0^i(t,x_s), \quad q = \mathcal{S}(t)x_s
\end{aligned}
\tag{6.50}
$$

where the subscript 0 in $(F_0, \mathcal{B}_0^i, u_0^i, \mathcal{W}_0, h_0^i)$ denotes that they are elements of the $O(\epsilon)$ approximation of the x_s-subsystem of Eq. 6.47.

We now state three assumptions on the system of Eq. 6.50, which are needed to synthesize a robust state feedback controller of the form of Eq. 6.48.

Assumption 6.7 *Referring to the system of Eq. 6.50, there exist a set of integers $(r_1, r_2, \ldots, r_l)$ and a coordinate transformation $(\zeta, \eta) = T(x_s, \theta)$ such that the representation of the system, in the coordinates (ζ, η), takes the form:*

$$
\begin{aligned}
\dot{\zeta}_1^{(1)} &= \zeta_2^{(1)} \\
&\vdots \\
\dot{\zeta}_{r_1-1}^{(1)} &= \zeta_{r_1}^{(1)} \\
\dot{\zeta}_{r_1}^{(1)} &= L_{F_0}^{r_1} h_0^1(t, x_s) + \sum_{i=1}^{l} L_{\mathcal{B}_0^i} L_{F_0}^{r_1-1} h_0^1(t, x_s) u_0^i + L_{\mathcal{W}_0} L_{F_0}^{r_1-1} h_0^1(t, x_s) \\
&\vdots \\
\dot{\zeta}_1^{(l)} &= \zeta_2^{(l)} \\
&\vdots \\
\dot{\zeta}_{r_l-1}^{(l)} &= \zeta_{r_l}^{(l)} \\
\dot{\zeta}_{r_l}^{(l)} &= L_{F_0}^{r_l} h_0^l(t, x_s) + \sum_{i=1}^{l} L_{\mathcal{B}_0^i} L_{F_0}^{r_l-1} h_0^l(t, x_s) u_0^i + L_{\mathcal{W}_0} L_{F_0}^{r_l-1} h_0^l(t, x_s) \\
\dot{\eta}_1 &= \Psi_1(t, \zeta, \eta, \theta, \dot{\theta}) \\
&\vdots \\
\dot{\eta}_{m-\sum_i r_i} &= \Psi_{m-\sum_i r_i}(t, \zeta, \eta, \theta, \dot{\theta}) \\
y^i &= \zeta_1^{(i)}, \quad i = 1, \ldots, l
\end{aligned}
\tag{6.51}
$$

where

$$
\begin{aligned}
x_s = T^{-1}(\zeta, \eta, \theta), \zeta &= [\zeta^{(1)} \quad \cdots \quad \zeta^{(l)}]^T \in \mathbb{R}^{\Sigma_i r_i}, \\
\eta &= \begin{bmatrix} \eta_1 & \cdots & \eta_{m-\Sigma_i r_i} \end{bmatrix}^T \in \mathbb{R}^{m-(\Sigma_i r_i)}.
\end{aligned}
$$

The above assumption is always satisfied for systems for which $r_i = 1$, for all $i = 1, \ldots, l$; this requirement can be easily achieved by selecting $b^i(z, t) \neq \phi_j$, for $i = 1, \ldots, l$, $j = 2, \ldots, \infty$.

Referring to the system of Eq. 6.51, we assume, for simplicity, that its characteristic matrix is nonsingular $\forall\, t \in [0, \infty)$, $\forall\, x_s \in \mathcal{H}_s(t)$.

Assumption 6.8 *The dynamical system:*

$$\begin{aligned} \dot{\eta}_1 &= \Psi_1(t, \zeta, 0, 0, 0) \\ &\vdots \\ \dot{\eta}_{m-\sum_i r_i} &= \Psi_{m-\sum_i r_i}(t, \zeta, 0, 0, 0) \end{aligned} \tag{6.52}$$

is locally exponentially stable.

Similar to assumption 5.3 imposed for the synthesis of robust nonlinear controllers for parabolic PDEs with fixed spatial domains, we now impose the counterpart of this assumption to allow the synthesis of robust nonlinear controllers for parabolic PDEs with time-dependent spatial domains.

Assumption 6.9 *There exists a known function $c_0(t, x_s)$ such that the following condition holds:*

$$\left|\left[L_{\mathcal{W}_0} L_{F_0}^{r_1-1} h_0^1(t, x_s) \quad \cdots \quad L_{\mathcal{W}_0} L_{F_0}^{r_l-1} h_0^l(t, x_s)\right]^T\right| \leq c_0(t, x_s) \tag{6.53}$$

for all $t \in [0, \infty)$, $x_s \in \mathcal{H}_s(t)$, $\theta \in \mathbb{R}^q$, $t \geq 0$.

In order to obtain estimates of the states x_s of the system of Eq. 6.50 from the measurements q^κ, $\kappa = 1, \ldots, p$, we impose the following assumption.

Assumption 6.10 *$p = m$ (i.e., the number of measurements is equal to the number of slow modes), and the inverse of the operator $Q(t)$ for all $t \in [0, \infty)$ exists so that $\hat{x}_s = Q^{-1}(t)q$, where $\hat{x}_s$ is an estimate of x_s.*

Theorem 6.2 provides an explicit formula for the robust controller, conditions that ensure boundedness of the state, and a precise characterization of the ultimate uncertainty attenuation level. The proof of the theorem is given in Appendix E.

Theorem 6.2 *Consider the parabolic infinite-dimensional system of Eq. 6.46 for which assumptions 6.1, 6.5, and 6.6 hold, and the finite-dimensional system of Eq. 6.50, for which assumptions 6.7, 6.8, 6.9, and 6.10 hold, under the robust output feedback controller:*

$$\begin{aligned} u_0 = a_0(x_s, x_f, \bar{v}, t) := [C_0(t, \hat{x}_s)]^{-1} &\Bigg\{ \sum_{i=1}^{l}\sum_{k=1}^{r_i} \frac{\beta_{ik}}{\beta_{ir_i}} (v_i^{(k)} - L_{F_0}^k h_0^i(t, \hat{x}_s)) \\ &+ \sum_{i=1}^{l}\sum_{k=1}^{r_i} \frac{\beta_{ik}}{\beta_{ir_i}} (v_i^{(k-1)} - L_{F_0}^{k-1} h_0^i(t, \hat{x}_s)) - \chi[c_0(t, \hat{x}_s)] \\ &\times \frac{\sum_{i=1}^{l}\sum_{k=1}^{r_i} \frac{\beta_{ik}}{\beta_{ir_i}} (L_{F_0}^{k-1} h_0^i(t, \hat{x}_s) - v_i^{(k-1)})}{\left|\sum_{i=1}^{l}\sum_{k=1}^{r_i} \frac{\beta_{ik}}{\beta_{ir_i}} (L_{F_0}^{k-1} h_0^i(t, \hat{x}_s) - v_i^{(k-1)})\right| + \phi} \Bigg\} \end{aligned} \tag{6.54}$$

where $\hat{x}_s = S^{-1}q$, $\frac{\beta_{ik}}{\beta_{ir_i}} = [\frac{\beta_{ik}^1}{\beta_{ir_i}^1} \cdots \frac{\beta_{ik}^l}{\beta_{ir_i}^l}]^T$ *are column vectors of parameters chosen so that the roots of the equation* $det(B(s)) = 0$, *where* $B(s)$ *is an* $l \times l$ *matrix, whose* (i, j)*-th element is of the form* $\sum_{k=1}^{r_i} \frac{\beta_{jk}^i}{\beta_{jr_i}^i} s^{k-1}$, *lie in the open left half of the complex plane, and* χ, ϕ *are adjustable parameters with* $\chi > 1$ *and* $\phi > 0$. *Then, there exist positive real numbers* (δ, ϕ^*) *such that for each* $\phi \leq \phi^*$, *there exists* $\epsilon^*(\phi)$, *such that if* $\phi \leq \phi^*, \epsilon \leq \epsilon^*(\phi)$ *and* $\max\{|x_s(0)|, \|x_f(0)\|_2, \|\theta\|, \|\dot{\theta}\|, \|\tilde{v}\|\} \leq \delta$,

(a) the state of the infinite-dimensional closed-loop system is bounded, and

(b) the outputs of the infinite-dimensional closed-loop system satisfy:

$$\limsup_{t \to \infty} |y^i - v_i| \leq d_0, \quad i = 1, \ldots, l \tag{6.55}$$

where $d_0 = O(\phi + \epsilon)$ *is a positive real number.*

Remark 6.11 The explicit formula of the controller of Eq. 6.54 was derived through combination of geometric control concepts and Lyapunov techniques; the derivation is similar to the one used in Chapter 5 for the synthesis of robust controllers for parabolic PDE systems with fixed spatial domains, and is omitted for brevity.

Remark 6.12 Similar to the case of robust control of parabolic PDEs with fixed spatial domains (see remark 5.8), the use of x_f-feedback in the controller of Eq. 6.54 does not lead to destabilization of the fast dynamics of the closed-loop systems because ϵ is sufficiently small and the controller of Eq. 6.54 does not include terms of the form $O(\frac{1}{\epsilon})$. Furthermore, it can be shown, using singular perturbations, that the controllers of Eqs. 6.37 and 6.54 possess a robustness property with respect to asymptotically stable unmodeled dynamics, provided that they are sufficiently fast.

Remark 6.13 Finally, we note that the nonlinear model reduction and control methods presented in this chapter can be readily generalized to parabolic PDE systems in which the manipulated inputs enter in a nonlinear fashion and can be represented by the following infinite-dimensional form:

$$\begin{aligned} \dot{x} &= \mathcal{A}(t)x + \mathcal{B}(t)g(u) + f(t, x, \theta), \quad x(0) = x_0 \\ y_c &= \mathcal{C}(t)x, \quad y_m = \mathcal{Q}(t)x \end{aligned} \tag{6.56}$$

where $g(\cdot)$ is a nonlinear smooth function. This can be done by using $g(u)$ as a "virtual manipulated input," computing a virtual control action, u_v, for $g(u)$, and then solving through Newton's method the nonlinear equation $g_u = u_v$ at each time instant to compute the value of u (the reader may refer to the next chapter for an application of this approach to the Czochralski crystal growth process where the manipulated input T_i (i-th heater temperature) enters the PDE in the form T_i^4 due to the nonlinear dependence of the radiation mechanism on temperature.

6.7 Application to a Catalytic Rod with Moving Boundary and Uncertainty

We consider the catalytic rod example of section 4.3 with moving boundary and uncertainty in the heat of reaction. The process is described by the following parabolic PDE:

$$\frac{\partial \bar{x}}{\partial t} = \frac{\partial^2 \bar{x}}{\partial z^2} + \beta_T e^{-\frac{\gamma}{1+\bar{x}}} + \beta_U(b(z,t)u(t) - \bar{x}) - \beta_{T,n} e^{-\gamma} \tag{6.57}$$

subject to the Dirichlet boundary conditions:

$$\bar{x}(0,t) = 0, \quad \bar{x}(l(t),t) = 0 \tag{6.58}$$

and the initial condition:

$$\bar{x}(z,0) = 0.5 \tag{6.59}$$

where $\bar{x}$ is the state, γ, β_u are dimensionless process parameters, β_T denotes a dimensionless heat of reaction (which is assumed to be unknown and time-varying; uncertain variable), $\beta_{T,n}$ denotes a *nominal* dimensionless heat of reaction, $b(z,t) = [b^1(z,t)\ b^2(z,t)]$ are the actuator distribution functions, and $u = [u^1\ u^2]^T$ is the manipulated input vector. The spatial domain is assumed to change according to the relation of Eq. 6.15 and the following typical values are given to the process parameters:

$$\beta_{T,n} = 60.0, \quad \beta_U = 2.0, \quad \gamma = 4.0. \tag{6.60}$$

For the system of Eq. 6.57, the differential operator is of the form:

$$\begin{aligned} \mathcal{A}(t)x &= \frac{\partial^2 \bar{x}}{\partial z^2} + \dot{l}(t)\frac{z}{l(t)}\frac{\partial \bar{x}}{\partial z}, \\ x \in D(\mathcal{A}) &= \{x \in \mathcal{H}([0,l(t)];\mathrm{I\!R});\quad \bar{x}(0,t) = 0;\quad \bar{x}(l(t),t) = 0\} \end{aligned} \tag{6.61}$$

and a countable, $\forall\, t \in [0,\infty)$, orthogonal basis of $\mathcal{H}(t)$ is:

$$\phi_j(z,t) = \sqrt{\frac{2}{l(t)}}\, sin\left(j\,\pi\,\frac{z}{l(t)}\right), \quad j = 1,\ldots,\infty. \tag{6.62}$$

Note that the above set of basis functions satisfies the conditions of assumption 6.2.

The control objective is to stabilize the system at the unstable steady state $\bar{x}(z,t) = 0$ in the presence of time-varying uncertainty in the dimensionless heat of the reaction β_T (i.e., $\beta_T = \beta_{T,n} + \theta(t)$ where $\theta(t) = 0.35\,\beta_{T,n}\, sin(0.524\,t)$) by employing a nonlinear output feedback controller that uses two point measurements of the state at $z = 0.25l(t)$ and $z = 0.625l(t)$ (i.e., moving sensors with $s(z,t) = \delta(z - 0.25l(t))$ and $s(z,t) = \delta(z - 0.625l(t))$, respectively, where $\delta(\cdot)$ is the standard Dirac function). We note that this selection for $\theta(t)$ satisfies the requirements

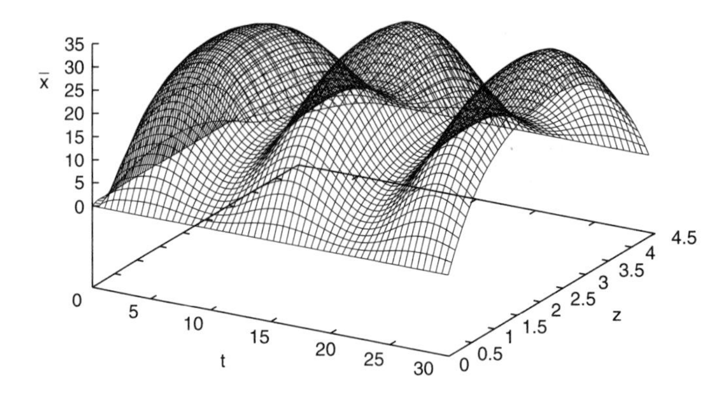

FIGURE 6.9. Open-loop profile of $\bar{x}$ in the presence of uncertainty.

of theorem 6.2 that $\theta(t), \dot{\theta}(t)$ should be sufficiently small (see closed-loop simulations below), while it leads to a very poor open-loop behavior for $\bar{x}(z,t)$ (see Figure 6.9). Since the maximum open-loop value of $\bar{x}(z,t)$ occurs at $z = l(t)/2$ and the first two modes of the process become unstable for some $l(t) \in [\pi, 1.4\pi]$, the controlled outputs are defined as:

$$
\begin{aligned}
y^1(t) &= \int_0^{l(t)} \phi_1(z,t)\bar{x}(z,t)\,dz, \\
y^2(t) &= \int_0^{l(t)} \phi_2(z,t)\bar{x}(z,t)\,dz.
\end{aligned} \tag{6.63}
$$

The actuator distribution functions are taken to be $b^1(z,t) = 1$ (uniform in space, distributed control action) and $b^2(z,t) = \delta(z - \frac{2}{3}l(t))$ (moving point-control actuation).

For the system of Eq. 6.57, we consider the first two modes as the dominant ones ($\epsilon = 0.11$) and use Galerkin's method to derive a two-dimensional ODE system that was used for the synthesis of a nonlinear robust output feedback controller through application of the formula of Eq. 6.54. The resulting controller takes the form:

$$
\begin{aligned}
u^1 &= \frac{1}{\sqrt{2\,l(t)}\frac{2\beta_u}{\pi}}\Bigg[\left(\beta_u - \left(\frac{\pi}{l(t)}\right)^2 - 1\right)(\phi_1, \hat{x}_s) - \left(\phi_1, e^{-\frac{\gamma}{1+\hat{x}_s}} - e^{-\gamma}\right) \\
&\quad - \beta_u\phi_1\left(\frac{2}{3}\pi, t\right)u_2(t) - \chi\theta_{max}\left|\left(\phi_1, e^{-\frac{\gamma}{1+\hat{x}_s}}\right)\right|\frac{(\phi_1, \hat{x}_s)}{(\phi_1, \hat{x}_s) + \phi}\Bigg] \\
u^2 &= \frac{1}{\beta_u\phi_2\left(\frac{2}{3}\pi, t\right)}\Bigg[\left(\beta_u - \left(\frac{2\pi}{l(t)}\right)^2 - 1\right)(\phi_2, \hat{x}_s) - \left(\phi_2, e^{-\frac{\gamma}{1+\hat{x}_s}} - e^{-\gamma}\right) \\
&\quad - \chi\theta_{max}\left|\left(\phi_2, e^{-\frac{\gamma}{1+\hat{x}_s}}\right)\right|\frac{(\phi_2, \hat{x}_s)}{(\phi_2, \hat{x}_s) + \phi}\Bigg]
\end{aligned} \tag{6.64}
$$

where θ_{max} is an upper bound of the disturbance $\theta_{max} \geq \sup_{t\in[0,\infty)}\{|\theta(t)|\}$, and for the values $\theta_{max} = 21$, $\chi = 1.2$, $\phi = 0.005$, the controller achieves

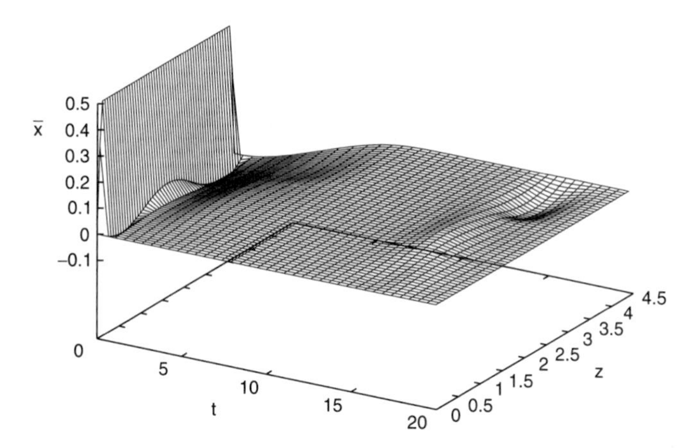

FIGURE 6.10. Closed-loop profile of $\bar{x}$ under nonlinear robust output feedback control.

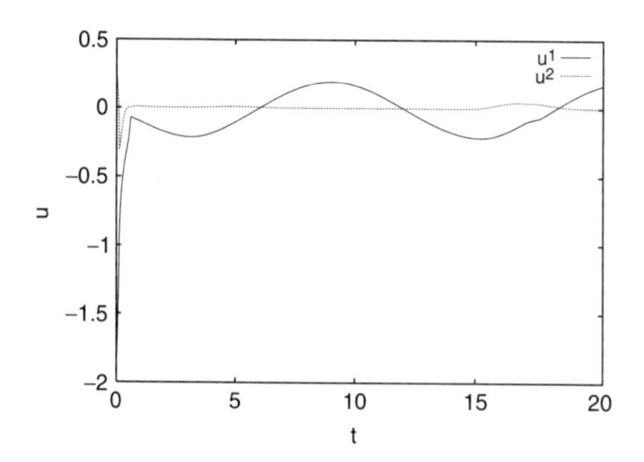

FIGURE 6.11. Manipulated input profiles for nonlinear robust output feedback controller.

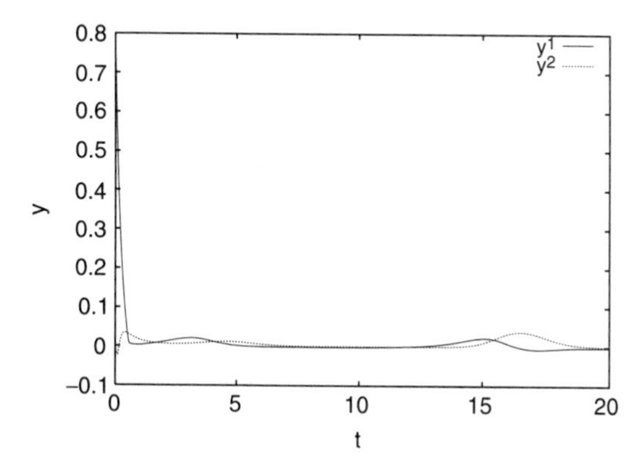

FIGURE 6.12. Closed-loop output profile under nonlinear robust output feedback control.

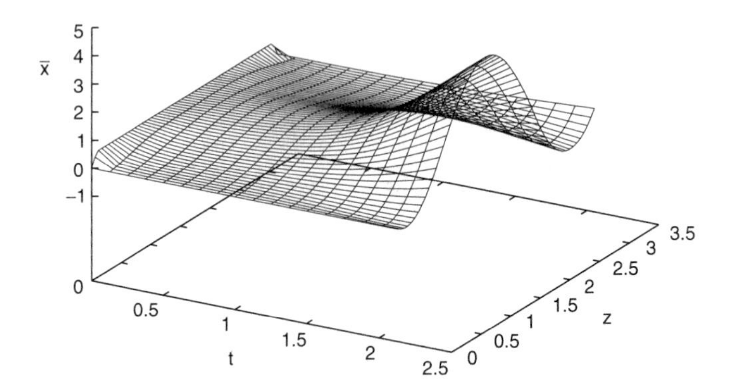

FIGURE 6.13. Closed-loop profile of $\bar{x}$ under nonlinear output feedback control (no uncertainty compensation).

an uncertainty attenuation level $d = 0.05$ (note that $0.05 = O(\epsilon + \phi) = O(0.11 + 0.005)$).

Figure 6.10 shows the evolution of the closed-loop rod temperature profile under the nonlinear robust output feedback controller, while Figure 6.11 shows the corresponding manipulated input profile. Clearly, the proposed controller regulates the temperature profile at $\bar{x}(z,t) = 0$, attenuating the effect of the uncertain variable (note that the requirement $\limsup_{t\to\infty}|y| \leq 0.05$ is enforced in the closed-loop system; Figure 6.12). For the sake of comparison, we also implement the same controller as before without the term which compensates for the effect of the uncertainty (i.e., $\chi = 0$). Figure 6.13 shows the evolution of the closed-loop rod temperature profile. It is clear that this controller cannot regulate the temperature profile at the desired steady state, $\bar{x}(z,t) = 0$ because it does not compensate for the effect of the uncertainty.

We finally note that nonlinear robust static output feedback controllers synthesized under the assumption of fixed spatial domain (i.e., $l(t) = \pi$ and $l(t) = 1.4\pi$) lead to closed-loop instability, as verified by simulations.

6.8 Conclusions

We presented general methods for the synthesis of nonlinear and robust time-varying output feedback controllers for systems of quasi-linear parabolic PDEs with time-dependent spatial domains, whose dynamics can be separated into slow and fast ones. Initially, a nonlinear model reduction procedure, based on combination of Galerkin's method with the concept of approximate inertial manifold, was employed for the derivation of ODE systems that yield solutions close, up to a desired accuracy, to the ones of the PDE system, for almost all times. Then, these ODE systems were used as the basis for the explicit construction of nonlinear and robust output feedback controllers via geometric control methods and Lyapunov techniques. Differences in the nature of the model reduction and control problems between parabolic PDE systems with fixed and moving spatial domains were identified and discussed. The control methods were successfully used to stabilize an unstable steady-state of a diffusion-reaction process with time-dependent spatial domain and were shown to outperform nonlinear and robust controller synthesis methods that do not account for the variation of the spatial domain.

Chapter 7

Case Studies

In this chapter, we present applications of the nonlinear control methods for parabolic PDE systems with fixed and moving spatial domains presented in Chapters 4 and 6 to the rapid thermal chemical vapor deposition process introduced in subsection 1.2.2 and a Czochralski crystal growth process, respectively. The results on nonlinear control of rapid thermal chemical vapor deposition were first presented in [10, 9, 11], and the results on nonlinear control of Czochralski crystal growth in [8].

7.1 Nonlinear Control of Rapid Thermal Chemical Vapor Deposition

7.1.1 Introduction

Rapid thermal chemical vapor deposition (RTCVD) is a rapidly growing technology in the microelectronics industry. The central idea of RTCVD is to use a series of lamps to radiatively heat a thin silicon wafer from room temperature to 1200 K at very high heating rates (more than 150 K/s), and then keep it at the high temperature for a short time. This sharp increase in the temperature of the wafer reduces significantly the overall thermal budget of the process (the overall processing time is usually less than a minute) and the diffusion length, thereby preserving dopant profiles from previous steps, and allows the fabrication of very small devices by using temperature as a switch in ending a process cycle. These features make RTCVD an attractive alternative over conventional furnace-based chemical vapor deposition processes employed in the fabrication of devices with submicron dimensional constraints. Even though RTCVD possesses many significant advantages, its widespread use is seriously limited by the lack of adequate wafer temperature control to achieve the tight requirements of uniformity and repeatability set by the industry. The main obstacles in achieving spatially uniform wafer temperature (and thus, uniform film deposition) are the highly nonlinear (owing to the radiative heat transfer mechanism) dynamic (owing to the very rapid heating of the wafer), and spatially varying nature of the RTCVD process that makes the development and implementation of effective model-based feedback controllers a very difficult task (see, for example, [24, 122, 131, 28, 93, 135] for previous results on control). The main challenge in the design of model-based feedback controllers for RTCVD processes is that the dynamic models of such processes consist of nonlinear parabolic partial differential equation (PDE) systems, which are distributed parameter (infinite-dimensional) systems, and thus, cannot be directly used for the design of practically implementable (low-dimensional) controllers [1].

In the remainder of this section, we focus on the development and implementation of a nonlinear control system on the RTCVD process introduced in subsection 1.2.2. Initially, a detailed mathematical model of the RTCVD process is presented, which consists of a nonlinear parabolic PDE describing the time evolution of the wafer temperature across the radius of the wafer, coupled with a set of nonlinear ODEs, which describe the time evolution of the concentrations of the various species. Then, the synthesis of a nonlinear output feedback controller, based on the RTCVD process model by following the control method introduced in Chapter 4, is discussed. The controller uses measurements of wafer temperature at four locations to manipulate the power of the top lamps in order to achieve uniform temperature, and thus, uniform deposition of the thin film on the wafer over the entire process cycle. The ability of the nonlinear output feedback controller to achieve a spatially uniform temperature profile (and therefore, final film thickness) in the presence of significant model uncertainty and disturbances is successfully tested through simulations and is shown to be superior to that of a proportional integral (PI) control scheme.

7.1.2 Modeling of RTCVD process

We consider the low-pressure RTCVD process introduced in subsection 1.2.1. We recall that the objective of the process is to deposit a $0.5\,\mu m$ film of polycrystalline silicon on a 6-inch wafer in 40 seconds. To achieve this objective, the reactor is fed with 10% SiH_4 in Ar at 5 $Torr$ pressure, and the heating lamps are used to heat the wafer from room temperature to $1200\,K$ (this is the temperature where the deposition reactions take place), at a heating rate of the order of $180\,K/sec$.

A detailed mathematical model for the RTCVD process has been developed in [134] under the following standard assumptions (see also [24] for similar assumptions):

1. Wafer temperature uniformity in the azimuthal direction due to wafer rotation and symmetric reactor design.

2. Negligible wafer temperature variations in the axial direction due to small thickness of the wafer.

3. Negligible heat transfer from the wafer to the reactant gases due to low-pressure conditions inside the chamber.

4. Negligible heat of deposition reactions compared to radiative heat transfer from the lamps to the wafer.

5. Constant optical properties of the wafer and the chamber.

6. Perfect mixing of the reacting mixture.
7. Spatially uniform quartz chamber-wall thermal dynamics.

Under the above assumptions, an energy balance on the wafer yields the following nonlinear parabolic PDE:

$$\rho_w T_{amb} \frac{\partial}{\partial t}(C_{p_w}(T)T) = \frac{T_{amb}}{R_w^2}\frac{1}{r}\frac{\partial}{\partial r}\left(\kappa(T) r \frac{\partial T}{\partial r}\right) - \frac{q_{rad}(T,r)}{\delta z} \tag{7.1}$$

subject to the boundary conditions

$$\left.\frac{\partial T}{\partial r}\right|_{r=0} = 0 \tag{7.2}$$

$$\left.\left(\frac{T_{amb}}{R_w}\kappa(T)\frac{\partial T}{\partial r}\right)\right|_{r=1} = -\sigma \epsilon_w T_{amb}^4 (T^4 - T_c^4) + q_{edge} u_b. \tag{7.3}$$

In the above equations, T_{amb} denotes the ambient temperature, $T = T'/T_{amb}$ denotes the dimensionless wafer temperature, ρ_w, C_{p_w}, R_w denote the density, heat capacity, and radius of the wafer, $r = r'/R_w$ denotes the dimensionless radial coordinate, q_{rad} is a term that accounts for radiative energy transfer between the wafer and its environment (see below for an explicit statement of the radiative phenomena contributing to this term), $T_c = T_c'/T_{amb}$ denotes the dimensionless temperature of the chamber, σ denotes the Stefan-Boltzmann constant, ϵ_w denotes the emissivity of the wafer, q_{edge} denotes the energy flux at the edge of the wafer, u_b denotes the percentage of the side lamp power that is used. The wafer heat capacity and thermal conductivity depend on temperature according to the following relations:

$$\begin{aligned} \kappa(T) &= 50.5\, ln(T T_{amb})^2 - 734.0\, ln(T T_{amb}) + 2.69 \times 10^3 \;\; W/m\;K \\ C_{p_w}(T) &= 1.06 \times 10^3 - 1.04 \times 10^5/(T T_{amb}) J/(Kg\;K). \end{aligned} \tag{7.4}$$

The radiative energy transfer term q_{rad} consists of two parts: the radiant energy absorbed from the heating lamps and the radiant energy exchanged between the wafer and reactor walls, that is:

$$q_{rad} = -Q_{lamps,w} \cdot u + q_{dw,t} + q_{dw,b} \tag{7.5}$$

where $Q_{lamps,w}$ is a vector of the total energy emitted from the three lamp banks and absorbed by the wafer, $u = [u_A u_B u_C]$ is the percentage of the lamp power that is used, $q_{dw,t}$ ($q_{dw,b}$) is the net radiative energy transfered to the wafer top (bottom) surface from sources other than the lamps (e.g., reflection from the quartz process chamber). The heating lamp energy flux to the wafer surface, $Q_{lamps,w}$, is computed by a ray-trace algorithm, which calculates the radiant energy flux distributions directly from the lamps to the wafer as well as the contribution of the reflected rays. The radiant energy flux distribution for each lamp bank as a function of wafer radial position can be found in [134]. The radiation exchange between the wafer and the walls (terms $q_{dw,t}$ and $q_{dw,b}$) is computed using the net-radiation method. We note that $q_{dw,t}$, $q_{dw,b}$ are highly nonlinear functions of the wafer and chamber temperatures, geometry of the reactor, and emissivity of wafer and chamber.

An energy balance on the quartz chamber yields the following ordinary differential equation.

$$T_{amb} M_c \frac{dT_c}{dt} = \epsilon_c Q_{lamps} \cdot u - A_{hem} q_h - A_{cyl} q_c - Q_{convect} - \sigma \epsilon_c A_c T_{amb}^4 (T_c^4 - 1) \tag{7.6}$$

where M_c denotes the chamber thermal mass, ϵ_c denotes the emissivity of the chamber, A_{hem} denotes the chamber hemispherical area, A_{cyl} denotes the chamber cylindrical area and A_c denotes the chamber outside area. q_h and q_c denote the net energy radiated from the hemispherical and cylindrical portions of the quartz chamber, respectively, and their expressions computed from the net-radiation method can be found in [134]. The term $Q_{lamps} \cdot u$ represents energy absorbed by the chamber directly from the heating lamps, while $Q_{convect}$ denotes the energy transfered from the quartz chamber to the cooling gas by forced convective cooling. The explicit computation of $Q_{convect}$ can be done by a simple cooling gas energy balance and is given in [134].

The assumption of perfect mixing of the reacting mixture allows us to derive the following set of ODEs which describe the time evolution of the mole fraction of SiH_4, X_{SiH_4}, and hydrogen, X_{H_2}:

$$\begin{aligned} \frac{dX_{SiH_4}}{dt} &= -\alpha \int_{A_w} R_s(T, X_{SiH4}, X_{H2})\, dA_w + \frac{1}{\tau} (X_{SiH_4}^{in} - X_{SiH_4}) \\ \frac{dX_{H_2}}{dt} &= 2\alpha \int_{A_w} R_s(T, X_{SiH4}, X_{H2})\, dA_w - \frac{1}{\tau} X'_{H_2} \end{aligned} \tag{7.7}$$

where α is the mole to mole conversion factor, A_w is the wafer area, τ is the residence time, $X_{SiH_4}^{in}$ is the mole fraction of SiH_4 in the inlet stream to the reactor, and R_s is the rate of the deposition reactions:

$$R_s(T, X_{SiH4}, X_{H2}) = \frac{k_0\, exp\left(\frac{-\gamma}{RTT_{amb}}\right) X_{SiH_4} P_{tot}}{1 + b X_{SiH_4} P_{tot} + \frac{\sqrt{X_{H_2} P_{tot}}}{c}} \tag{7.8}$$

where k_0 is the pre-exponential constant, γ is the activation energy for deposition, P_{tot} is the total pressure and b, c are constants. The deposition rate of Si onto the wafer surface is governed by the following expression:

$$\frac{dS}{dt} = \frac{MW_{Si}}{\rho_{Si}} R_s(T, X_{SiH_4}, X_{H_2}) \tag{7.9}$$

where MW_{Si} and ρ_{Si} denote the molecular weight and density of Si, respectively. Referring to the expression of the deposition rate, we note the Arrhenius dependence of the deposition rate on wafer temperature which clearly shows that nonuniform temperature results in nonuniform deposition, thereby implying the need to develop and implement a nonlinear feedback controller on the process in order to achieve radially uniform wafer temperature.

7.1.3 Nonlinear controller synthesis—Closed-loop simulations

The objective of this section is to design and implement a nonlinear low-order output feedback controller on the RTCVD process, and test its robustness with respect to significant model uncertainty and disturbances. The values of the process parameters used in the simulations are given in Table 7.1. Initially, a second-order finite difference scheme with 100 discretization points was used to compute an accurate solution of the model of the RTCVD process (Eqs. 7.1, 7.6, and 7.7). For time integration, the Euler method was used. Following [134], the nonlinear boundary condition of Eq. 7.3 was solved simultaneously at each time step using a Gauss–Newton method. Several 40 second simulation runs of the process were performed with the following initial conditions: $T = 1$, $T_c = 1$, $S = 0$, $X_{SiH_4} = 0.1$ and $X_{H_2} = 0$. "Snapshots" of the temperature profiles were used as data for determining the dominant spatial temperature modes (wafer temperature empirical eigenfunctions) through Karhunen–Loéve expansion (See Appendix F for a detailed presentation of this method). Then, a collocation formulation of Galerkin's method was used to obtain a low-order model that describes the wafer temperature. The wafer temperature was expanded in terms of the first four empirical eigenfunctions. The three roots of the fourth empirical eigenfunction were used as collocation points, thereby forcing the residual to be orthogonal, and therefore zero, at these points. Additional collocation points were added at $r = 0$ and $r = 1$ to satisfy the boundary conditions. Therefore, the dimension of the constructed low-order model was five. In order to test the validity of the low-order model, simulations of the detailed and low-order models were performed by using an open-loop recipe (obtained from a trial and error procedure) to manipulate the power of the top lamps in order to heat the wafer from room temperature to 1200 K. The wafer temperature profiles obtained by the detailed and low-order models, and their difference at all positions and times, are shown in Figure 7.1. Clearly, the predictive capabilities of the low-order model are excellent. The fifth-order

TABLE 7.1. RTCVD process parameters.

A_w	=	182.41×10^{-4}	m^2
A_c	=	1217.31×10^{-4}	m^2
A_{cyl}	=	794.83×10^{-4}	m^2
A_{hem}	=	422.48×10^{-4}	m^2
L_c	=	15.43×10^{-2}	m
L_s	=	10×10^{-2}	m
R_c	=	8.2×10^{-2}	m
R_w	=	7.62×10^{-2}	m
δz	=	0.05×10^{-2}	m
M_c	=	1422.6	$J\,K^{-1}$
q_{edge}	=	49×10^4	$J\,s^{-1}\,m^{-2}$
b	=	78.95	$torr^{-1}$
c	=	0.38	$torr^{1/2}$
e_w	=	0.7	
e_c	=	0.37	
ρ_c	=	2.6433×10^3	$kg\,m^{-3}$
V_c	=	638.25×10^6	m^3
ρ_w	=	2.3×10^3	$kg\,m^{-3}$
$X_{SiH_4}^{in}$	=	0.1	$kmol_{SiH_4}\,kmol_{feed}^{-1}$
P_{tot}	=	5.0	$torr$
MW_{Si}	=	28.086	$kg\,kmol^{-1}$
ρ_{Si}	=	2.3×10^3	$kg\,m^{-3}$
R	=	8.314×10^3	$J\,kmol^{-1}\,K^{-1}$
T_{amb}	=	300.0	K
α	=	12.961×10^6	$kmol^{-1}$
k_0	=	263.158×10^1	$kmol\,m^{-2}\,s^{-1}\,torr^{-1}$
γ	=	153.809×10^6	$J\,kmol^{-1}$
σ	=	5.6705×10^{-8}	$J\,s^{-1}\,m^{-2}\,K^{-4}$
τ	=	0.380	s

model was used to synthesize a nonlinear multivariable output feedback controller using the controller synthesis formula of theorem 4.2. The controller uses measurements of the wafer temperature at five locations across the wafer and adjusts the powers of the four top lamps (see Figure 7.2 for a schematic of the control configuration). A simulation run was performed to evaluate the performance of the nonlinear controller for a 40 *sec* cycle with initial conditions $T = 1$, $T_c = 1$, $S = 0$, $X_{SiH_4} = 0.1$, and $X_{H_2} = 0$. Figure 7.3 shows the spatiotemporal evolution of the wafer (top figure), the temperature distribution along the radius of the wafer at $t = 40$ *sec* (middle figure), and the thickness of the deposition (bottom figure). The performance of the nonlinear controller is excellent, achieving an almost uniform (less than 1% variation) thin film deposition. For the sake of comparison, we also implemented on the process four proportional integral controllers with the following tuning parameters (these values were computed through extensive trial and error to obtain the best possible response).

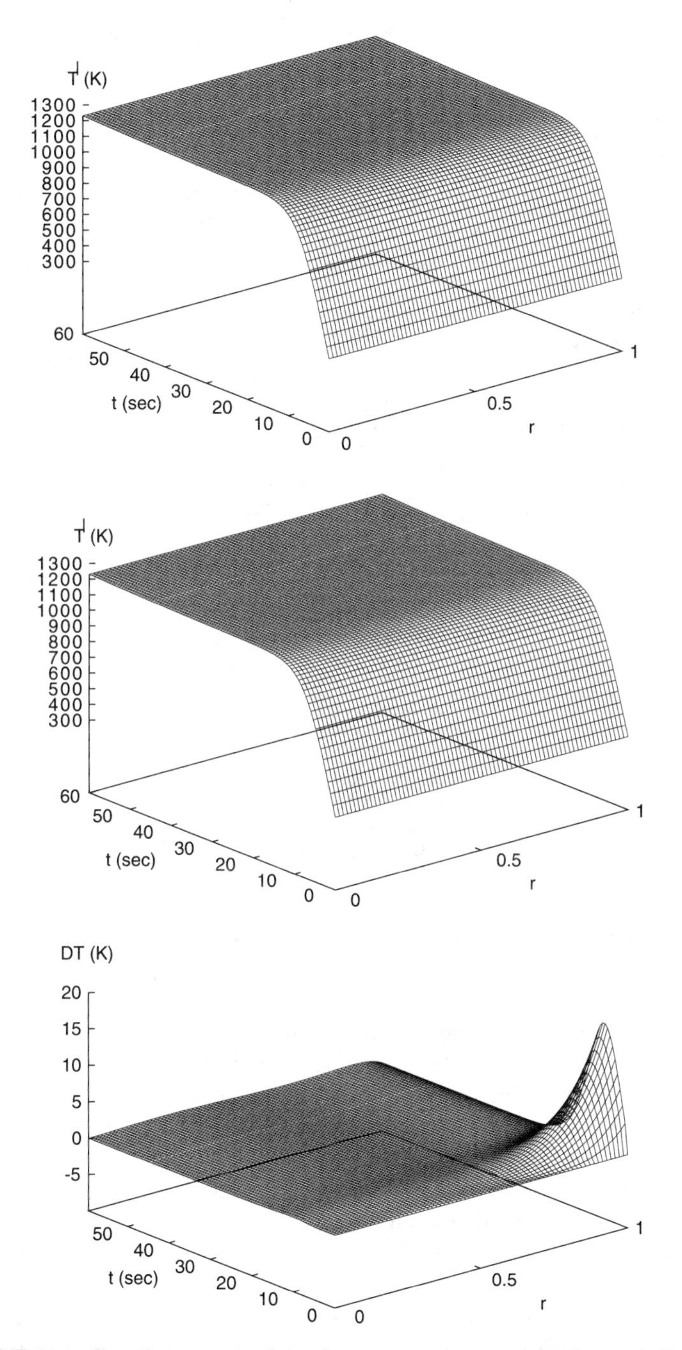

FIGURE 7.1. Spatiotemporal wafer temperature profiles—full model (top plot), reduced-order model (middle plot), temperature difference (DT) between profiles predicted by the full-order and reduced-order models (bottom plot).

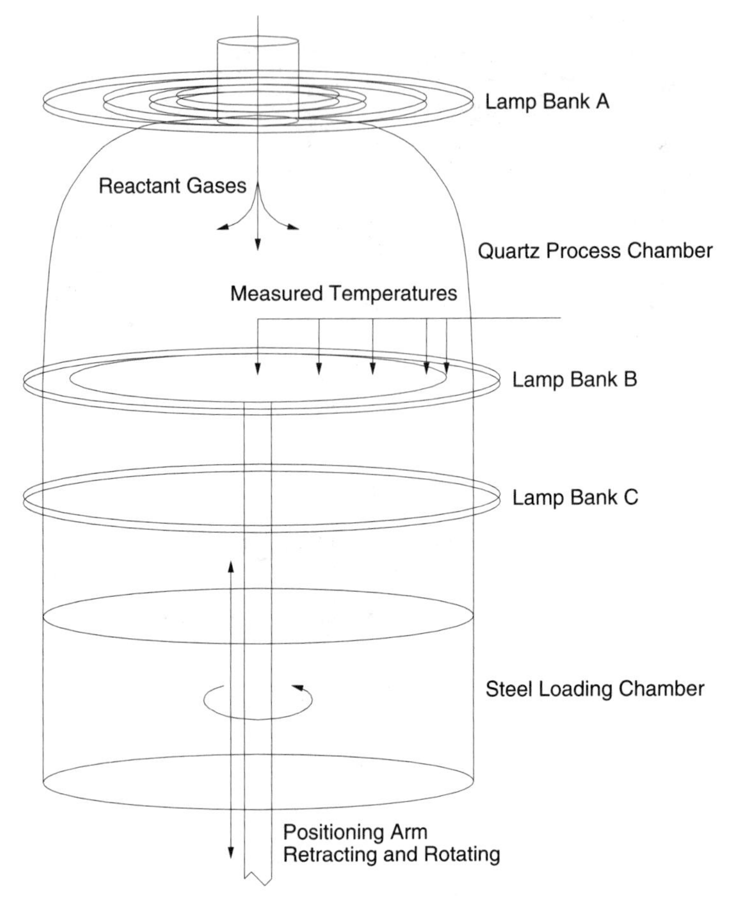

FIGURE 7.2. Control problem specification for RTCVD process.

The final thickness of the deposition achieved by PI control is displayed in Figure 7.3 (bottom plot; dashed line). The performance of the four proportional integral controllers is significantly inferior (more than 5% variation in thin film thickness) to the one obtained by the nonlinear output feedback controller.

Finally, we performed two sets of simulation runs to evaluate the robustness properties of the nonlinear controller in the presence of significant model uncertainty and disturbances. In the first set of simulation runs, we consider a 40 *sec* cycle for the RTCVD process with initial conditions $T = 1$, $T_c = 1$, $S = 0$, $X_{SiH_4} = 0.1$, and $X_{H_2} = 0$. A 20 K decrease in the ambient temperature at $t = 15$ *sec* and a 20% reduction in the power of the side lamp bank B at $t = 30$ *sec* are considered as unmeasured disturbances. Figure 7.4 shows the spatiotemporal evolution of the wafer temperature under nonlinear control, while Figure 7.5 shows the thickness of the deposition

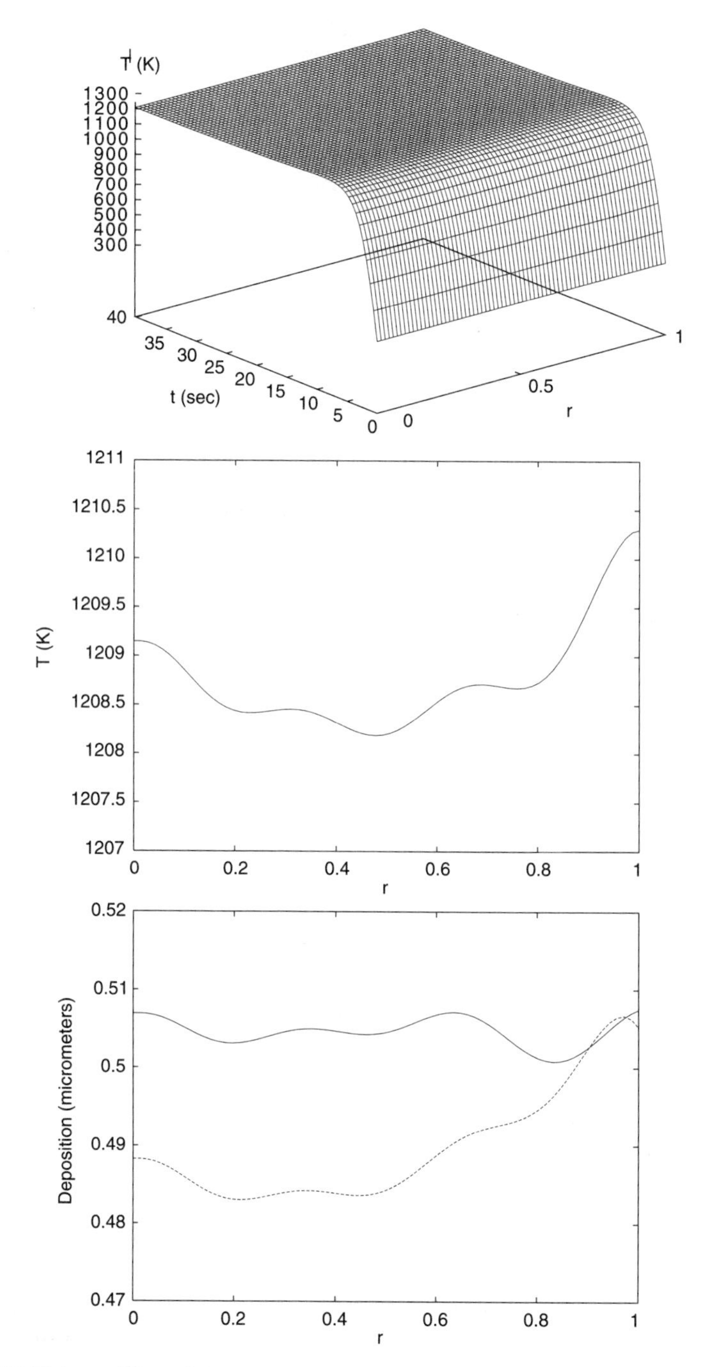

FIGURE 7.3. Closed-loop spatiotemporal wafer temperature profile (top figure), wafer temperature profile at $t = 40$ *sec* (middle figure), and deposition thickness profile at $t = 40$ *sec* (bottom figure) under nonlinear control (solid line) and PI control (dashed line).

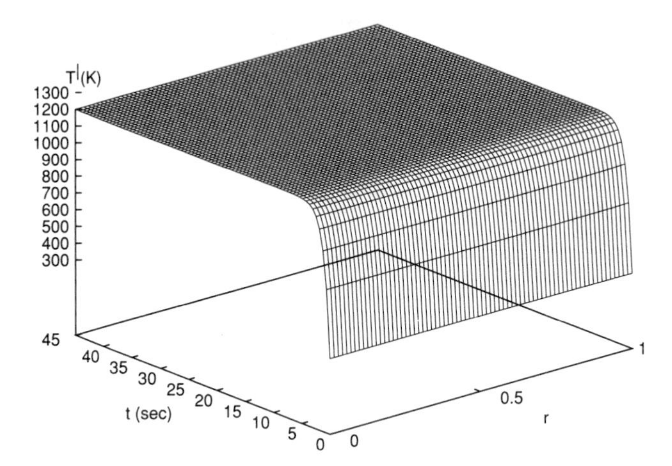

FIGURE 7.4. Spatiotemporal wafer temperature profile under nonlinear control in the presence of disturbances in the ambient temperature and lamb bank B.

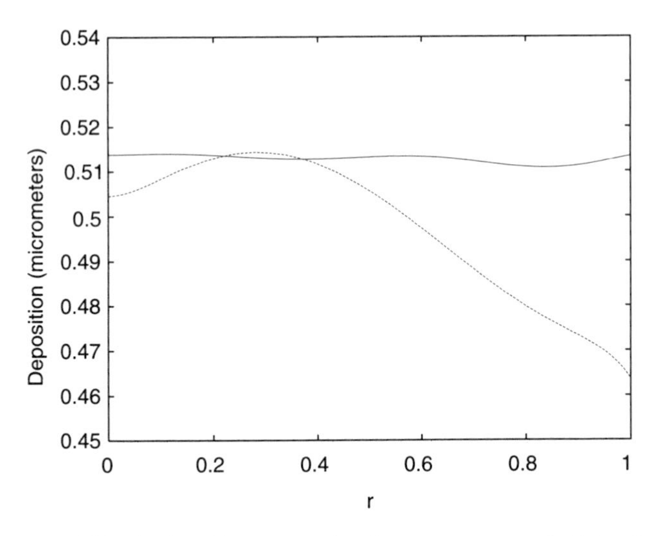

FIGURE 7.5. Final film thickness in the presence of disturbances in the ambient temperature and lamb bank B under nonlinear control (solid line) and PI control (dashed line).

at $t = 40.8$ *sec* under nonlinear (solid line) and PI (dashed line) control. It is clear that the nonlinear controller achieves uniform deposition of the thin film on the wafer (less than 1% nonuniformity) and outperforms the PI control scheme (more than 8% nonuniformity). In the second set of simulation runs, we consider a 40 *sec* cycle for the RTCVD process with initial conditions $T = 1$, $T_c = 1$, $S = 0$, $X_{SiH_4} = 0.1$, and $X_{H_2} = 0$. In this case,

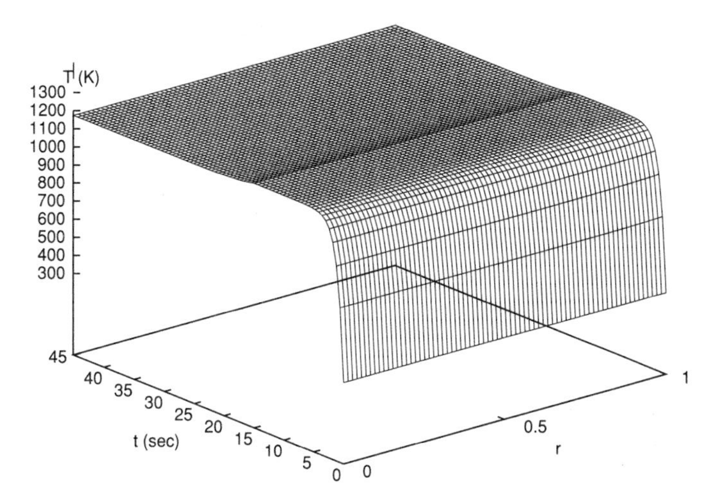

FIGURE 7.6. Spatiotemporal wafer temperature profile under nonlinear control in the presence of parametric model uncertainty under nonlinear control.

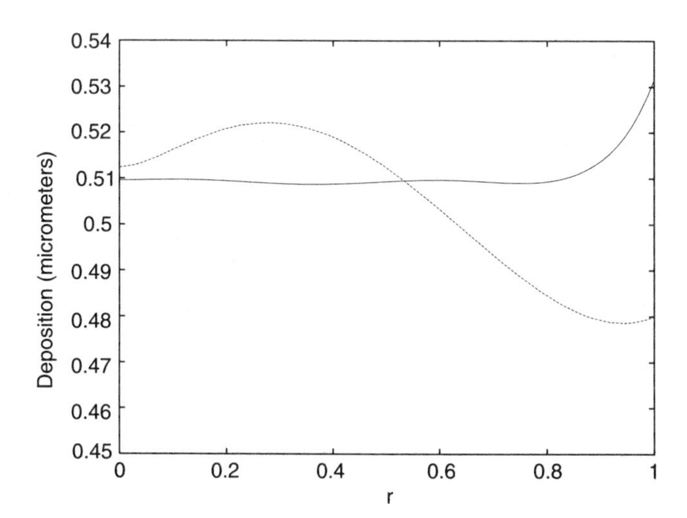

FIGURE 7.7. Final film thickness in the presence of parametric model uncertainty under nonlinear (solid line) and PI control (dashed line).

a 50% error in the emissivity of the wafer was introduced at $t = 15$ *sec*. Figure 7.6 shows the spatiotemporal evolution of the wafer temperature under nonlinear control, while Figure 7.7 shows the thickness of the deposition at $t = 40.8$ *sec* under nonlinear (solid line) and PI (dashed line) control. Again, the nonlinear controller achieves almost uniform deposition of the thin film on the wafer (less than 4% nonuniformity), while the PI control scheme leads to a significantly nonuniform response (more than 10% variation in final film thickness).

7.2 Nonlinear Control of Czochralski Crystal Growth

7.2.1 Introduction

Czochralski crystal growth (CZ) is a well-established industrial process used for the production of single crystals like silicon (Si) and gallium arsenide ($GaAs$). Such crystals are widely used for the construction of wafers employed in the production of microelectronic chips. The central idea of the CZ process is to grow a crystal from a melt by pulling a seed crystal very slowly within a well-regulated thermal environment in a furnace. For the subsequent processing steps, it is important to form a cylindrical crystal with a desired radius and length which includes very low concentrations of impurities and dislocations, as well as a uniform dopant distribution.

The current practice in achieving a constant crystal radius is to use a proportional integral derivative (PID) controller that manipulates the pulling rate of the crystal from the melt to adjust the crystal radius, while the regulation of the crystal temperature is addressed by adjusting the heat transfered to the melt and crystal by manipulation (typically via PID) of the heater power. The tuning of such PID controllers based on fundamental lumped parameter models that describe the dominant dynamic characteristics of the CZ process has been addressed in a series of papers published by Gevelber and Stephanopoulos [74, 72, 73, 70, 71]. Even though these works provide valuable insights into the nature of the crystal radius and thermal dynamics, identify natural control objectives and variables, as well as structural limitations on the best achievable closed-loop performance, they do not account for the presence of spatial variations of the temperature inside the crystal that constitute the main cause for dislocations and defects [132, 70]. Furthermore, PID controllers do not account for the fact that CZ processes exhibit nonlinear and time-varying behavior and involve coupling of variables evolving in widely different time-scales. In an effort to overcome the limitations of conventional controllers, the problems of controlling the radius of the crystal and reducing the thermal strain in the interface between crystal and melt were recently addressed within a model predictive control framework [84, 85]. However, in these papers, the key practical issue of deriving accurate low-order approximations of the distributed process model that can be used in real-time control implementation was not addressed. The main challenge in the design of model-based feedback controllers for

CZ processes is the fact that the dynamic models of such processes are typically in the form of nonlinear parabolic partial differential equations (PDE) with time-dependent spatial domains. These are distributed parameter (infinite-dimensional) systems, and therefore, they cannot be directly used for the design of practically implementable (low-dimensional) controllers.

We use the method proposed in Chapter 6 for nonlinear control of parabolic PDE systems with time-dependent spatial domains to synthesize a nonlinear controller for temperature control in the Czochralski crystal growth process. Specifically, we propose a control configuration and a nonlinear multivariable model-based controller for the reduction of thermal gradients inside the crystal in the Czochralski crystal growth process after the crystal radius has reached its final value. Even though the process model is a distributed system, the proposed controller is of low order, and therefore, it can be readily implemented in real time.

The rest of this chapter is structured as follows: A fundamental mathematical model that describes the evolution of the temperature inside the crystal in the radial and axial directions and accounts for radiative heat exchange between the crystal and its surroundings and motion of the crystal boundary is initially presented. This model is numerically solved by using Galerkin's method, and the behavior of the crystal temperature is studied to obtain valuable insights that lead to the precise formulation of the control problem and the derivation of a simplified one-dimensional in a space PDE model with moving domain which is used for controller synthesis. The method of Chapter 6 is then employed to construct a fourth-order model that describes the dominant thermal dynamics of the Czochralski process and synthesize a fourth-order nonlinear controller that can be readily implemented in practice. The proposed control scheme is successfully implemented on a Czochralski process used to produce a 0.7 m long silicon crystal with a radius of 0.05 m and is shown to significantly reduce the axial and radial thermal gradients inside the crystal compared to the open-loop operation and to the case of using a single control actuator. The robustness of the proposed controller with respect to parametric model uncertainty, melt and chamber temperature disturbances, and unmodeled actuator and sensor dynamics is demonstrated through simulations.

7.2.2 Czochralski crystal growth process description and modeling

We focus on the Czochralski crystal growth process shown in Figure 7.8 used to produce a 0.7 m long silicon crystal with a radius of 0.05 m. The process comprises of a cylindrical chamber that includes a rotating pedestal that can move in the axial direction. A crucible containing silicon (Si) crystals is placed on the pedestal and heaters (placed on the sides of the

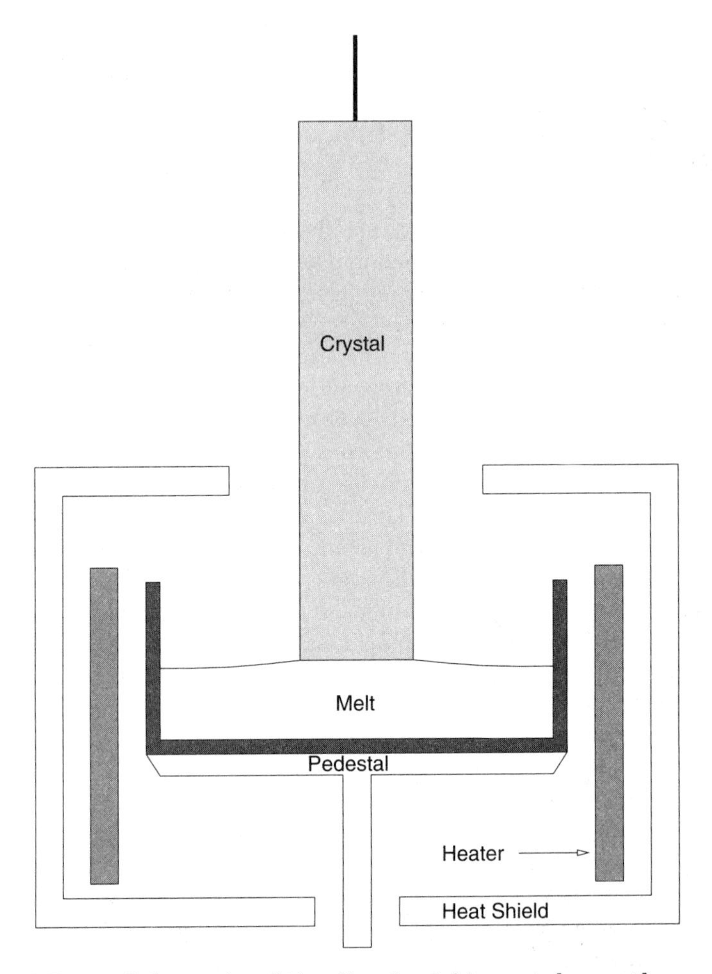

FIGURE 7.8. Schematic of the Czochralski crystal growth process.

chamber and under the pedestal) are used to increase the temperature of the *Si* crystals inside the crucible (through radiation) above the melting point of *Si*. A *Si* seed crystal comes in contact with the melt, and the temperature of the melt is adjusted until the meniscus is supported by the end of the seed. Once the meniscus has been stabilized, the seed crystal is pulled away from the melt and new crystal is formed [83]. The interface between the crystal and the melt is maintained at a constant position during the operation of the process by moving the position of the pedestal higher with time. As the length of the crystal becomes larger, part of it leaves the chamber and starts cooling, at which point the thermal gradients inside the crystal become large and may cause thermal strain inside the crystal [132].

If the cooling conditions are not properly regulated, large thermal strain may cause microdefects (e.g., dislocations) inside the crystal [70], and may even lead to fracture. As a result, the cooling process should be carefully regulated. Finally, the process is terminated when the crystal-melt interface reaches the crucible bottom.

The development of detailed mathematical models for the Czochralski crystal growth process is an area that has received significant attention (see, for example, the book [83]), and at this point comprehensive models are available [58, 59, 56, 8, 60, 57, 151, 138, 139]. Since the objective of our work is to develop a control configuration and a model-based feedback controller that will smoothly regulate the cooling process of the crystal as it leaves the chamber, our modeling effort focuses on the development of a mathematical model that describes the spatiotemporal evolution of the crystal temperature, after the crystal radius has obtained its final value, and accounts for radiative heat exchange between the crystal, heater shield, crucible, melt surface, and the environment. Moreover, our control objective allows making the following simplifying assumptions in the model development: a) the crystal radius and the meniscus height are assumed to be constant; this is typically achieved by using a controller that manipulates the pulling rate and the chamber temperature to maintain these variables constant and allows neglecting the detailed dynamics of the crystal/melt interface in our analysis (see [4] for a detailed discussion on crystal radius control), b) the temperature distribution inside the crystal is assumed to be axisymmetric owing to the constant rotation of the crucible, c) radiation is assumed to be the dominant heat transfer mechanism; this assumption is justified from the fact that the temperatures of the chamber, melt, and crystal surfaces are very high (1000–1700 K), d) secondary radiation is not taken into account since it has a smaller effect on the crystal temperature profile compared to the primary radiation and the temperatures of the surrounding surfaces are kept constant at the desired set points using control, e) the melt and chamber temperature and the pulling rate are assumed to be constant, and the melt/crystal interface is assumed to be flat; this allows us to neglect the melt dynamics, f) the solidification front remains in a specified region of the heater as the melt level drops; this is achieved in practice by raising the crucible through movement of the pedestal [58], g) the concentration of dopant, oxygen, and carbon are not explicitly included in the model; this is done to simplify the controller synthesis task, and special consideration is taken in the tuning of the controller (see Subsection 7.2.4 below) to ensure that large and abrupt heater temperature (manipulated input) changes, which could cause large variations in the concentration fields, do not occur.

Under these assumptions, an application of an energy balance to a differential element of the crystal yields the following two-dimensional parabolic PDE:

$$\frac{\partial T_c}{\partial t} + v_p \frac{\partial T_c}{\partial z} = \frac{k}{\rho c_p}\left[\frac{1}{r}\frac{\partial}{\partial r}\left(r\frac{\partial T_c}{\partial r}\right) + \frac{\partial^2 T_c}{\partial z^2}\right] \tag{7.10}$$

subject to the following boundary conditions:

$$T_c(r,0,t) = T_{mp}, \quad 0 \le r \le R \tag{7.11}$$

$$\left.\frac{\partial T_c}{\partial r}\right|_0 = 0, \quad 0 \le z \le l(t) \tag{7.12}$$

$$\begin{aligned}\left.\frac{k}{\sigma}\frac{\partial T_c}{\partial r}\right|_R &= \epsilon_{w_{cr}}\epsilon_{w_m}F_{cr\to m}(R,z)\big[T_m^4 - T_c^4(R,z,t)\big] + \epsilon_{w_{cr}}\epsilon_{w_{ch}}F_{cr\to ch}(R,z)\\ &\times\big[T_{ch}^4 - T_c^4(R,z,t)\big] + \epsilon_{w_{cr}}\epsilon_{w_{amb}}F_{cr\to amb}(R,z)\\ &\times\big[T_{amb}^4 - T_c^4(R,z,t)\big], \quad 0 \le z \le l(t)\end{aligned} \tag{7.13}$$

$$\begin{aligned}\left.\frac{k}{\sigma}\frac{\partial T_c}{\partial z}\right|_{l(t)} &= \epsilon_{w_{cr}}\epsilon_{w_m}F_{cr\to m}(r,l(t))\big[T_m^4 - T_c^4(r,l(t),t)\big]\\ &+\epsilon_{w_{cr}}\epsilon_{w_{ch}}F_{cr\to ch}(r,l(t))\big[T_{ch}^4 - T_c^4(r,l(t),t)\big]\\ &+\epsilon_{w_{cr}}\epsilon_{w_{amb}}F_{cr\to amb}(r,l(t))\big[T_{amb}^4 - T_c^4(r,l(t),t)\big],\\ &\qquad 0 \le r \le R.\end{aligned} \tag{7.14}$$

In the above equations, T_c is the temperature of the crystal, t is the time, r is the radial direction, z is the axial direction, v_p is the pulling speed, T_{ch} is the chamber temperature, T_{amb} is the ambient temperature, T_{mp} is the melting point temperature of silicon, T_m is the temperature of the melt, σ is the Stefan–Boltzmann constant, $\epsilon_{w_{cr}}$, ϵ_{w_m}, $\epsilon_{w_{ch}}$, $\epsilon_{w_{amb}}$ denote the emissivities of the crystal, melt, chamber, and ambient, respectively, $F_{cr\to j}$ is the view factor from the surface of a differential element of the crystal at $r = R$, cr, to surface j, and $l(t)$ is the total height of the crystal at time t. The model of Eqs. 7.10–7.14 constitutes a parabolic PDE system with moving boundary owing to the variation of the length, $l(t)$, of the crystal in the axial direction ($l(t) = \int_0^t v_p(s)ds$, where $v_p(t)$ is the pulling rate).

In Eq. 7.10, the terms $\frac{\partial T_c}{\partial t}$ and $v_p\frac{\partial T_c}{\partial z}$ describe the rate of change of crystal temperature and the convection effect due to the motion of the crystal, respectively, while the terms $\frac{k}{\rho c_p}[\frac{1}{r}\frac{\partial}{\partial r}(r\frac{\partial T_c}{\partial r}) + \frac{\partial^2 T_c}{\partial z^2}]$ account for heat conduction inside the crystal. On the other hand, the boundary condition of Eq. 7.11 states that the crystal temperature in the crystal/melt interface is equal to the Si melting point temperature and the boundary condition of Eq. 7.12 is based on the assumption of axisymmetric crystal. Furthermore, the boundary condition of Eq. 7.13 accounts for radiative heat exchange between the *differential surface at the side of the crystal* and melt (term $\epsilon_{w_{cr}}\epsilon_{w_m}F_{cr\to m}(R,z)[T_m^4 - T_c^4(R,z,t)]$), chamber (term $\epsilon_{w_{cr}}\epsilon_{w_{ch}}F_{cr\to ch}(R,z)[T_{ch}^4 - T_c^4(R,z,t)]$) and ambient (term

TABLE 7.2. Physical properties of Si.

Melting point	T_{mp}	1683	K
Crystal specific heat	c_p	1000	$J\ kg^{-1}\ K^{-1}$
Crystal density	ρ	2420	$kg\, m^{-1}$
Crystal thermal conductivity	k	22	$J\,(s\ m\ K)^{-1}$
Crystal emissivity	$\epsilon_{w_{cr}}$	0.7	
Melt emissivity	ϵ_{w_m}	0.7	

$\epsilon_{w_{cr}}\epsilon_{w_{amb}}F_{cr\to amb}(R,z)[T_{amb}^4 - T_c^4(R,z,t)]$), and the boundary condition of Eq. 7.14 accounts for radiative heat exchange between the *differential surface at the top of the crystal* and melt (term $\epsilon_{w_{cr}}\epsilon_{w_m}F_{cr\to m}(r,l(t))[T_m^4 - T_c^4(r,l(t),t)]$), chamber (term $\epsilon_{w_{cr}}\epsilon_{w_{ch}}F_{cr\to ch}(r,l(t))[T_{ch}^4 - T_c^4(r,l(t),t)]$) and ambient (term $\epsilon_{w_{cr}}\epsilon_{w_{amb}}F_{cr\to amb}(r,l(t))[T_{amb}^4 - T_c^4(r,l(t),t)]$). The computation of the various view factors through decomposition of the corresponding complex geometries into simple geometries for which view factors can be computed analytically is discussed in the Appendix.

Finally, the values of the physical properties of Si are given in Table 7.2, and the values of the parameters of the process are given in Table 7.3. We note that: a) the Si crystal properties are assumed to be independent of the temperature and concentrations of dopant, carbon, and oxygen, and b) even though our study does not focus on a specific experimental or industrial Czochralski crystallizer, the values of the process parameters in Table 7.3 are within the range of values normally employed in industrial crystallizers.

Remark 7.1 Regarding the development of the above model, we must note that the focus of our control effort is on the reduction of the magnitude of the thermal gradients inside the body of the crystal, and thus we focus on the development of a model that describes the spatiotemporal evolution of the crystal temperature, once the crystal radius has obtained its final value, and accounts for the interactions (radiative heat transfer) between the crystal and its surroundings. There is an extensive literature on the

TABLE 7.3. Process parameters.

Chamber height	h_{ch}	0.18	m
Chamber radius	R_{ch}	0.15	m
Chamber emissivity	$\epsilon_{w_{ch}}$	0.3	
Chamber wall temperature	T_{ch}	1500	K
Pulling velocity	u_p	$1.66\ 10^{-5}$	$m\ sec^{-1}$
Melt temperature	T_m	1705	K
Initial length of crystal	l_0	0.05	m
Final length of crystal	l_f	0.70	m
Radius of crystal	R	0.05	m
Ambient temperature	T_{amb}	600	K
Steady-state heater temperature	T_{sp_i}	1000	K

development of integrated (global) models for the Czochralski process that include the crystal/melt interface shape and melt temperature dynamics (see, e.g., [58, 59, 56, 151] for details). We finally note that the technique for order reduction and controller design that is presented in the next section can be also applied to these integrated crystal growth models to design controllers for crystal temperature regulation, but such an application is outside the scope of this work.

7.2.3 Control-relevant analysis of the Czochralski crystal growth

The objective of this section is to study the thermal gradients inside the crystal in the Czochralski crystal growth process to obtain insights that will be used to formulate a meaningful control problem and derive an appropriate model for controller synthesis. The mathematical model of the process of Figure 7.8 consisting of the two-dimensional parabolic PDE of Eq. 7.10 and the boundary conditions of Eqs. 7.11–7.14 was solved by using Galerkin's method. Specifically, 30 global eigenfunctions (that is, trigonometric functions that cover the entire domain) in the axial direction and 30 global (Bessel) eigenfunctions in the radial direction were used as basis functions in Galerkin's method to discretize the process model in space and reduce it into a large set (900 equations) of ODEs. The time integration of the large set of ODEs was performed by utilizing explicit Euler. It was verified that further increase in the number of eigenfunctions in both r and z directions as well as reduction in the step of time-integration results in negligible improvements in the accuracy of the computed solution. In all the simulation runs, the crystal is initially assumed to be at $l(0) = 0.05\ m$ and $T_c(r, z, 0) = 0.99\ T_{mp}$.

Figures 7.9 and 7.10a show the contour plots of the crystal temperature as a function of the axial and radial directions at four different time instants during the operation of the process and the temporal evolution of the crystal temperature in the axial direction at the center of the crystal, respectively, when the temperature of the ambient (void space in the furnace surrounding the crystal) is set at $T_{amb} = 600\ K$. We observe that the temperature drop inside the crystal in the axial direction is much larger (almost 700 K) than the temperature drop inside the crystal in the radial direction (less than 20 K; see also Figure 7.10b). This can also be seen in Figures 7.10c and 7.10d that show the maximum thermal gradient in the radial and axial direction, respectively, as a function of the axial parameter and time. We observe that the maximum thermal gradient in the axial direction is 5 times larger than the maximum thermal gradient in the radial direction for all times. To further investigate this observation, we show in Figures 7.12 and 7.14 the temporal evolution of the crystal temperature when the temperature of the ambient is set at $T_{amb} = 1000\ K$ and $T_{amb} = 1400\ K$, respectively. Again, it is clear that the variation of the crystal temperature in the radial direction is negligible compare to the variation

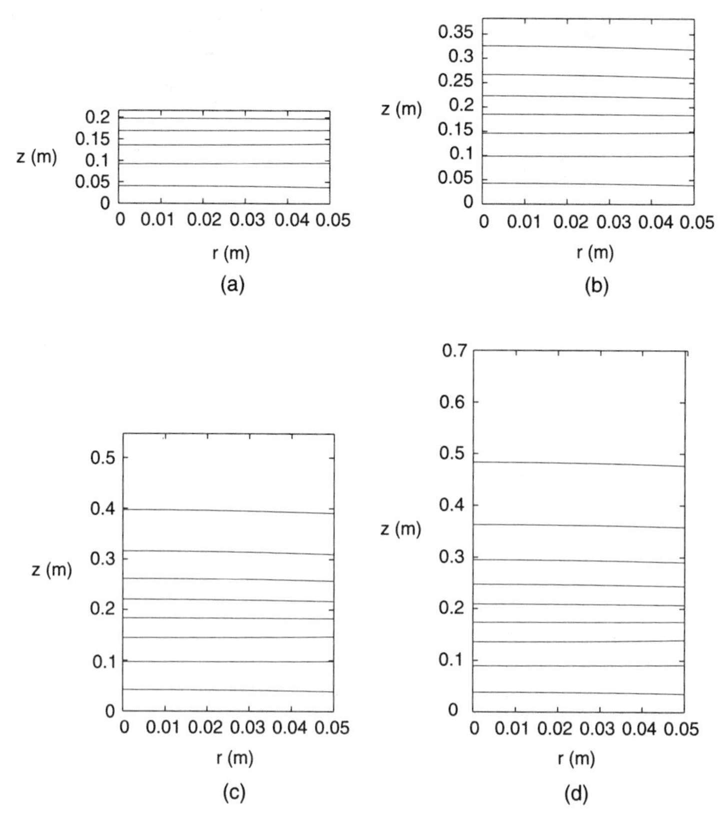

FIGURE 7.9. Temperature of crystal as a function of radial and axial coordinates for $T_{amb} = 600\,K$. (a) $t = 1 \times 10^4\ s$, (b) $t = 2 \times 10^4\ s$, (c) $t = 3 \times 10^4\ s$, and (d) $t = 4 \times 10^4\ s$. Each contour represents a 100 K temperature difference.

in the axial one. This can also be seen in Figures 7.11 and 7.13 presenting contour plots of the crystal temperature, when the ambient temperature is $T_{amb} = 1000\ K$ and $T_{amb} = 1400\ K$, respectively. This conclusion is very important because it allows neglecting the radial dependence of the crystal temperature in the process model that will be used as the basis for the synthesis of a nonlinear feedback controller, thereby yielding the following one-dimensional parabolic PDE with moving boundary:

$$\begin{aligned}
\frac{\partial T_c}{\partial t} + v_p \frac{\partial T_c}{\partial z} &= \frac{k}{\rho c_p} \frac{\partial^2 T_c}{\partial z^2} + \frac{2\sigma \epsilon_{w_{cr}} \epsilon_{w_m}}{\rho c_p R} F_{cr \to m}(R, z)\left[T_m^4 - T_c^4(z, t)\right] \\
&\quad + \frac{2\sigma \epsilon_{w_{cr}} \epsilon_{w_{ch}}}{\rho c_p R} F_{cr \to ch}(R, z)\left[T_{ch}^4 - T_c^4(z, t)\right] \\
&\quad + \frac{2\sigma \epsilon_{w_{cr}} \epsilon_{w_{amb}}}{\rho c_p R} F_{cr \to amb}(R, z)\left[T_{amb}^4 - T_c^4(z, t)\right] \quad (7.15)
\end{aligned}$$

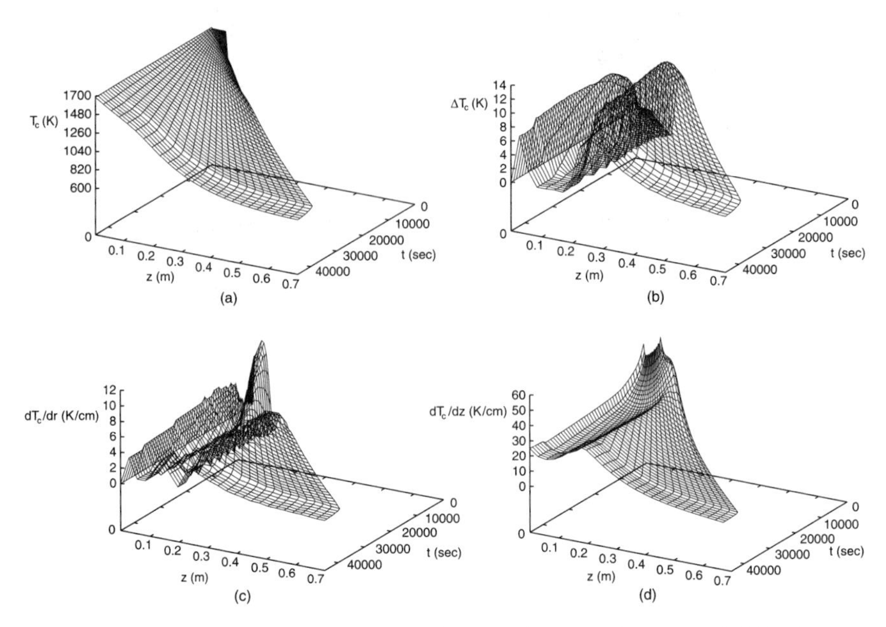

FIGURE 7.10. (a) Temperature of crystal at $r = 0.0\ m$, (b) Maximum temperature difference inside the crystal in the radial direction, (c) Maximum thermal gradient inside the crystal in the radial direction, and (d) Maximum thermal gradient inside the crystal in the axial direction as a function of time and axial coordinate for $T_{amb} = 600\ K$.

subject to the boundary conditions:

$$T_c(0,t) = T_{mp} \tag{7.16}$$

$$\begin{aligned} \frac{k}{\sigma}\frac{\partial T_c}{\partial z}\bigg|_{l(t)} &= \epsilon_{w_{cr}}\epsilon_{w_m}F_{cr\to m}(l(t))\big[T_m^4 - T_c^4(l(t),t)\big] + \epsilon_{w_{cr}}\epsilon_{w_{ch}}F_{cr\to ch}(l(t)) \\ &\quad \times \big[T_{ch}^4 - T_c^4(l(t),t)\big] + \epsilon_{w_{cr}}\epsilon_{w_{amb}}F_{cr\to amb}(l(t)) \\ &\quad \times \big[T_{amb}^4 - T_c^4(l(t),t)\big]. \end{aligned} \tag{7.17}$$

An additional conclusion that follows from the study of Figures 7.10, 7.12, and 7.14 is that the ambient temperature T_{amb} has a very significant effect on the axial profile of the crystal temperature. This suggests that T_{amb} is a meaningful choice for manipulated variable in order to enforce a desired temperature drop inside the crystal. Therefore, we formulate the control problem as the one of controlling the temperature gradient of the crystal in the axial direction by manipulating the temperature of extra heaters placed above the chamber at equispaced intervals. The use of extra heaters to control the crystal temperature is also motivated by the realization that

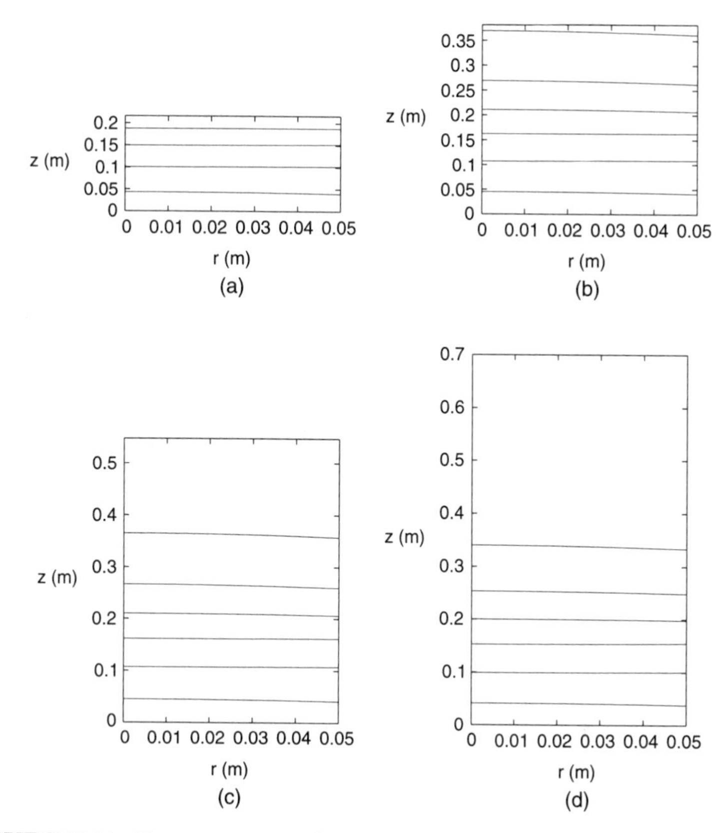

FIGURE 7.11. Temperature of crystal as a function of radial and axial coordinates for $T_{amb} = 1000\ K$. (a) $t = 1 \times 10^4\ s$, (b) $t = 2 \times 10^4\ s$, (c) $t = 3 \times 10^4\ s$, and (d) $t = 4 \times 10^4\ s$. Each contour represents a 100 K temperature difference.

the regulation of thermal gradients in the axial direction requires the use of a manipulated variable that is distributed along the length of the crystal. We will show later on that the use of extra heaters to control axial thermal gradients will also lead to significant reduction of the radial thermal gradients inside the crystal (which cannot be directly controlled) owing to the well-regulated thermal environment in which the crystal grows in such a case.

Our attention now turns to the development of the control configuration that will be used to regulate the thermal gradients inside the crystal; this entails the computation of the number of separate components of the heaters needed to achieve the desired regulation of the thermal gradients. To this end, we initially assumed that the temperature of the heaters is spatially uniform and used a single proportional integral controller to

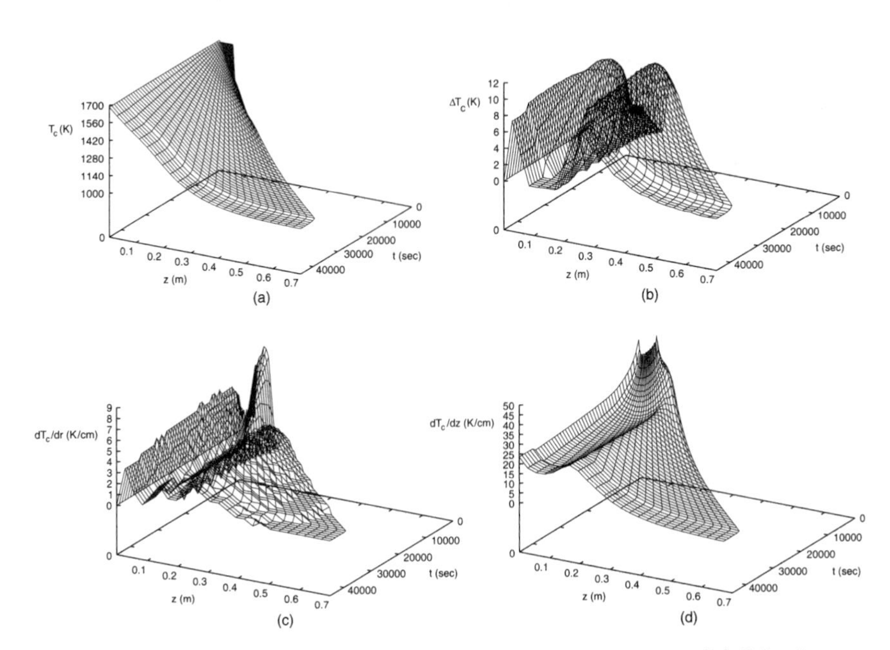

FIGURE 7.12. (a) Temperature of crystal at $r = 0.0\ m$, (b) Maximum temperature difference inside the crystal in the radial direction, (c) Maximum thermal gradient inside the crystal in the radial direction, and (d) Maximum thermal gradient inside the crystal in the axial direction, as a function of time and axial coordinate for $T_{amb} = 1000\ K$.

manipulate this temperature of the form:

$$\begin{aligned} \dot{\xi} &= y_{sp} - y, \quad \xi(0) = 0 \\ u &= K(y_{sp} - y) + \frac{1}{\tau}\xi. \end{aligned} \tag{7.18}$$

The objective of the controller was to keep the surface temperature of the crystal at position $z = 0.48\ m$ at a constant value of 1250 K and enforce a smooth temperature inside the crystal below the critical value of 33 K/cm [70]. The parameters of the controller are given in Table 7.4.

Figure 7.16a shows the temporal evolution of the temperature at the center of the crystal in the axial direction in the closed-loop system. Clearly, the controller drives the temperature of the crystal to its new set point, and the temperature gradients in the radial direction remain small for

TABLE 7.4. PI controller parameters.

$z_s[m]$	$z_f[m]$	$z_m[m]$	K	τ	$y_{sp}[K]$
0.18	0.71	0.48	3	1.0	1250

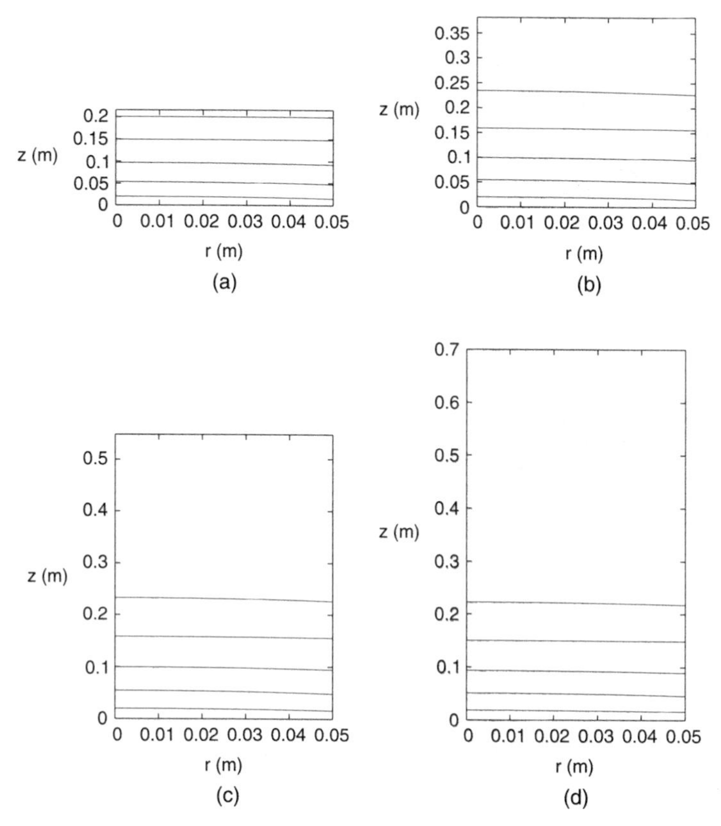

FIGURE 7.13. Temperature of crystal as a function of radial and axial coordinates for $T_{amb} = 1400$ *K*. (a) $t = 1 \times 10^4$ *s*, (b) $t = 2 \times 10^4$ *s*, (c) $t = 3 \times 10^4$ *s*, and (d) $t = 4 \times 10^4$ *s*. Each contour represents a 50 *K* temperature difference.

almost all times (see Figure 7.16c and compare with 7.10c). However, the controller fails to establish a smooth temperature drop inside the crystal in the axial direction, enforcing a large temperature drop close to the crystal/melt interface (even if small in absolute values) and achieving an almost uniform crystal temperature far away from the interface; this can be seen in Figure 7.15, which shows the contour plots of the crystal temperature as a function of the axial and radial directions at four different time instants during the operation of the process. Figure 7.17 shows the corresponding profile for the manipulated input, which changes smoothly with time to achieve the control objective (the initial sharp change of the manipulated input is actually a smooth change that occurs over a time period of 250 *s* and is due to the fact that the time axis covers the entire process cycle). Note that the controller is activated (i.e., $u(t) \neq 1$) when the crystal enters in the zone in which the control actuator operates. Regarding Figure 7.16,

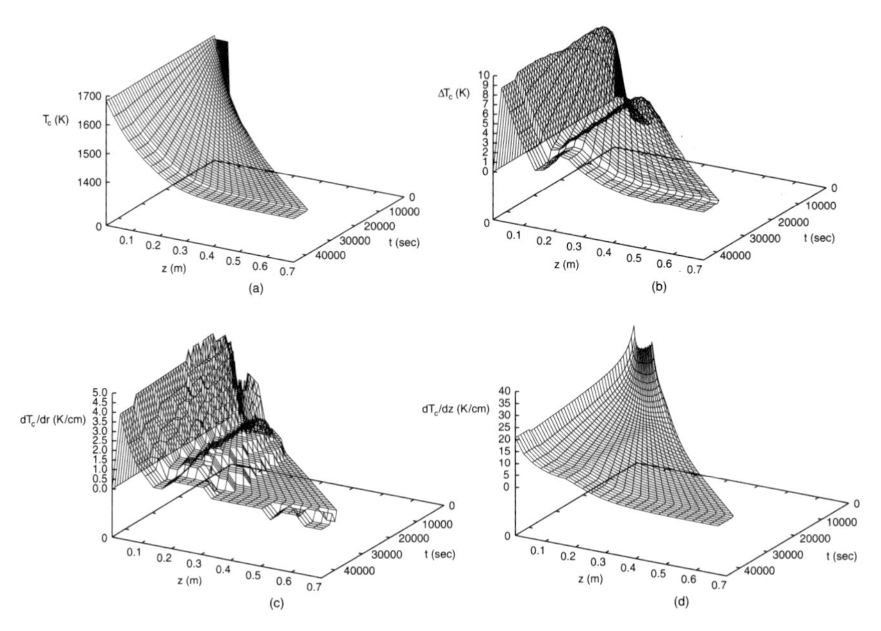

FIGURE 7.14. (a) Temperature of crystal at $r = 0.0\ m$, (b) Maximum temperature difference inside the crystal in the radial direction, (c) Maximum thermal gradient inside the crystal in the radial direction, and (d) Maximum thermal gradient inside the crystal in the axial direction as a function of time and axial coordinate for $T_{amb} = 1400\ K$.

it is worth noting that the radial temperature nonuniformity remains small (less than 10 K) for the whole process time, which further validates the conclusion that we drawn on the basis of the open-loop behavior that radial thermal gradients are much smaller than axial thermal gradients.

On the basis of the simulation results for the case of using a single control actuator and driven by our desire to achieve a smooth temperature drop inside the crystal, we formulate the control problem as the one of controlling the temperature gradient of the crystal in the axial direction by manipulating the temperature of three extra heaters placed at three equispaced intervals as shown in Figure 7.18.

To account for the heaters used for crystal temperature control in the two-dimensional process model, we substitute the boundary conditions of Eqs. 7.13 and 7.14 with the following boundary conditions:

$$\begin{aligned}\frac{k}{\sigma}\frac{\partial T_c}{\partial r}\bigg|_R &= \epsilon_{w_{cr}}\epsilon_{w_m}F_{cr\to m}(R,z)\big[T_m^4 - T_c^4(R,z,t)\big] + \epsilon_{w_{cr}}\epsilon_{w_{ch}}F_{cr\to ch}(R,z)\\ &\quad\times\big[T_{ch}^4 - T_c^4(R,z,t)\big] + \epsilon_{w_{cr}}\sum_{i=1}^{n}\epsilon_{w_i}F_{cr\to i}(R,z)\\ &\quad\times\big[T_i^4(t) - T_c^4(R,z,t)\big],\quad 0\le z\le l(t)\end{aligned} \tag{7.19}$$

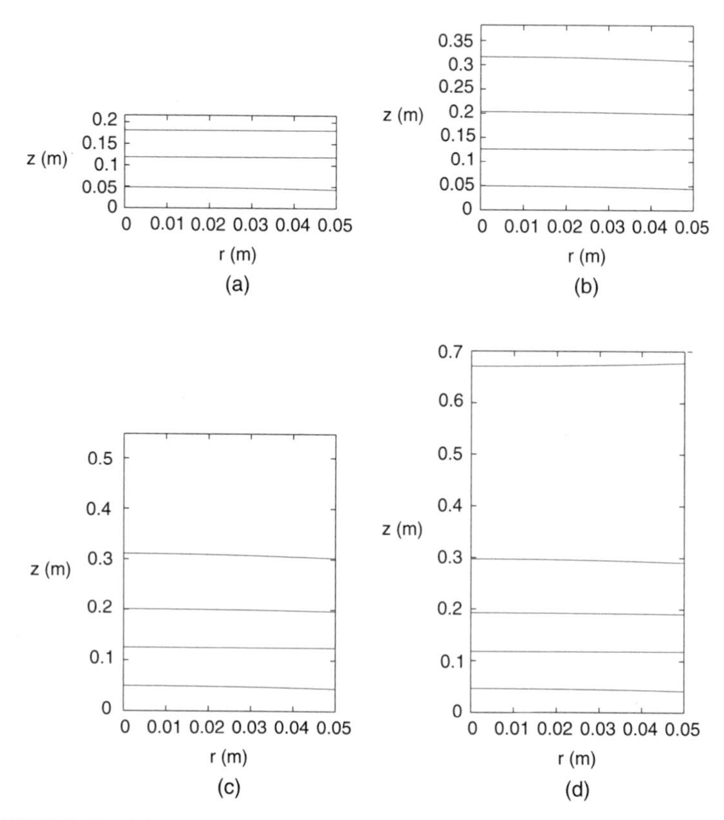

FIGURE 7.15. Temperature of crystal as a function of radial and axial coordinates—one PI controller. (a) $t = 1 \times 10^4$ *s*, (b) $t = 2 \times 10^4$ *s*, (c) $t = 3 \times 10^4$ *s*, and (d) $t = 4 \times 10^4$ *s*. Each contour represents a 100 *K* temperature difference.

$$
\begin{aligned}
\frac{k}{\sigma}\frac{\partial T_c}{\partial z}\bigg|_{l(t)} &= \epsilon_{w_{cr}}\epsilon_{w_m}F_{cr\to m}(r,l(t))\big[T_m^4 - T_c^4(r,l(t),t)\big] \\
&+ \epsilon_{w_{cr}}\epsilon_{w_{ch}}F_{cr\to ch}(r,l(t))\big[T_{ch}^4 - T_c^4(r,l(t),t)\big] \\
&+ \epsilon_{w_{cr}}\sum_{i=1}^{n}\epsilon_{w_i}F_{cr\to i}(r,l(t))\big[T_i^4(t) - T_c^4(r,l(t),t)\big], \quad 0 \le r \le R
\end{aligned}
\tag{7.20}
$$

where n is the number of heaters used for control, $T_i(t)$ is the temperature of the i-th heater, and $F_{cr\to i}$ is the view factor of a differential crystal surface element to the heater surface. Moreover, with the addition of the heaters, the one-dimensional model that will be used for

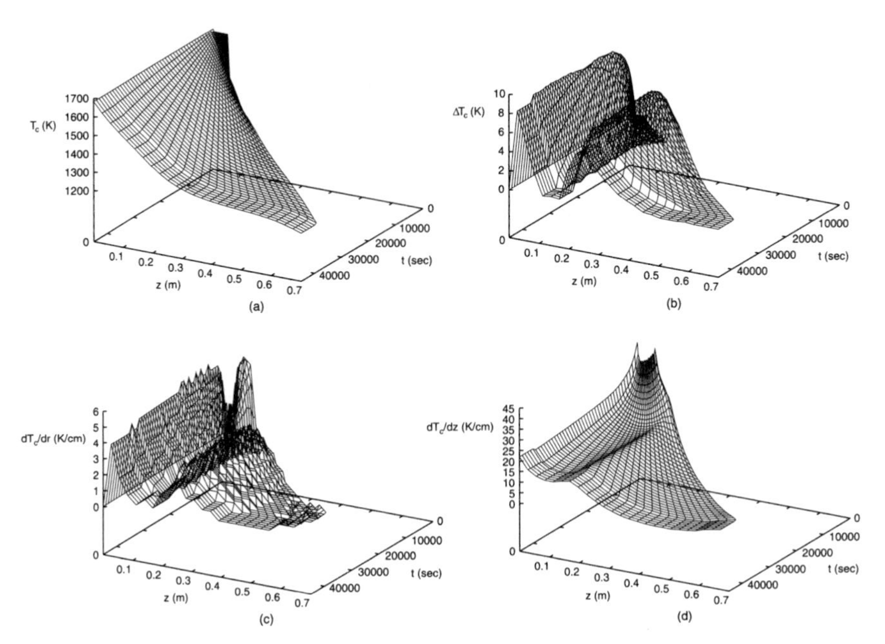

FIGURE 7.16. (a) Temperature of crystal at $r = 0.0\,m$, (b) Maximum temperature difference inside the crystal in the radial direction, (c) Maximum thermal gradient inside the crystal in the radial direction, and (d) Maximum thermal gradient inside the crystal in the axial direction as a function of time and axial coordinate—one PI controller.

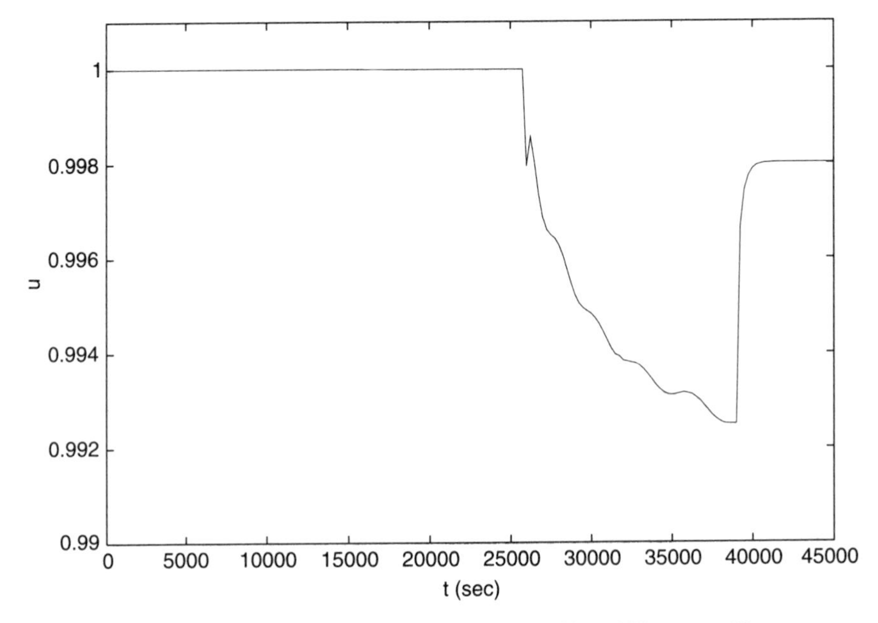

FIGURE 7.17. Manipulated input profile—PI controller.

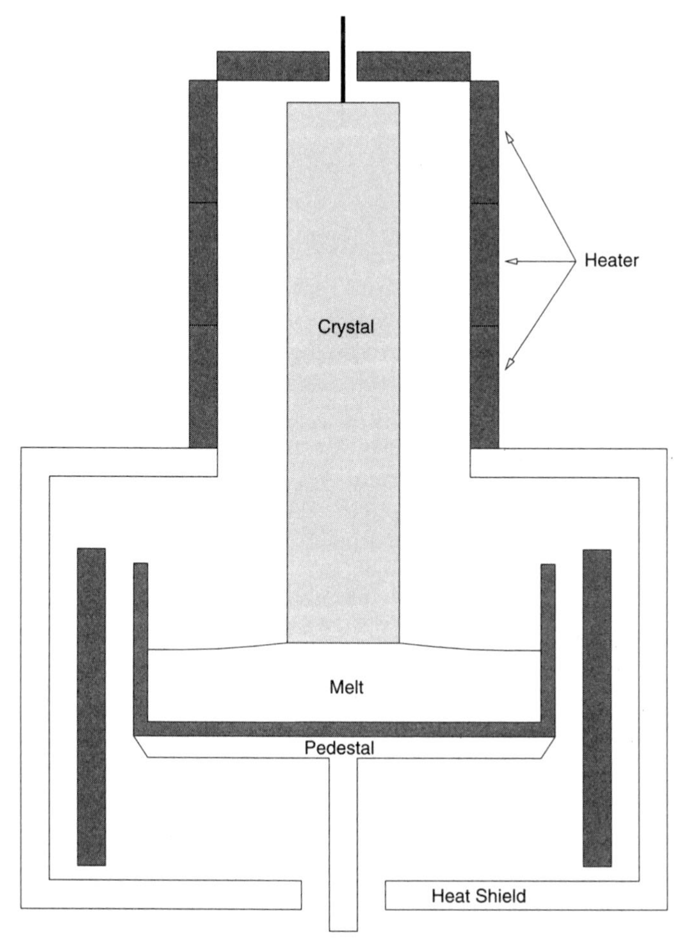

FIGURE 7.18. Control configuration for crystal temperature control in the Czochralski crystal growth.

the controller design takes the form:

$$\begin{aligned}\frac{\partial T_c}{\partial t} + v_p \frac{\partial T_c}{\partial z} &= \frac{k}{\rho c_p} \frac{\partial^2 T_c}{\partial z^2} + \frac{2\sigma \epsilon_{w_{cr}} \epsilon_{w_m}}{\rho c_p R} F_{cr \to m}(R, z)\left[T_m^4 - T_c^4(z,t)\right] \\ &+ \frac{2\sigma \epsilon_{w_{cr}} \epsilon_{w_{ch}}}{\rho c_p R} F_{cr \to ch}(R, z)\left[T_{ch}^4 - T_c^4(z,t)\right] \\ &+ \frac{2\sigma \epsilon_{w_{cr}}}{\rho c_p R} \sum_{i=1}^{n} \epsilon_{w_i} F_{cr \to i}(R, z)\left[T_i^4(t) - T_c^4(z,t)\right] \end{aligned} \quad (7.21)$$

subject to the following boundary conditions,

$$
\begin{aligned}
T_c(0,t) &= T_{mp} \\
\frac{k}{\sigma}\frac{\partial T_c}{\partial z}\bigg|_{l(t)} &= \epsilon_{w_{cr}}\epsilon_{w_m}F_{cr\to m}(l(t))\left[T_m^4 - T_c^4(l(t),t)\right] + \epsilon_{w_{cr}}\epsilon_{w_{ch}}F_{cr\to ch}(l(t)) \\
&\quad \times\left[T_{ch}^4 - T_c^4(l(t),t)\right] + \epsilon_{w_{cr}}\sum_{i=1}^{n}\epsilon_{w_i}F_{cr\to i}(l(t)) \\
&\quad \times\left[T_i^4(t) - T_c^4(l(t),t)\right].
\end{aligned}
\tag{7.22}
$$

Referring to Eqs. 7.21–7.22, note the lack of the terms corresponding to heat transfer between the crystal and the ambient (compare with Eqs. 7.19–7.20); these terms have been substituted by the terms that account for heat transfer between the crystal and the extra heaters that are used as control actuators.

The performance of the proposed control configuration that employs three heaters to regulate the thermal gradients inside the crystal compared to the open-loop system and the closed-loop system in the case of using one heater will be evaluated in section 7.2.4.

Remark 7.2 We note that another approach for the simulation of the crystal temperature concerns the elimination of the thermal dynamics of the process on the basis that they are slower than the dynamics of the crystal radius and solves for the temperature profile inside the growing crystal as a series of solutions of a steady state thermal model for different lengths (see, for example, [56, 60]). However, in the present study, we have decided to include the thermal dynamics in the model of the process that is used for the implementation of the controller because the use of feedback control or the presence of disturbances (e.g., variations in the pulling rate, and melt temperature) may significantly modify the dynamic behavior of the process compared to the open-loop behavior.

7.2.4 Nonlinear controller synthesis—Closed-loop simulations

In this subsection, our objective is to synthesize and implement a low-order nonlinear output feedback controller on the crystal growth process that enforces a desired smooth temperature profile in the axial direction inside the crystal. We begin with the reduction of the one-dimensional parabolic PDE model of Eq. 7.21–7.22 into a small set of nonlinear ODEs utilizing the model reduction algorithm of subsection 4.2. This set of ODEs will be subsequently used for controller design.

The one-dimensional model of Eq. 7.21–7.22 can be formulated in the general form of Eq. 5.1–5.3 with:

$$
A = 0.0, \qquad B = \frac{k}{\rho c_p}, \qquad w = \frac{\sigma\epsilon_{w_{cr}}}{\rho c_p},
$$

$$
g(u) = \left[(T_1^4/T_{sp_1}^4) \quad \cdots \quad (T_l^4/T_{sp_l}^4)\right]
$$

$$
b(z,t) = \left[\epsilon_{w_1} \left(\frac{2}{R} F_{cr\to 1}(R,z) + \frac{\delta(z-l(t))}{kl(t)} F_{cr\to 1}(z) \right) T_{sp1}^4 \quad \cdots \right.
$$
$$
\left. \epsilon_{w_l} \left(\frac{2}{R} F_{cr\to l}(R,z) + \frac{\delta(z-l(t))}{kl(t)} F_{cr\to l}(z) \right) T_{spl}^4 \right],
$$
$$
f(z,t,T_c) = -u_p \frac{\partial T_c}{\partial z} + \frac{2\sigma\epsilon_{w_{cr}}}{\rho c_p R} \left(\epsilon_{w_m} F_{cr\to m}(R,z)[T_m^4 - T_c^4] \right.
$$
$$
\left. + \epsilon_{w_{ch}} F_{cr\to ch}(R,z)[T_{ch}^4 - T_c^4] - \sum_{i=1}^{l} \epsilon_{w_i} F_{cr\to i}(R,z) T_c^4 \right)
$$
$$
+ \delta(z-l(t)) \frac{\sigma\epsilon_{w_{cr}}}{k\rho c_p l(t)} \left(\epsilon_{w_m} F_{cr\to m}(z)[T_m^4 - T_c^4] \right.
$$
$$
\left. + \epsilon_{w_{ch}} F_{cr\to ch}(z)[T_{ch}^4 - T_c^4] - \sum_{i=1}^{l} \epsilon_{w_i} F_{cr\to i}(z) T_c^4 \right) \tag{7.23}
$$

where $\delta(\cdot)$ is the standard Dirac function. Defining the new variable $x = (T_c - T_{mp})/T_{mp}$ and nondimensionalizing the temperature for this parabolic PDE system, the spatial differential operator $\mathcal{A}$ can be defined as:

$$
\mathcal{A}(t)x = \frac{k}{\rho c_p} \frac{\partial^2 \bar{x}}{\partial z^2},
$$
$$
x \in D(\mathcal{A}(t)) = \left\{ x \in \mathcal{H}([0, l(t)]; \mathbb{R}); \quad \bar{x}(0,t) = 0; \quad \frac{\partial \bar{x}}{\partial z}(l(t),t) = 0 \right\}. \tag{7.24}
$$

Note that $\mathcal{A}(t)$ includes the higher-order spatial derivative (conduction term) included in the PDE of Eq. 7.21, while, owing to the time-dependent nature of the pulling rate v_p (in general), it does not include the first-order spatial derivative (convective term). The eigenvalue problem of $\mathcal{A}(t)$ can be solved analytically and yields:

$$
\lambda_j(t) = -\frac{k}{\rho c_p} \left[\frac{\pi(2j+1)}{2l(t)} \right]^2,
$$
$$
\phi_j(t,z) = \sqrt{\frac{2}{l(t)}}\, sin\left(\frac{\pi(2j+1)}{2l(t)} z \right), \quad j = 0, \ldots, \infty \tag{7.25}
$$

where the eigenfunctions $\phi_j(t,z)$ form a countable, $\forall t \in [0,\infty)$, and orthogonal basis of $\mathcal{H}(t)$.

We used the nonlinear model reduction method discussed in subsection 4.2, that is, combination of Galerkin's method with approximate inertial manifolds to derive a fourth-order ODE model that uses a 4-th order approximation for x_f (i.e., $m = 4$ and $\tilde{m} = 4$). The eigenfunctions $\phi_j(t,z)$ were used as basis functions in Galerkin's method. Figure 7.19 shows the temporal evolution of the crystal temperature in the axial direction at $r = 0.034$ computed by the two-dimensional model (Figure 7.19a), the

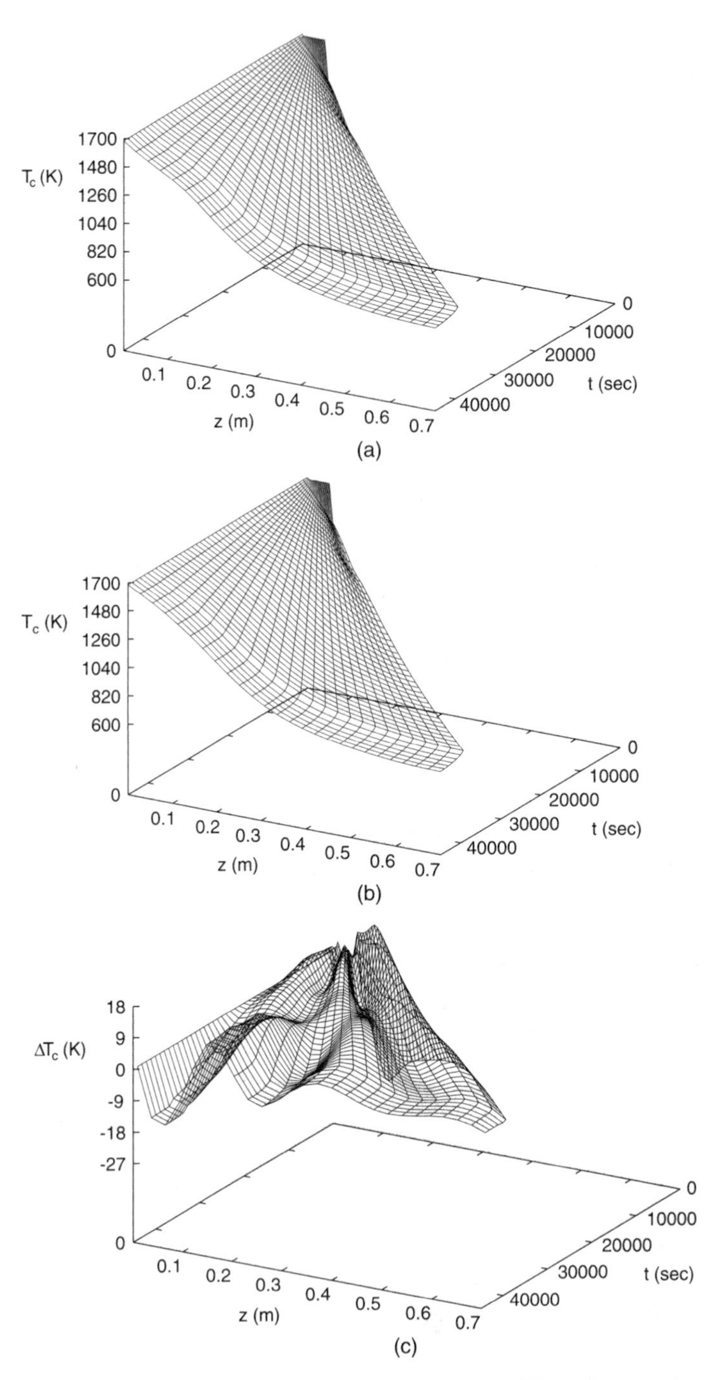

FIGURE 7.19. Comparison of temperature profiles of open loop system at $T_{amb} = 600\,K$. (a) Two-dimensional model at $r = 0.034\,m$, (b) reduced-order model and (c) temperature difference between the two-dimensional model at $r = 0.034\,m$ and the reduced-order model.

temporal evolution of the crystal temperature in the axial direction computed by the fourth-order model (Figure 7.19b), and the difference between the crystal temperatures predicted by the two-dimensional model ($r = 0.034$) and the fourth-order model (Figure 7.19c), for $T_{amp} = 600$. It is clear that the predictive capabilities of the reduced-order model are very good for a wide range of operating conditions, and thus, it makes sense to use this fourth-order model as a basis for the synthesis of a nonlinear output feedback controller.

The control objective is to enforce an almost linear axial temperature profile in the crystal by manipulating the temperature of the heaters and using point measurements of the surface temperature of the crystal. Specifically, we consider the control configuration shown in Figure 7.18. The heater was assumed to be divided into three equispaced regions, and the temperature in each one of these regions is adjusted by the controller. The region where the ith controller operates is determined by the variables z_{si}, z_{f_i}, which represent the minimum and maximum distances of the controller region from the melt/crystal interface, respectively (note that the distribution function of the κth control actuator is $b^i(z) = H(z - z_{si}) - H(z - z_{f_i})$, where $H(\cdot)$ is the standard Heavyside function). The controlled and measured outputs were taken to be identical and defined as:

$$
\begin{aligned}
y^1(t) = q^1(t) &= \int_0^{l(t)} \delta(z - z_{m1})\, T_c(R, z, t)\, dz \\
y^2(t) = q^2(t) &= \int_0^{l(t)} \delta(z - z_{m2})\, T_c(R, z, t)\, dz \qquad (7.26)\\
y^3(t) = q^3(t) &= \int_0^{l(t)} \delta(z - z_{m3})\, T_c(R, z, t)\, dz
\end{aligned}
$$

where $T_c(R, z, t)$ is the surface temperature of the crystal. The point measurements of the temperature on the surface of the crystal can be readily obtained in practice with optimal pyrometers. Finally, the fourth-order model obtained through Galerkin's method with approximate inertial manifolds with the specifications for manipulated inputs and controlled and measured outputs was used for the synthesis of a nonlinear output feedback controller by utilizing the formula of theorem 6.1. In order to incorporate integral action in this controller, the term $v_\kappa - h_0^i(\eta)$ was substituted by the term $v_i - y^i$. All the parameters used in the control problem are given in Table 7.5.

TABLE 7.5. Nonlinear controller parameters.

Controller i	$z_{si}[m]$	$z_{fi}[m]$	$z_{mi}[m]$	β_i	$v_i[K]$
1	0.18	0.38	0.28	1.0	1400
2	0.38	0.58	0.48	1.0	1250
3	0.58	0.71	0.68	1.0	1100

Several simulations runs were performed to evaluate: (a) the ability of the nonlinear controller to enforce a linear temperature profile in the axial direction inside the crystal, (b) the robustness properties of the nonlinear controller with respect to parametric model uncertainty, disturbances, and unmodeled actuator and sensor dynamics. In all the simulation runs, the crystal was initially ($t = 0$ sec) assumed to be at $T_c(z, r, 0) = 1670$ K and have a length of $l(0) = 0.05$ m. The initial conditions of the fourth-order observer included in the nonlinear output controller were computed by using the initial condition assumed for the crystal. Moreover, in all the simulation runs, the objective of the controller is to regulate the three controlled outputs at $y^1 = 1400$ K, $y^2 = 1250$ K, $y^3 = 1100$ K; this motivated our objective to enforce a linear temperature drop inside the crystal of 8 K/cm. As we will see in our results below, these set points allow enforcing a smooth linear temperature drop inside the crystal at all times during the growth. Furthermore, we note that in order to avoid disturbing the concentration profiles of the dopant and oxygen through unnecessarily large variations of the manipulated inputs, a lower bound (constraint) is implemented on the control action that does not allow $u^i(t)$ to become smaller than 0.82 (note that $u^i(t) = T_i/T_{sp_i}$ where $T_{sp_i} = 1000$ K for all $i = 1, 2, 3$ is a reference temperature for the ith heater). Finally, the tuning parameters of the nonlinear controller are chosen so that the computed variations in the control action with respect to time are small.

Initially, the set-point tracking capability of the nonlinear controller was evaluated under nominal conditions. Figures 7.20 and 7.21a show the contour plots of the crystal temperature as a function of the axial and radial directions at four different time instants during the operation of the process and the temporal evolution of the crystal temperature in the axial direction at the center of the crystal, respectively, in the closed-loop system. Clearly, the nonlinear controller drives the controlled outputs to their new set points, while enforcing the desired temperature drop inside the crystal at all radial positions (note that the radial temperature nonuniformity is less than 20 K, which is clearly smaller to the one computed for the open-loop system; compare Figure 7.21b and Figure 7.10b), without creating large radial temperature gradients at any time (the radial temperature gradients remain small for all times; see Figure 7.21c and compare with 7.10c). As can be seen from Figure 7.21d, the temperature gradient in the axial direction remains bounded throughout the length of the crystal, and is much smaller than in the open loop case (Figure 7.10d). Figure 7.22 shows the corresponding profiles for the three manipulated inputs, which change smoothly with time to achieve the control objective. Note also that each one of the three controllers is activated (i.e., $u^i(t) \neq 1$) when the crystal enters in the zone in which the corresponding control actuator operates and the initial sharp profiles are actually the slow activation of the controllers over a period of 250 s. Figure 7.23a shows the crystal temperature at $r = 0.05$ m, and Figure 7.23b shows the maximum thermal gradient

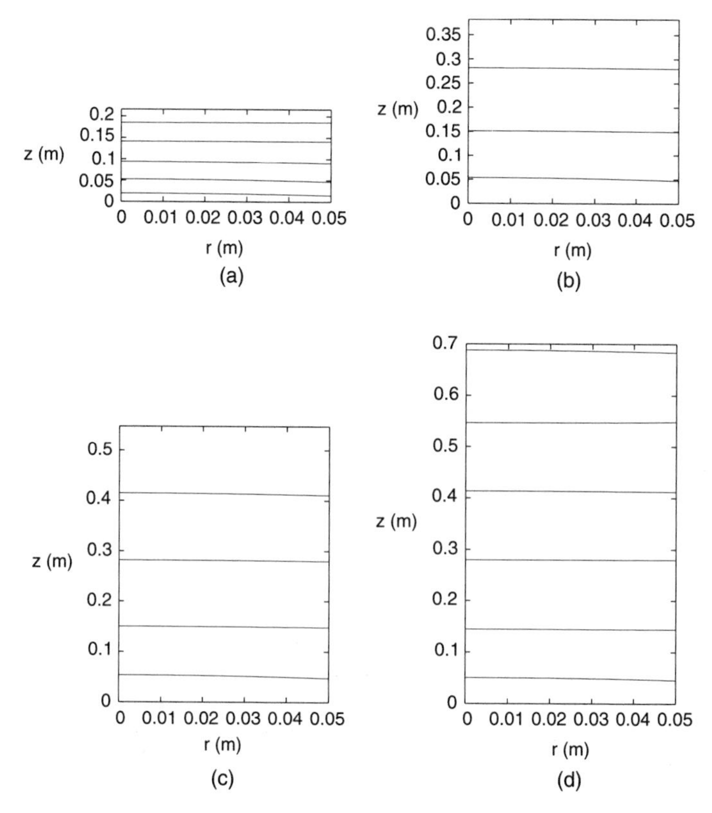

FIGURE 7.20. Temperature of crystal as a function of radial and axial coordinates under nonlinear control—nominal case. (a) $t = 1 \times 10^4$ s, (b) $t = 2 \times 10^4$ s, (c) $t = 3 \times 10^4$ s, and (d) $t = 4 \times 10^4$ s. Each contour represents a 100 K temperature difference.

inside the crystal in the axial direction, at the end of the process cycle under the proposed nonlinear control system, the PI controller, and the case of constant ambient temperature; it is clear that the proposed control scheme reduces the thermal gradients in the axial direction compared to the other approaches.

Next, the robustness properties of the nonlinear controller in the presence of disturbances, parametric uncertainties, and unmodeled actuator and sensor dynamics were investigated. Figure 7.25 shows the closed-loop crystal temperature contours at four different time instants, and Figure 7.26a shows the closed-loop crystal temperature in the axial direction at the center of the crystal in the presence of a 10% disturbance in the melt temperature and 20% disturbance in the temperature of the chamber. Figure 7.27 shows the manipulated input profiles under the nonlinear controller. The nonlinear controller enforces the desired linear drop in the crystal temperature in the axial direction, attenuating the effect of the

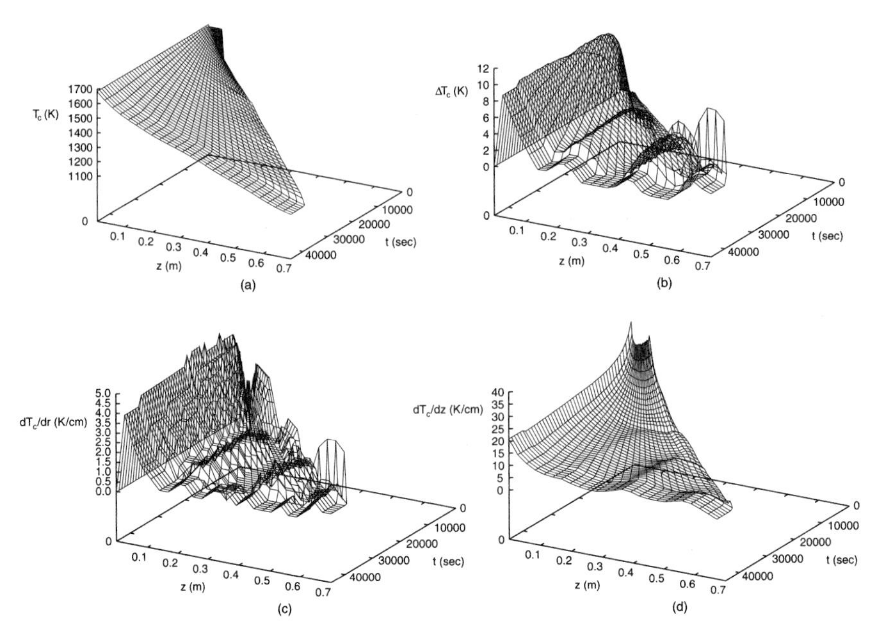

FIGURE 7.21. (a) Temperature of crystal at $r = 0.0\ m$, (b) Maximum temperature difference inside the crystal in the radial direction, (c) Maximum thermal gradient inside the crystal in the radial direction, (d) Maximum thermal gradient inside the crystal in the axial direction as a function of time and axial coordinate under nonlinear control—nominal case.

disturbances (compare the closed-loop temperature profiles of Figure 7.26 with the open-loop temperature profiles of Figure 7.24 ($T_{amb} = 1000\ K$), under the same disturbances), while keeping the temperature gradients in the radial (Figure 7.26c) and axial (Figure 7.26d) directions small. We also successfully studied the robustness of the nonlinear controller with respect to 10% error in the emissivity of the crystal $e_{w_{cr}}$. The nonlinear controller was found to drive the outputs to their set points with small temperature gradients in the radial and axial directions (the detailed results are given in [4] and are omitted here for brevity). Furthermore, we studied the robustness properties of the nonlinear controller in the presence of unmodeled actuator and sensor dynamics. To account for the actuator dynamics, the process model of Eq. 7.10 was augmented with the dynamical system $\epsilon_c \dot{z}_{1i} = -z_{1i} + z_{2i}$, $\epsilon_c \dot{z}_{2i} = -z_{2i} + u^i$, $i = 1, 2, 3$, where $z_{1i}, z_{2i} \in \mathbb{R}$ are the actuator states, z_{1i} is the actuator output and ϵ_c is a small parameter characterizing how fast are the actuator dynamics. To account for the sensor dynamics, the process model of Eq. 7.10 was augmented with the dynamical system $\epsilon_m \dot{z}_{3\kappa} = -z_{3\kappa} + z_{4\kappa}$, $\epsilon_m \dot{z}_{4\kappa} = -z_{4\kappa} + q^\kappa$, $\kappa = 1, 2, 3$, where $z_{3\kappa}, z_{4\kappa} \in \mathbb{R}$ are the sensor states, $z_{3\kappa}$ is the sensor output, and ϵ_m is a small parameter characterizing how fast are the sensor dynamics. We

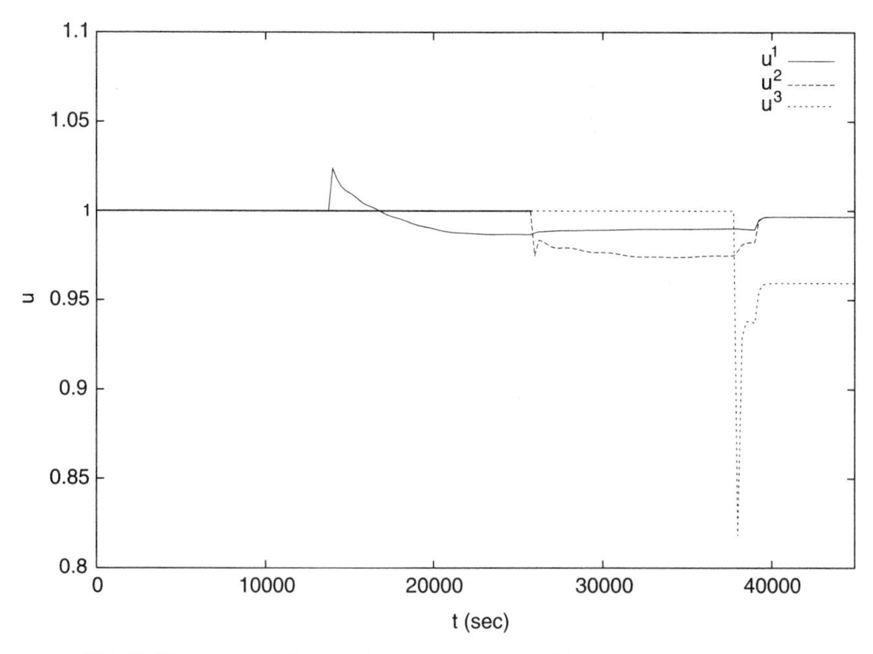

FIGURE 7.22. Manipulated input profiles—nominal case.

found that the maximum values of ϵ_c and ϵ_m for which the stability of the closed-loop system is guaranteed are $\epsilon_c = 0.02$ and $\epsilon_m = 0.02$, thereby implying that the nonlinear controller is robust with respect to stable and sufficiently fast unmodeled sensor and actuators dynamics.

Finally, it is important to point out another advantage of the proposed control configuration. Even though there is no explicit way to reduce the thermal gradients in the radial direction inside the crystal through direct feedback control, reduction of such gradients is accomplished indirectly during the regulation of the thermal gradients in the axial direction owing to the well-regulated environment in which the crystal grows in such a case. This point is made clear by comparing Figures 7.10c, 7.12c, 7.14c, and 7.24c with Figures 7.21c and 7.26c that present the evolution of the radial thermal gradients inside the crystal in the open- and closed-loop systems, respectively, and show that a significant reduction of the thermal gradients in the radial direction is achieved in the case of using the proposed control scheme compared to the open-loop system.

Remark 7.3 The decision to pick the three set-point values so that an almost linear axial temperature drop is enforced inside the crystal was motivated by our objective to reduce the axial thermal gradient $\partial T_c/\partial z$ for all $z \in [0, l(t)]$ and all $t \in [0, \infty)$, thereby reducing the possibilities for crystal dislocation and defects due to temperature nonuniformity (the reader may refer to [70, 132] for more information). We note that the nonlinear

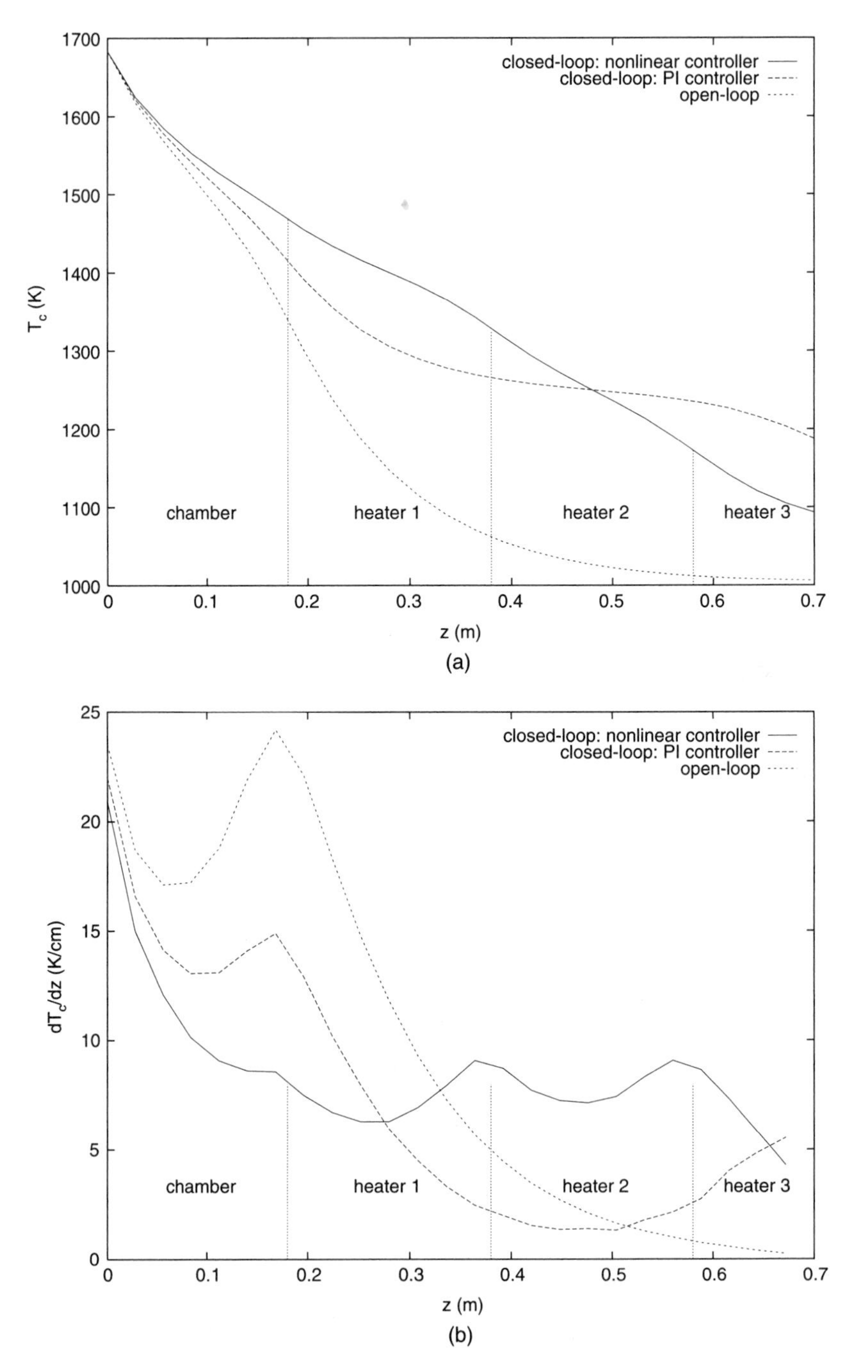

FIGURE 7.23. (a) Temperature of crystal at $r = 0.05\ m$, and (b) Maximum thermal gradient inside the crystal in the axial direction at the end of the process—nominal case.

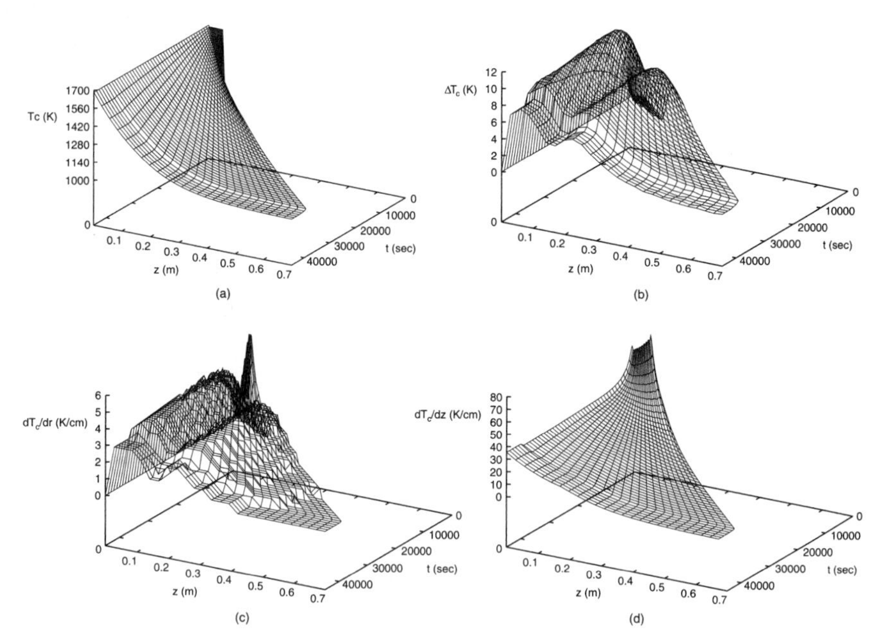

FIGURE 7.24. (a) Temperature of crystal at $r = 0.0\,m$, (b) Maximum temperature difference inside the crystal in the radial direction, (c) Maximum thermal gradient inside the crystal in the radial direction, and (d) Maximum thermal gradient inside the crystal in the axial direction as a function of time and axial coordinate for $T_{amb} = 1000\,K$—exogenous disturbances.

controller can be used to enforce any other desired smooth temperature drop inside the crystal by appropriate choice of the values of the three set points.

Remark 7.4 It is important to point out that even though the fourth-order model obtained through Galerkin's method and approximate inertial manifolds was used to synthesize a nonlinear geometric controller that can be readily implemented in practice; this fourth-order model can be also used for the design of optimization-based controllers including model predictive controllers because it is of very low order and, therefore, it leads to optimization programs that can be rapidly solved on-line.

Remark 7.5 The variation of the crystal size with respect to time is a moving boundary characteristic that should be included in the context of controlling the thermal gradients. Of course, there are other moving boundary issues in the Czochralski process that should be included in the development of an integrated simulation of the process; however, these are not important in the context of controlling thermal gradients inside the crystal, when the crystal radius has reached a constant value.

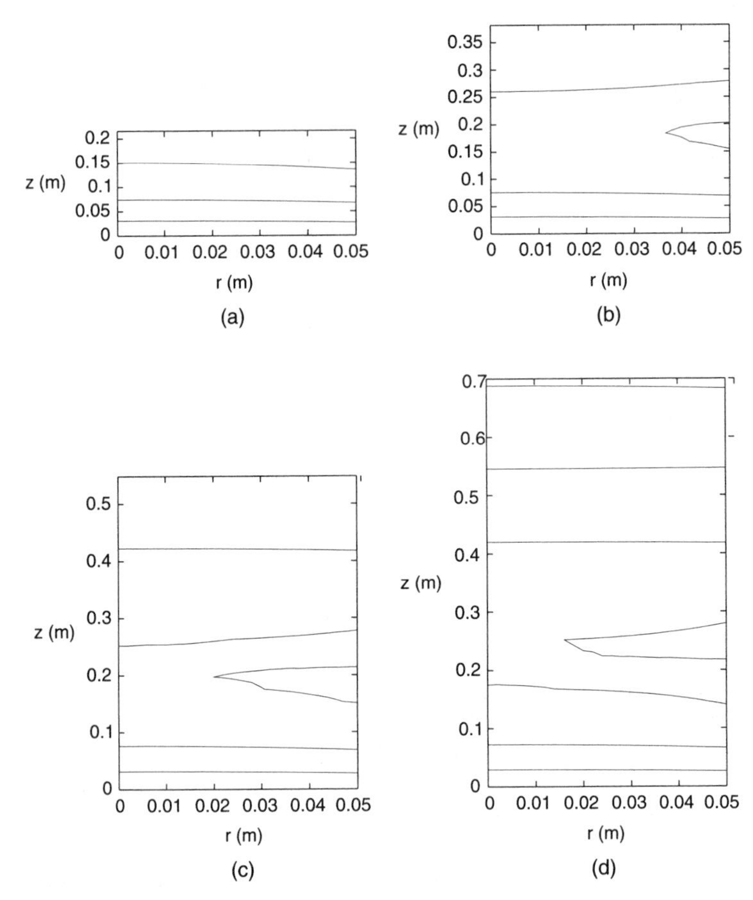

FIGURE 7.25. Temperature of crystal as a function of radial and axial coordinates under nonlinear contro—exogenous disturbances. (a) $t = 1 \times 10^4$ s, (b) $t = 2 \times 10^4$ s, (c) $t = 3 \times 10^4$ s, and (d) $t = 4 \times 10^4$ s. Each contour represents a 100 K temperature difference.

Remark 7.6 We note that even though our study on control of the Czochralski process focused on the regulation of the temperature profile in the axial direction inside the crystal under the assumption that the crystal radius is constant, the proposed nonlinear controller can be *directly* coupled with another nonlinear controller that manipulates the pulling velocity and chamber temperature to regulate the crystal radius. This is possible because the time scale of operation of the pulling velocity-crystal radius control loop is much faster than the time scale of operation of the heater temperature-crystal temperature control loop, while the manipulation of the chamber temperature, T_{ch}, keeps the pulling velocity within specified limits, which allows decoupling the synthesis of the two controllers (see also [84] for similar conclusions). The synthesis of a nonlinear controller

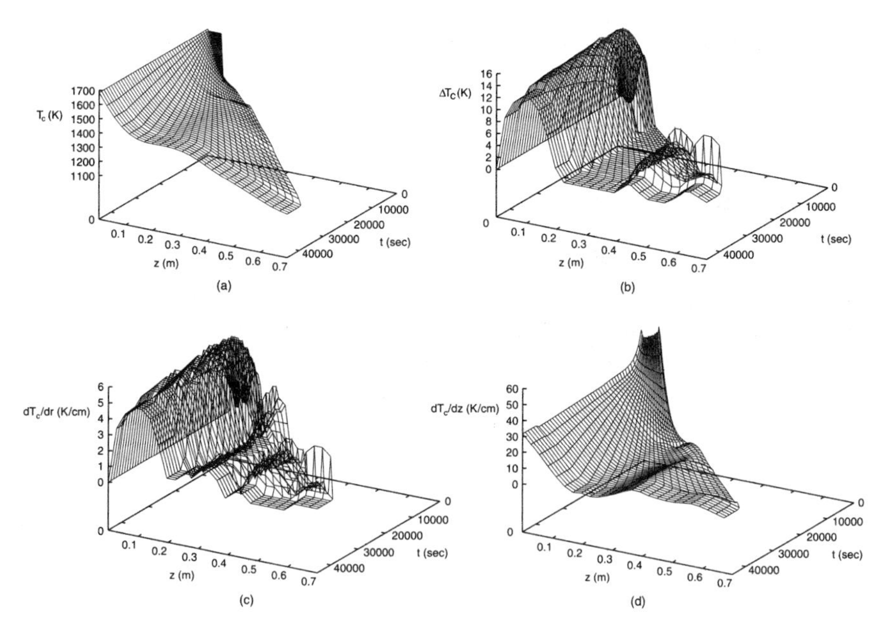

FIGURE 7.26. (a) Temperature of crystal at $r = 0.0\,m$, (b) Maximum temperature difference inside the crystal in the radial direction, (c) Maximum thermal gradient inside the crystal in the radial direction, and (d) Maximum thermal gradient inside the crystal in the axial direction as a function of time and axial coordinate under nonlinear control—exogenous disturbances.

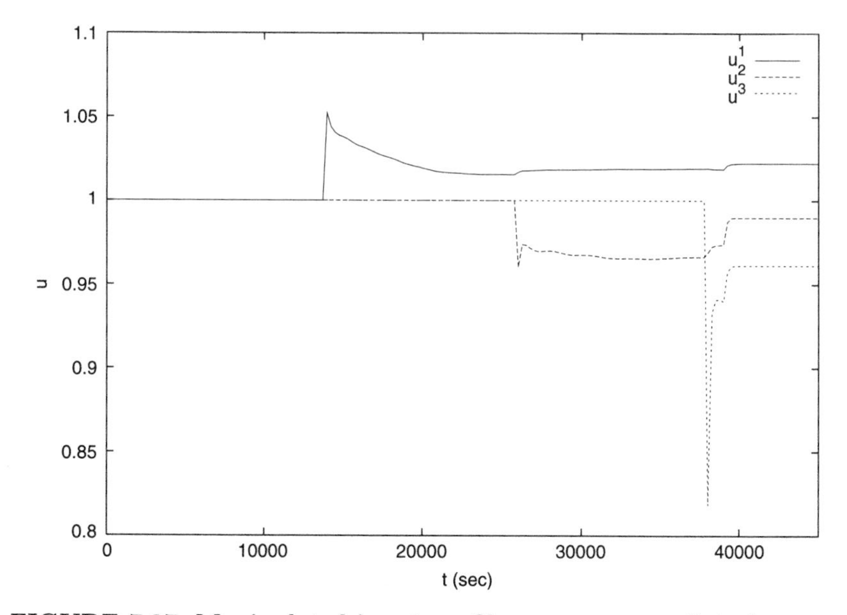

FIGURE 7.27. Manipulated input profiles—exogenous disturbances.

on the basis of an ODE model that describes the rate of change of the crystal radius as a function of the pulling velocity is presented in the Appendix.

7.3 Conclusions

In this chapter, we presented two successful applications of the nonlinear control methods for parabolic PDE systems with fixed and moving spatial domains presented in Chapters 4 and 6 to a rapid thermal chemical vapor deposition process and a Czochralski crystal growth process, respectively. Nonlinear low-dimensional feedback controllers were synthesized and implemented on detailed fundamental models of these processes. The performance of the controllers was successfully tested through simulations.

Appendix A

Proofs of Chapter 2

The proofs of theorem 2.2 and proposition 2.4, concerning the case of systems of quasi-linear hyperbolic PDEs, are conceptually similar to the ones given for the linear counterparts of these results, and will therefore be omitted for brevity.

Proof of Proposition 2.2. Consider the closed-loop system of Eq. 2.26. Differentiating the output of this system with respect to time, we obtain the following set of equations:

$$\begin{aligned}
y &= \mathcal{C}kx \\
\frac{dy}{dt} &= \mathcal{C}k\left(A\frac{\partial}{\partial z} + B + wb(z)\mathcal{S}\right)x \\
\frac{d^2y}{dt^2} &= \mathcal{C}k\left(A\frac{\partial}{\partial z} + B + wb(z)\mathcal{S}\right)^2 x \\
&\;\;\vdots \\
\frac{d^{\bar{\sigma}}y}{dt^{\bar{\sigma}}} &= \mathcal{C}k\left(A\frac{\partial}{\partial z} + B + wb(z)\mathcal{S}\right)^{\bar{\sigma}} x + \mathcal{C}k\left(A\frac{\partial}{\partial z} + B + wb(z)\mathcal{S}\right)^{\bar{\sigma}-1} \\
&\quad \times wb(z)sv
\end{aligned} \tag{A.1}$$

It is sufficient to show that $\bar{\sigma} = \sigma$. Note that

$$\begin{aligned}
\mathcal{C}k\left(A\frac{\partial}{\partial z} + B + wb(z)\mathcal{S}\right) &= \mathcal{C}k\left(A\frac{\partial}{\partial z} + B\right) + \mathcal{C}kwb(z)\mathcal{S} \\
&= \mathcal{C}k\left(A\frac{\partial}{\partial z} + B\right),
\end{aligned}$$

because $\mathcal{C}kwb(z) = 0$. Similarly, it can be shown, by induction, that

$$\mathcal{C}k\left(A\frac{\partial}{\partial z} + B + wb(z)\mathcal{S}\right)^i = \mathcal{C}k\left(A\frac{\partial}{\partial z} + B\right)^i, \quad \text{for } i = 1, \ldots, \sigma - 1.$$

Using this simplification in the expressions for the time derivatives (Eq. A.1), it follows directly that:

$$\begin{aligned}
y &= \mathcal{C}kx \\
\frac{dy}{dt} &= \mathcal{C}k\left(A\frac{\partial}{\partial z} + B\right)x
\end{aligned}$$

$$\frac{d^2y}{dt^2} = \mathcal{C}k\left(A\frac{\partial}{\partial z} + B\right)^2 x$$

$$\vdots$$

$$\frac{d^\sigma y}{dt^\sigma} = \mathcal{C}k\left(A\frac{\partial}{\partial z} + B\right)^\sigma x + \mathcal{C}k\left(A\frac{\partial}{\partial z} + B\right)^{\sigma-1} wb(z)\mathcal{S}x$$
$$+ \mathcal{C}k\left(A\frac{\partial}{\partial z} + B\right)^{\sigma-1} wb(z)sv. \tag{A.2}$$

This completes the proof of the proposition. □

Proof of Theorem 2.1. Under the controller of Eq. 2.28, the closed-loop system takes the form:

$$\frac{\partial x}{\partial t} = A\frac{\partial x}{\partial z} + Bx + wb(z)\left[\gamma_\sigma \mathcal{C}k\left(A\frac{\partial}{\partial z} + B\right)^{\sigma-1} wb(z)\right]^{-1}$$
$$\times\left\{v - \mathcal{C}kx - \sum_{\nu=1}^{\sigma}\gamma_\nu \mathcal{C}k\left(A\frac{\partial}{\partial z} + B\right)^\nu x\right\} \tag{A.3}$$
$$y = \mathcal{C}kx.$$

From the result of proposition 2.2, it follows that a differentiation of the output of the system of Eq. A.3 yields the following equations:

$$y = \mathcal{C}kx$$
$$\frac{dy}{dt^2} = \mathcal{C}k\left(A\frac{\partial}{\partial z} + B\right)x$$
$$\frac{d^2y}{dt^2} = \mathcal{C}k\left(A\frac{\partial}{\partial z} + B\right)^2 x$$
$$\vdots \tag{A.4}$$
$$\frac{d^\sigma y}{dt^\sigma} = \mathcal{C}k\left(A\frac{\partial}{\partial z} + B\right)^\sigma x + \mathcal{C}k\left(A\frac{\partial}{\partial z} + B\right)^{\sigma-1} wb(z)$$
$$\times\left[\gamma_\sigma \mathcal{C}k\left(A\frac{\partial}{\partial z} + B\right)^{\sigma-1} wb(z)\right]^{-1}$$
$$\times\left\{v - \mathcal{C}kx - \sum_{\nu=1}^{\sigma}\gamma_\nu \mathcal{C}k\left(A\frac{\partial}{\partial z} + B\right)^\nu x\right\}$$

Substituting the above relations into Eq. 2.27, one can easily show that the result of the theorem holds. □

Proof of Proposition 2.3. Utilizing the expressions for the output derivatives of Eq. A.2, the closed-loop system of Eq. A.3 takes the form:

$$\frac{\partial x}{\partial t} = A\frac{\partial x}{\partial z} + Bx + wb(z)\left[\mathcal{C}k\left(A\frac{\partial}{\partial z} + B\right)^{\sigma-1} wb(z)\right]^{-1}$$

$$\times\left\{\frac{1}{\gamma_\sigma}(v-y)-\sum_{\nu=1}^{\sigma-1}\frac{\gamma_\nu}{\gamma_\sigma}\frac{d^\nu y}{dt^\nu}-Ck\left(A\frac{\partial}{\partial z}+B\right)^\sigma x\right\} \tag{A.5}$$

$$y=Ckx.$$

Defining the state vectors $\zeta_i=\frac{d^{i-1}y}{dt^{i-1}}$, $i=1,\ldots,\sigma$, the system of Eq. A.5 can be equivalently written in the form of the following interconnection:

$$\begin{aligned}
\dot{\zeta}_1 &= \zeta_2 \\
&\vdots \\
\dot{\zeta}_{\sigma-1} &= \zeta_\sigma \\
\dot{\zeta}_\sigma &= -\frac{1}{\gamma_\sigma}\zeta_1-\frac{\gamma_1}{\gamma_\sigma}\zeta_2-\cdots-\frac{\gamma_{\sigma-1}}{\gamma_\sigma}\zeta_\sigma+\frac{1}{\gamma_\sigma}v \\
\frac{\partial x}{\partial t} &= A\frac{\partial x}{\partial z}+Bx+wb(z)\left[Ck\left(A\frac{\partial}{\partial z}+B\right)^{\sigma-1}wb(z)\right]^{-1} \\
&\quad\times\left\{\bar{\zeta}-Ck\left(A\frac{\partial}{\partial z}+B\right)^\sigma x\right\}
\end{aligned} \tag{A.6}$$

where $\bar{\zeta}=\frac{1}{\gamma_\sigma}(v-\zeta_1)-\sum_{\nu=1}^{\sigma-1}\frac{\gamma_\nu}{\gamma_\sigma}\zeta_{\nu+1}$. Condition 1 of the proposition guarantees that the ζ-subsystem of the above interconnection is exponentially stable, and thus, the following condition holds:

$$|\bar{\zeta}|\leq K_1|\bar{\zeta}_0|e^{-a_1t} \tag{A.7}$$

where $|\cdot|$ denotes the standard Eucledian norm, K_1, a_1 are positive real numbers, with $K_1\geq 1$, and $\bar{\zeta}_0$ is the value of the variable $\bar{\zeta}$ at time $t=0$, (i.e. $\bar{\zeta}_0=\frac{1}{\gamma_\sigma}(v(0)-\zeta_{(1)0})-\sum_{\nu=1}^{\sigma-1}\frac{\gamma_\nu}{\gamma_\sigma}\zeta_{(\nu+1)0})$. From condition 2, we have that the differential operator of the system of Eq. 2.35 generates an exponentially stable semigroup $\bar{U}$, that is $\|\bar{U}\|_2\leq K_2e^{-a_2t}$, where K_2, a_2 are positive real numbers, with $K_2\geq 1$. Utilizing Eq. 2.15, the following estimate can be written for the state x of the system of Eq. A.5:

$$\|x\|_2\leq K_2\|x_0\|_2e^{-a_2t}+K_2\int_0^t e^{-a_2(t-\tau)}\|\bar{W}(z)\|_2|\bar{\zeta}(\tau)|d\tau \tag{A.8}$$

where

$$\bar{W}(z)=wb(z)\left[Ck\left(A\frac{\partial}{\partial z}+B\right)^{\sigma-1}wb(z)\right]^{-1}. \tag{A.9}$$

Substituting Eq. A.7 into Eq. A.8, we have that:

$$\begin{aligned}
\|x\|_2 &\leq K_2\|x_0\|_2e^{-a_2t}+K_2K_1\int_0^t e^{-a_2(t-\tau)}\|\bar{W}(z)\|_2|\bar{\zeta}_0|e^{-a_1\tau}d\tau \\
&= K_2\|x_0\|_2e^{-a_2t}+K_2K_1\|\bar{W}(z)\|_2|\bar{\zeta}_0|e^{-a_2t}\int_0^t e^{(a_2-a_1)\tau}d\tau.
\end{aligned} \tag{A.10}$$

Let $\delta = a_2 - a_1$. If $\delta = 0$, $\|x\|_2 \leq K_2\|x_0\|_2 e^{-a_2 t} + K_2 K_1 \|\bar{W}(z)\|_2 |\bar{\zeta}_0| e^{-a_2 t} t \leq K_2\|x_0\|_2 e^{-a_2 t} + K_2 K_1 \|\bar{W}(z)\|_2 \frac{|\bar{\zeta}_0|}{a_2 - a_3} e^{-a_3 t}$, where $0 < a_3 < a_2$. Clearly, in this case, the closed-loop system is exponentially stable. Furthermore, If $\delta > 0$, $\|x\|_2 \leq K_2\|x_0\|_2 e^{-a_2 t} + K_2 K_1 \|\bar{W}(z)\|_2 \frac{|\bar{\zeta}_0|}{\delta} e^{-a_1 t}(1 - e^{-(a_2 - a_1)t})$, while if $\delta < 0$, $\|x\|_2 \leq K_2\|x_0\|_2 e^{-a_2 t} + K_2 K_1 \|\bar{W}(z)\|_2 \frac{|\bar{\zeta}_0|}{|\delta|} e^{-a_2 t}(1 - e^{(a_2 - a_1)t})$. In either case ($\delta > 0, \delta < 0$), we have that:

$$\|x\|_2 \leq K_2\|x_0\|_2 e^{-\tilde{a}t} + K_2 K_1 \|\bar{W}(z)\|_2 \frac{|\bar{\zeta}_0|}{|\delta|} e^{-\tilde{a}t} \tag{A.11}$$

where $\tilde{a} = \min\{a_1, a_2\}$. From Eqs. A.7–A.11, the exponential stability of the closed-loop system of Eq. A.5 follows directly. □

Proof of Theorem 2.3.

Part 1: Stability analysis. Substituting the controller of Eq. 2.39 in the system of Eq. 2.9, we have:

$$\begin{aligned}
\frac{\partial x}{\partial t} &= A\frac{\partial x}{\partial z} + Bx + wb(z)\left[\gamma_\sigma Ck\left(A\frac{\partial}{\partial z} + B\right)^{\sigma-1} wb(z)\right]^{-1} \\
&\quad \times \left\{v - Ck\eta - \sum_{\nu=1}^{\sigma}\gamma_\nu Ck\left(A\frac{\partial}{\partial z} + B\right)^{\nu}\eta\right\} \\
\frac{\partial \eta}{\partial t} &= A\frac{\partial \eta}{\partial z} + B\eta + wb(z)\left[\gamma_\sigma Ck\left(A\frac{\partial}{\partial z} + B\right)^{\sigma-1} wb(z)\right]^{-1} \\
&\quad \times \left\{v - Ck\eta - \sum_{\nu=1}^{\sigma}\gamma_\nu Ck\left(A\frac{\partial}{\partial z} + B\right)^{\nu}\eta\right\} + \mathcal{P}(q - \mathcal{Q}p\eta).
\end{aligned} \tag{A.12}$$

Introducing the error coordinate $\bar{e} = x - \eta$, the above closed-loop system can be written as:

$$\begin{aligned}
\frac{\partial x}{\partial t} &= A\frac{\partial x}{\partial z} + Bx + wb(z)\left[\gamma_\sigma Ck\left(A\frac{\partial}{\partial z} + B\right)^{\sigma-1} wb(z)\right]^{-1} \\
&\quad \times \left\{v - Ckx - \sum_{\nu=1}^{\sigma}\gamma_\nu Ck\left(A\frac{\partial}{\partial z} + B\right)^{\nu}x\right\} + \bar{\mathcal{X}}\bar{e} \\
\frac{\partial \bar{e}}{\partial t} &= (\mathcal{L} - \mathcal{P}\mathcal{Q}p)\bar{e}
\end{aligned} \tag{A.13}$$

where

$$\bar{\mathcal{X}}\bar{e} = wb(z)\left[\gamma_\sigma Ck\left(A\frac{\partial}{\partial z} + B\right)^{\sigma-1} wb(z)\right]^{-1}\left\{Ck\bar{e} + \sum_{\nu=1}^{\sigma}\gamma_\nu Ck\left(A\frac{\partial}{\partial z} + B\right)^{\nu}\bar{e}\right\}. \tag{A.14}$$

Because the operator $\mathcal{P}$ is designed such that the operator $\mathcal{L} - \mathcal{P}\mathcal{Q}p$ generates an exponentially stable semigroup, the following estimate can be written for the evolution of the state $\bar{e}$ of the above system:

$$\|\bar{e}\|_2 \leq K_3\|\bar{e}_0\|_2 e^{-\bar{a}t} \tag{A.15}$$

where $\bar{e}_0$ denotes the vector of initial conditions and K_3, $\bar{a}$ are positive real numbers, with $K_3 \geq 1$. Furthermore, utilizing Eq. 2.15 and the fact that the conditions of proposition 2.3 hold, the following inequality can be written for the state x of the system of Eq. A.13:

$$\|x\|_2 \leq K_4\|x_0\|_2 e^{-\hat{a}t} + K_4 \int_0^t e^{-\hat{a}(t-\tau)} \|\bar{\mathcal{X}}\bar{e}(\tau)\|_2 d\tau \tag{A.16}$$

where K_4, $\hat{a}$ are positive real numbers, with $K_4 \geq 1$. Substituting the inequality of Eq. A.15 into Eq. A.16, and performing similar calculations as in the proof of proposition 2.3, we can show that the closed-loop system is exponentially stable.

Part 2: Input/output response. Under consistent initialization of the states x and η, that is, $x(z,0) = \eta(z,0)$, it follows that $\bar{e}(z,0) = 0$. From the dynamical system for $\bar{e}$, it is clear that if $\bar{e}(z,0) = 0$, then $\bar{e}(z,t) = 0$ for all $t \geq 0$. Thus, the system of Eq. A.13 reduces to:

$$\begin{aligned} \frac{\partial x}{\partial t} &= A\frac{\partial x}{\partial z} + Bx + wb(z)\left[\gamma_\sigma Ck\left(A\frac{\partial}{\partial z} + B\right)^{\sigma-1} wb(z)\right]^{-1} \\ &\quad \times \left\{v - Ckx - \sum_{\nu=1}^{\sigma} \gamma_\nu Ck\left(A\frac{\partial}{\partial z} + B\right)^{\nu} x\right\}. \end{aligned} \tag{A.17}$$

A direct application of theorem 2.1 completes the proof of the theorem. □

Proof of Theorem 2.4.

Part 1: Stability analysis. Substituting the controller of Eq. 2.43 in the system of Eq. 2.7, we have:

$$\begin{aligned} \frac{\partial x}{\partial t} &= A\frac{\partial x}{\partial z} + f(x) + g(x)b(z)\left[\gamma_\sigma C L_g\left(\sum_{j=1}^{n}\frac{\partial \eta_j}{\partial z}L_{a_j} + L_f\right)^{\sigma-1} h(\eta)b(z)\right]^{-1} \\ &\quad \times \left\{v - Ch(\eta) - \sum_{\nu=1}^{\sigma}\gamma_\nu C\left(\sum_{j=1}^{n}\frac{\partial \eta_j}{\partial z}L_{a_j} + L_f\right)^{\nu} h(\eta)\right\} \\ \frac{\partial \eta}{\partial t} &= A\frac{\partial \eta}{\partial z} + f(\eta) + g(\eta)b(z)\left[\gamma_\sigma C L_g\left(\sum_{j=1}^{n}\frac{\partial \eta_j}{\partial z}L_{a_j} + L_f\right)^{\sigma-1} h(\eta)b(z)\right]^{-1} \\ &\quad \times \left\{v - Ch(\eta) - \sum_{\nu=1}^{\sigma}\gamma_\nu C\left(\sum_{j=1}^{n}\frac{\partial \eta_j}{\partial z}L_{a_j} + L_f\right)^{\nu} h(\eta)\right\} + \bar{\mathcal{P}}(q - \mathcal{Q}p(\eta)). \end{aligned} \tag{A.18}$$

In order to perform a local analysis of the stability properties of the above system, we consider its linearization:

$$\frac{\partial x}{\partial t} = A\frac{\partial x}{\partial z} + B(z)x + w(z)b(z)\left[\gamma_\sigma Ck(z)\left(A\frac{\partial}{\partial z} + B(z)\right)^{\sigma-1} w(z)b(z)\right]^{-1}$$

$$\times \left\{ v - \mathcal{C}k(z)\eta - \sum_{\nu=1}^{\sigma} \gamma_\nu \mathcal{C}k(z) \left(A\frac{\partial}{\partial z} + B(z) \right)^{\nu} \eta \right\} + \bar{\mathcal{P}}(q - \mathcal{Q}p(z)\eta) \tag{A.19}$$

$$\frac{\partial \eta}{\partial t} = A\frac{\partial \eta}{\partial z} + B(z)\eta + w(z)b(z) \left[\gamma_\sigma \mathcal{C}k(z) \left(A\frac{\partial}{\partial z} + B(z) \right)^{\sigma-1} w(z)b(z) \right]^{-1}$$

$$\times \left\{ v - \mathcal{C}k(z)\eta - \sum_{\nu=1}^{\sigma} \gamma_\nu \mathcal{C}k(z) \left(A\frac{\partial}{\partial z} + B(z) \right)^{\nu} \eta \right\} + \bar{\mathcal{P}}(q - \mathcal{Q}p(z)\eta).$$

Introducing the error coordinate $\bar{e} = x - \eta$, we can write the above closed-loop system as:

$$\frac{\partial x}{\partial t} = A\frac{\partial x}{\partial z} + B(z)x + w(z)b(z) \left[\gamma_\sigma \mathcal{C}k(z) \left(A\frac{\partial}{\partial z} + B(z) \right)^{\sigma-1} w(z)b(z) \right]^{-1}$$

$$\times \left\{ v - \mathcal{C}k(z)x - \sum_{\nu=1}^{\sigma} \gamma_\nu \mathcal{C}k(z) \left(A\frac{\partial}{\partial z} + B(z) \right)^{\nu} x \right\} + \tilde{\mathcal{X}}\bar{e} \tag{A.20}$$

$$\frac{\partial \bar{e}}{\partial t} = (\bar{\mathcal{L}} - \bar{\mathcal{P}}\mathcal{Q}p(z))\bar{e}$$

where

$$\tilde{\mathcal{X}}\bar{e} = w(z)b(z) \left[\gamma_\sigma \mathcal{C}k(z) \left(A\frac{\partial}{\partial z} + B(z) \right)^{\sigma-1} w(z)b(z) \right]^{-1}$$

$$\times \left\{ \mathcal{C}k(z)\bar{e} + \sum_{\nu=1}^{\sigma} \gamma_\nu \mathcal{C}k(z) \left(A\frac{\partial}{\partial z} + B(z) \right)^{\nu} \bar{e} \right\}. \tag{A.21}$$

From the conditions of the theorem, we have that the x-subsystem (with $\bar{e} = 0$) of the above interconnection is exponentially stable and the $\bar{e}$-subsystem is also exponentially stable, because the operator $\bar{\mathcal{P}}$ is designed such that the operator $\bar{\mathcal{L}} - \bar{\mathcal{P}}\mathcal{Q}p(z)$ generates an exponentially stable semi-group. Following an approach analogous to the one used in the proof of proposition 2.3, one can show that the system of Eq. A.20 is exponentially stable. Utilizing the result of proposition 2.1, we have that the closed-loop system of Eq. A.18 is locally exponentially stable, because the linearized system of Eq. A.20 is exponentially stable.

Part 2: Input/output response. Under consistent initialization of the states x and η, that is, $x(z,0) = \eta(z,0)$, it follows that $\bar{e}(z,0) = 0$. From the dynamical system for $\bar{e}$, it is clear that if $\bar{e}(z,0) = 0$, then $\bar{e}(z,t) = 0$ for all $t \geq 0$. Thus, the system of Eq. A.20 reduces to:

$$\frac{\partial x}{\partial t} = A\frac{t\partial x}{\partial z} + f(x) + g(x)b(z) \left[\gamma_\sigma C L_g \left(\sum_{j=1}^{n} \frac{\partial x_j}{\partial z} L_{a_j} + L_f \right)^{\sigma-1} h(x)b(z) \right]^{-1}$$

$$\times \left\{ v - Ch(x) - \sum_{\nu=1}^{\sigma} \gamma_\nu C \left(\sum_{j=1}^{n} \frac{\partial x_j}{\partial z} L_{a_j} + L_f \right)^{\nu} h(x) \right\} \tag{A.22}$$

$$y = Ch(x).$$

Since the characteristic index between y and v is σ, a differentiation of the output of the system of Eq. A.22 yields the following expressions:

$$\begin{aligned}
y &= Ch(x) \\
\frac{dy}{dt^2} &= C \left(\sum_{j=1}^{n} \frac{\partial x_j}{\partial z} L_{a_j} + L_f \right) h(x) \\
\frac{d^2 y}{dt^2} &= C \left(\sum_{j=1}^{n} \frac{\partial x_j}{\partial z} L_{a_j} + L_f \right) \left(\sum_{j=1}^{n} \frac{\partial x_j}{\partial z} L_{a_j} + L_f \right) h(x) \\
&\vdots \\
\frac{d^\sigma y}{dt^\sigma} &= C \left(\sum_{j=1}^{n} \frac{\partial x_j}{\partial z} L_{a_j} + L_f \right)^{\sigma} + C L_g \left(\sum_{j=1}^{n} \frac{\partial x_j}{\partial z} L_{a_j} + L_f \right)^{\sigma-1} h(x) b(z) \\
&\quad \times \left[\gamma_\sigma C L_g \left(\sum_{j=1}^{n} \frac{\partial x_j}{\partial z} L_{a_j} + L_f \right)^{\sigma-1} h(x) b(z) \right]^{-1} \\
&\quad \times \left\{ v - Ch(x) - \sum_{\nu=1}^{\sigma} \gamma_\nu C \left(\sum_{j=1}^{n} \frac{\partial x_j}{\partial z} L_{a_j} + L_f \right)^{\nu} h(x) \right\}.
\end{aligned} \tag{A.23}$$

Substituting the above relation into Eq. 2.27, one can easily show that the result of the theorem holds. □

Appendix B

Proofs of Chapter 3

Proof of Proposition 3.1.
Part 3.1: The proof of the first part of the proposition is given in the proof of proposition 2.2.

Part 2: Referring to the open-loop system of Eq. 3.1, let $\delta \leq \sigma$. We will first show that the result holds in the case $\delta < \sigma$. Consider the closed-loop system of Eq. 3.8. From part one of the proposition, we have that the characteristic index of the output y with respect to v in the closed-loop system is equal to σ. From the definition of characteristic index of y with respect to θ, we have that the following relations hold for the open-loop system:

$$C^i L_{W_k} \left(\sum_{j=1}^{n} \frac{\partial x_j}{\partial z} L_{a_j} + L_f \right)^{\mu-1} h(x) r_k^i(z) \equiv 0,$$
$$\forall \mu = 1, \ldots, \delta - 1, \; i = 1, \ldots, l. \quad \text{(B.1)}$$

Differentiating the output of the closed-loop system with respect to time and using the relations of Eq. B.1, we get:

$$\begin{aligned}
y &= Ch(x) \\
\frac{dy}{dt} &= C \left(\sum_{j=1}^{n} \frac{\partial x_j}{\partial z} L_{a_j} + L_f \right) h(x) \\
\frac{d^2y}{dt^2} &= C \left(\sum_{j=1}^{n} \frac{\partial x_j}{\partial z} L_{a_j} + L_f \right) \left(\sum_{j=1}^{n} \frac{\partial x_j}{\partial z} L_{a_j} + L_f \right) h(x) \\
&\vdots \\
\frac{d^\delta y}{dt^\delta} &= C \left(\sum_{j=1}^{n} \frac{\partial x_j}{\partial z} L_{a_j} + L_f \right)^{\delta-1} \left(\sum_{j=1}^{n} \frac{\partial x_j}{\partial z} L_{a_j} + L_f \right) h(x) \\
&\quad + C \sum_{k=1}^{q} L_{W_k} \left(\sum_{j=1}^{n} \frac{\partial x_j}{\partial z} L_{a_j} + L_f \right)^{\delta-1} h(x) r_k(z) \theta_k(t).
\end{aligned} \quad \text{(B.2)}$$

From the above equations, it is clear that the result of the second part of the proposition holds if $\delta < \sigma$. The same argument can be used to show that the result is also true for the case $\delta = \sigma$. The details in this case are omitted for brevity. □

Proof of Theorem 3.2

Part 1: Uncertainty decoupling. Necessity. We will proceed by contradiction. Consider the system of Eq. 3.1 and assume that $\delta \leq \sigma$. Referring to the closed-loop system of Eq. 3.8, we have, from proposition 1, that the characteristic indices σ, δ are preserved, and the condition $\delta \leq \sigma$ holds. This fact implies that θ affects directly the δ-th time derivative of y in the closed-loop system, and thus, y, which yields a contradiction.

Sufficiency. With reference to the system of Eq. 3.1, suppose that $\sigma < \delta$. In this case, a time differentiation of y up to σ-th order yields the following expressions:

$$
\begin{aligned}
y &= Ch(x) \\
\frac{dy}{dt} &= C\left(\sum_{j=1}^{n} \frac{\partial x_j}{\partial z} L_{a_j} + L_f\right) h(x) \\
\frac{d^2y}{dt^2} &= C\left(\sum_{j=1}^{n} \frac{\partial x_j}{\partial z} L_{a_j} + L_f\right)\left(\sum_{j=1}^{n} \frac{\partial x_j}{\partial z} L_{a_j} + L_f\right) h(x) \\
&\vdots \\
\frac{d^\sigma y}{dt^\sigma} &= C\left(\sum_{j=1}^{n} \frac{\partial x_j}{\partial z} L_{a_j} + L_f\right)^{\sigma-1}\left(\sum_{j=1}^{n} \frac{\partial x_j}{\partial z} L_{a_j} + L_f\right) h(x) \\
&\quad + CL_g\left(\sum_{j=1}^{n} \frac{\partial x_j}{\partial z} L_{a_j} + L_f\right)^{\sigma-1} h(x)b(z)u.
\end{aligned}
\tag{B.3}
$$

From the expression of the σ-th derivative of y, it is clear that there exists a control law of the form of Eq. 3.7 (e.g., the controller of Eq. 3.10) which guarantees that y is independent of θ in the closed-loop system, for all times. Finally, one can easily show, utilizing the expressions of Eq. B.3, that the controller of Eq. 3.10 enforces the input/output response of Eq. 3.9 in the closed-loop system.

Part 2: Boundedness. First, we note that whenever $\theta(t) \equiv 0$, the conditions (i) and (ii) of the theorem guarantee that the nominal closed-loop system is locally exponentially stable (this follows from proposition 2.3). Since the nominal closed-loop system is locally exponential stable, we have from theorem 3.1 that there exists a smooth Lyapunov functional $V : \mathcal{H}^n \times [\alpha, \beta] \rightarrow R$ of the form of Eq. 3.4, and a set of positive real numbers a_1, a_2, a_3, a_4, a_5, such that the following properties hold:

$$
\begin{aligned}
a_1\|x\|_2^2 \leq\ & V(t) \leq a_2\|x\|_2^2 \\
\frac{dV}{dt} =\ & \frac{\partial V}{\partial x}\bar{\mathcal{L}}(x) \leq -a_3\|x\|_2^2 \\
\left\|\frac{\partial V}{\partial x}\right\|_2 \leq\ & a_4\|x\|_2
\end{aligned}
\tag{B.4}
$$

if $\|x\|_2 \leq a_5$. Whenever $(\theta(t) \neq 0)$ the closed-loop system, under the control law of Eq. 3.10, takes the form:

$$\frac{\partial x}{\partial t} = \bar{\mathcal{L}}(x) + W(x)r(z)\theta(t) \tag{B.5}$$

where $\bar{\mathcal{L}}(x)$ is a nonlinear operator. Computing the time derivative of the functional $V : \mathcal{H}^n \times [\alpha, \beta] \to R$ along the trajectories of the uncertain closed-loop system of Eq. B.5, using Eq. B.4, and using the fact that, if $\|x\|_2 \leq a_5$, there exists a positive real number a_6 such that $\|W(x)\|_2 \leq a_6$, we get:

$$\begin{aligned} \frac{dV}{dt} &= \frac{\partial V}{\partial x}(\bar{\mathcal{L}}(x) + W(x)r(z)\theta(t)) \\ &\leq -a_3\|x\|_2^2 + a_4 a_6 \|x\|_2 \|r(z)\|_2 |\theta(t)|. \end{aligned} \tag{B.6}$$

From the last inequality of the above equation, we have that if $|\theta(t)| \leq a_3 a_5 / 4 a_4 a_6 \|r(z)\|_2 \leq \hat{\delta}, \forall\, t \geq 0$, then

$$\frac{dV}{dt} \leq -\frac{a_3}{2}\|x\|_2^2 \tag{B.7}$$

if $a_5/2 \leq \|x\|_2 \leq a_5$. From the above inequality, we have [92] that $\|x\|_2$ is a bounded quantity, which implies that the state of the closed-loop system is bounded. □

Proof of Theorem 3.3. First, we define the state vectors $\zeta_\nu^i = \frac{d^{\nu-1}y^i}{dt^{\nu-1}}$, $i = 1, \ldots, l$, $\nu = 1, \ldots, \sigma$, $\zeta_\nu = [\zeta_\nu^1 \zeta_\nu^2 \cdots \zeta_\nu^l]^T$, $\bar{\zeta}_1 = \zeta_1 - v$, $\tilde{\zeta}_\sigma^i = \zeta_\sigma^i + \frac{\gamma_1}{\gamma_\sigma}\bar{\zeta}_1^i + \frac{\gamma_2}{\gamma_\sigma}\zeta_2^i + \cdots + \frac{\gamma_{\sigma-1}}{\gamma_\sigma}\zeta_{\sigma-1}^i$, $\tilde{\zeta}_\sigma = [\tilde{\zeta}_\sigma^1 \ \tilde{\zeta}_\sigma^2 \ \cdots \ \tilde{\zeta}_\sigma^l]^T$ and for ease of notation, we also set:

$$\bar{s}(x) = \left[\mathcal{C} L_g \left(\sum_{j=1}^{n} \frac{\partial x_j}{\partial z} L_{a_j} + L_f \right)^{\sigma-1} h(x) b(z) \right]^{-1}. \tag{B.8}$$

Using the above notation the controller of Eq. 3.16 takes the form:

$$u = \bar{s}(x) \left\{ -\gamma_0 \tilde{\zeta}_\sigma - \sum_{\nu=1}^{\sigma-1} \frac{\gamma_\nu}{\gamma_\sigma} \zeta_{\nu+1} - 2K(t)\Delta(\tilde{\zeta}_\sigma, \phi)\tilde{\zeta}_\sigma \right\} \tag{B.9}$$

where $\Delta(\tilde{\zeta}_\sigma, \phi)$ is an $l \times l$ diagonal matrix, whose (i, i)-th element is of the form $(|\tilde{\zeta}_\sigma^i| + \phi)^{-1}$.

Part 1: Asymptotic output tracking. Using the definition for the state vectors ζ_ν, $\nu = 1, \ldots, \sigma$, and $\bar{\zeta}_1$, $\tilde{\zeta}_\sigma$, the closed-loop system

$$\begin{aligned} \frac{\partial x}{\partial t} &= A\frac{\partial x}{\partial z} + f(x) + g(x)b(z)u + W(x)r(z)\theta(t) \\ y &= \mathcal{C}h(x). \end{aligned} \tag{B.10}$$

can be equivalently written in the following form:

$$
\begin{aligned}
\dot{\bar{\zeta}}_1 &= \zeta_2 \\
&\vdots \\
\dot{\zeta}_{\sigma-1} &= -\frac{\gamma_1}{\gamma_\sigma}\bar{\zeta}_1 - \frac{\gamma_2}{\gamma_\sigma}\zeta_2 - \cdots - \frac{\gamma_{\sigma-1}}{\gamma_\sigma}\zeta_{\sigma-1} + \tilde{\zeta}_\sigma \\
\dot{\tilde{\zeta}}_\sigma &= \sum_{\nu=1}^{\sigma-1}\frac{\gamma_\nu}{\gamma_\sigma}\zeta_{\nu+1} - \gamma_0\tilde{\zeta}_\sigma - \sum_{\nu=1}^{\sigma-1}\frac{\gamma_\nu}{\gamma_\sigma}\zeta_{\nu+1} - 2K(t)\Delta(\tilde{\zeta}_\sigma, \phi)\tilde{\zeta}_\sigma + K(t) \\
\frac{\partial x}{\partial t} &= \mathcal{L}x + g(x)b(z)\bar{s}(x)\left\{-\gamma_0\tilde{\zeta}_\sigma - \sum_{\nu=1}^{\sigma-1}\frac{\gamma_\nu}{\gamma_\sigma}\zeta_{\nu+1} - 2K(t)\Delta(\tilde{\zeta}_\sigma, \phi)\tilde{\zeta}_\sigma\right\} \\
&\quad + W(x)r(z)\theta.
\end{aligned}
\tag{B.11}
$$

To establish the relation of Eq. 3.18, we will first work with the $\tilde{\zeta}_\sigma$-subsystem of Eq. B.11 and establish a bound for the state of this system in terms of the initial condition and the parameter ϕ (in order to simplify the presentation we will focus on the i-th input/output pair). To this end, we consider the following smooth functions $V^i : \mathbb{R} \to \mathbb{R}_{\geq 0}$, $i = 1, \ldots, l$:

$$
V^i = \frac{1}{2}(\tilde{\zeta}_\sigma^i)^2 \tag{B.12}
$$

to show that the time derivative of V^i is negative definite outside of a region that includes the steady state in the space and this region can be made arbitrarily small by picking ϕ sufficiently small. Calculating the time derivative of V^i along the trajectories of the $\tilde{\zeta}_\sigma^i$-subsystem of Eq. B.10, we have:

$$
\begin{aligned}
\frac{dV^i}{dt} &= \tilde{\zeta}_\sigma^i[-\gamma_0\tilde{\zeta}_\sigma^i - 2K(t)(|\tilde{\zeta}_\sigma^i| + \phi)^{-1}\tilde{\zeta}_\sigma^i + K(t)] \\
&\leq -\gamma_0(\tilde{\zeta}_\sigma^i)^2 - 2K(t)(|\tilde{\zeta}_\sigma^i| + \phi)^{-1}(\tilde{\zeta}_\sigma^i)^2 + K(t)|\tilde{\zeta}_\sigma^i| \\
&\leq -\gamma_0(\tilde{\zeta}_\sigma^i)^2 - 2K(t)(|\tilde{\zeta}_\sigma^i| + \phi)^{-1}(\tilde{\zeta}_\sigma^i)^2 \\
&\quad + K(t)(|\tilde{\zeta}_\sigma^i| + \phi)^{-1}[(\tilde{\zeta}_\sigma^i)^2 + |\tilde{\zeta}_\sigma^i|\phi] \\
&\leq -\gamma_0(\tilde{\zeta}_\sigma)^2 - K(t)(|\tilde{\zeta}_\sigma^i| + \phi)^{-1}|\tilde{\zeta}_\sigma|(|\tilde{\zeta}_\sigma^i| - \phi).
\end{aligned}
\tag{B.13}
$$

Clearly, when $|\tilde{\zeta}_\sigma^i| \geq \phi$, $\dot{V}^i$ is negative definite, which implies [86] that there exist positive real numbers $\hat{K} \geq 1, \bar{a}, \gamma$, such that the following bound holds for the norm of the state of the $\tilde{\zeta}_\sigma^i$-subsystem:

$$
|\tilde{\zeta}_\sigma^i| \leq \hat{K}|(\tilde{\zeta}_\sigma^i)_0|e^{-\bar{a}t} + \gamma\phi \tag{B.14}
$$

where $(\tilde{\zeta}_\sigma^i)_0$ denotes the value of $\tilde{\zeta}_\sigma^i$ at time $t = 0$. Consider now the following subsystem:

$$\begin{aligned} \dot{\bar{\zeta}}_1^i &= \zeta_2^i \\ &\vdots \\ \dot{\zeta}_{\sigma-1}^i &= -\frac{\gamma_1}{\gamma_\sigma}\bar{\zeta}_1^i - \frac{\gamma_2}{\gamma_\sigma}\zeta_2^i - \cdots - \frac{\gamma_{\sigma-1}}{\gamma_\sigma}\zeta_{\sigma-1}^i + \tilde{\zeta}_\sigma^i \end{aligned} \tag{B.15}$$

where $\tilde{\zeta}_\sigma^i$ can be thought of as an external input. Since the parameters γ_k are chosen so that the polynomial $1 + \gamma_1 s + \cdots + \gamma_\sigma s^\sigma = 0$ is Hurwitz, we have that there exists positive real numbers $K_\zeta, a_\zeta, \gamma_{\tilde{\zeta}_\sigma}$ such that the following bound can be written for the state vector $\zeta^i = [\bar{\zeta}_1^i \ \zeta_2^i \ \cdots \ \zeta_{\sigma-1}^i]$ of the system of Eq. B.15:

$$|\zeta^i| \leq K_\zeta \left|\zeta_0^i\right| e^{-a_\zeta t} + \gamma_{\tilde{\zeta}_\sigma} \left\|\tilde{\zeta}_\sigma^i\right\|. \tag{B.16}$$

Taking the limit as $t \to \infty$, using the property that $\lim_{t\to\infty}\sup_{t\geq 0}\{|\tilde{\zeta}_\sigma^i|\} = \sup_{t\geq 0}\{\lim_{t\to\infty}|\tilde{\zeta}_\sigma^i|\}$, and using Eq. B.14 we have:

$$\lim_{t\to\infty}|\zeta^i| \leq \gamma_{\tilde{\zeta}_\sigma^i}\gamma\phi. \tag{B.17}$$

Picking $\phi^* = d/\gamma_{\tilde{\zeta}_\sigma}\gamma$, the relation of Eq. 3.18 follows directly from the fact $\lim_{t\to\infty}|\bar{\zeta}_1^i| \leq \lim_{t\to\infty}|\zeta^i|$.

Part 2: Boundedness of the state. The proof that the state is bounded for all times, whenever the conditions of the theorem hold, can be obtained by using contradiction argument. It starts by assuming that there exists a maximal time T such that for all $t \in [0, T)$, the states $(x, \zeta, \tilde{\zeta}_\sigma)$ of the system of Eq. B.11 are bounded (note that T always exists because the system starts from bounded initial conditions). Then, the bounds that hold for the states $(x, \zeta, \tilde{\zeta}_\sigma)$ for $t \in [0, T)$ are derived and shown to continue to hold for $t \in [0, T + k_T]$, where k_T is some positive number. However, this contradicts the assumption that T is the maximal time in which the state is bounded, which implies that $T = \infty$, and thus the state of the closed-loop system is bounded for all times. □

Proof of Theorem 3.4. Using L_{ij} as defined in subsection 3.7.1, the representation of the system of Eq. 3.20, with $u = 0$, in the coordinates (x, η_f) takes the form:

$$\begin{aligned} \frac{\partial x}{\partial t} &= \mathcal{L}x + W(x)r(z)\theta(t) + \mathcal{L}_{12}\eta_f \\ \epsilon\frac{\partial \eta_f}{\partial t} &= \mathcal{L}_{22}\eta_f + \epsilon\left(\frac{\partial z_s}{\partial x}\dot{x} + \frac{\partial z_s}{\partial \theta}\dot{\theta}\right). \end{aligned} \tag{B.18}$$

Since from assumption 3.6, the system:

$$\epsilon \frac{\partial \eta_f}{\partial t} = \mathcal{L}_{22} \eta_f \tag{B.19}$$

is exponentially stable, we have that if $(x, \theta, \dot{\theta})$ are bounded, then there exists a set of positive real numbers $K_f, \bar{a}_f, \gamma_{\eta_f}$ such that the following bound can be written for the state of the η_f-subsystem of Eq. B.18, for all $t \geq 0$:

$$\|\eta_f\|_2 \leq K_f \|\eta_{f0}\|_2 e^{-\bar{a}_f \frac{t}{\epsilon}} + \gamma_{\eta_f} \epsilon. \tag{B.20}$$

Using assumption 3.5 and Eq. 2.15, we have that the following bound holds for the state of the x-subsystem of Eq. B.18, for all $t \geq 0$:

$$\begin{aligned} \|x\|_2 \leq & K_s \|x_0\|_2 e^{-a_s t} + K_s \int_0^t e^{-a_s(t-\tau)} \|W(x) r(z)\|_2 |\theta(\tau)| d\tau \\ & + K_s \int_0^t e^{-a_s(t-\tau)} \|\mathcal{L}_{12}\|_2 \|\eta_f\|_2 d\tau \\ \leq & K_s \|x_0\|_2 e^{-a_s t} + M_2 \|\theta\| + M_3 \sup_{t \geq 0}\{\|\eta_f\|_2\} \end{aligned} \tag{B.21}$$

where $M_2 = \frac{K_s \bar{M}_2 \|r(z)\|_2}{a_s}$, and $M_3 = \frac{K_s \|\mathcal{L}_{12}\|_2}{a_s}$, provided that the initial condition ($\|x_0\|_2$) and the inputs ($\|\theta\|, \|\dot{\theta}\|, \sup\{\|\eta_f\|_2\}$) are sufficiently small. Note that since we do not know a-priori that the states (x, η_f) of the system of Eq. B.18 are bounded, we have to work with truncations and exploit causality in order to prove boundedness. Let $\tilde{\delta}$ be as given in the statement of the theorem, so that $\max\{\|x_0\|_2, \|\eta_{f_0}\|_2, \|\theta\|, \|\dot{\theta}\|\} \leq \tilde{\delta}$, and let δ_x be a positive real number that satisfies

$$\delta_x > K_s \tilde{\delta} + M_2 \tilde{\delta} + d \tag{B.22}$$

where d is a positive real number specified in the statement of the theorem. Note that since $K_s \geq 1$, $\delta_x > \tilde{\delta}$, and using continuity with respect to initial conditions, we define $[0, T)$ to be the maximal interval in which $\|x_t\|_2 \leq \delta_x$, for all $t \in [0, T)$ and suppose that T is finite. We will now show by contradiction that $T = \infty$, provided that ϵ is sufficiently small.

First, notice that from Eq. B.19 we have that for all $t \in [0, T)$, $\|\eta_{f_t}\| \leq K_f \tilde{\delta} + \gamma_\eta \epsilon$. Let ϵ_0 be a positive real number so that $\epsilon \in (0, \epsilon_0]$ and define $\delta_{\eta_f} := K_f \tilde{\delta} + \gamma_{\eta_f} \epsilon_0$. Note that $\delta_{\eta_f} > \tilde{\delta}$. The following lemma will be used in our development (this lemma can be proven by using similar arguments as in the proof of lemma 3 in [45]). □

Lemma *Referring to the x-subsystem of Eq. B.18, let Eq. B.21 hold. Then, for each pair of positive real numbers $\bar{\delta}, \bar{d}$, there exists a positive real number ρ^* such that for each $\rho \in [0, \rho^*]$, if* $\max\{\|x_0\|_2, \|\eta_f\|_2, \|\theta\|, \|\dot{\theta}\|\} \leq \bar{\delta}$, *then* *the solution of the x-subsystem of Eq. B.18 with $x(0) = x_0$ exists for each $t \geq 0$ and satisfies:*

$$\|x\|_2 \leq K_s \|x_0\|_2 e^{-a_s t} + M_2 \|\theta\| + M_3 \sup_{t \geq \rho} \{\|\eta_f\|_2\} + \bar{d}. \tag{B.23}$$

Applying the result of the above lemma, we have that there exists a positive real number ρ (assume without loss of generality that $\rho < T$), such that if $\max\{\|x_0\|_2, \|\eta_f\|, \|\theta\|, \|\dot{\theta}\|\} \leq \delta_{\eta_f}$, then the solution of the x-subsystem of Eq. B.18 with $x(0) = x_0$ exists for each $t \in [0, T)$ and satisfies:

$$\|x\|_2 \leq K_s \|x_0\|_2 e^{-a_s t} + M_2 \|\theta\| + M_3 \sup_{t \geq \rho} \{\|\eta_f\|_2\} + \frac{d}{2}. \tag{B.24}$$

Substituting Eq. B.19 into Eq. B.24, if $\epsilon \in (0, \epsilon_0]$, we have for all $t \in [0, T)$:

$$\|x\|_2 \leq K_s \|x_0\|_2 e^{-a_s t} + M_2 \|\theta\| + \frac{d}{2} + M_3 K_f \|\eta_{f0}\|_2 \left(e^{-\bar{a}_f \frac{\rho}{\epsilon}} + \gamma_{\eta_f} \epsilon\right). \tag{B.25}$$

From the fact that the last term of the above equation vanishes as $\epsilon \to 0$, we have that there exists an $\epsilon_1 \in (0, \epsilon_0]$ such that if $\epsilon \in (0, \epsilon_1]$, then for all $t \in [0, T)$:

$$\|x\|_2 \leq K_s \|x_0\|_2 e^{-a_s t} + M_2 \|\theta\| + d. \tag{B.26}$$

From the definition of δ_x, the assumption that T is finite and continuity of x, there must exist some positive real number k such that $\sup_{t \in [0, T+k]} \{\|x\|_2\} < \delta_x$. This contradicts that T is maximal. Hence, $T = \infty$ and the inequality Eq. 3.32 holds for all $t \geq 0$.

Finally, letting ϵ_3 be such that $\gamma_{\eta_f} \epsilon \leq d$ for all $\epsilon \in [0, \epsilon_3]$ it follows that both inequalities of Eqs. 3.32 and 3.33 hold for $\epsilon \in (0, \epsilon^*]$, where $\epsilon^* = \min\{\epsilon_1, \epsilon_2, \epsilon_3\}$.

Proof of Theorem 3.6. Substituting the controller of Eq. 3.10 into the system of Eq. 3.20 and using the expression for the output derivatives of the closed-loop slow system of Eq. B.3, we have:

$$\begin{aligned}
\frac{\partial x}{\partial t} &= \mathcal{L}x + g(x)b(z)\left[\gamma_\sigma C L_g \left(\sum_{j=1}^{n} \frac{\partial x_j}{\partial z} L_{a_j} + L_f\right)^{\sigma-1} h(x)b(z)\right]^{-1} \\
&\quad \times \left\{\frac{1}{\gamma_\sigma}(v - y) - \sum_{\nu=1}^{\sigma-1} \frac{\gamma_\nu}{\gamma_\sigma} \frac{d^\nu y}{dt^\nu} - C\left(\sum_{j=1}^{n} \frac{\partial x_j}{\partial z} L_{a_j} + L_f\right)^{\sigma} h(x)\right\} \\
&\quad + W_1(x)r(z)\theta(t) + \mathcal{L}_{12}\eta_f \\
\epsilon\frac{\partial \eta_f}{\partial t} &= \mathcal{L}_{22}\eta_f + g_2(x)b(z)\left[\gamma_\sigma C L_g \left(\sum_{j=1}^{n} \frac{\partial x_j}{\partial z} L_{a_j} + L_f\right)^{\sigma-1} h(x)b(z)\right]^{-1}
\end{aligned}$$

$$\times\left\{\frac{1}{\gamma_\sigma}(v-y)-\sum_{\nu=1}^{\sigma-1}\frac{\gamma_\nu}{\gamma_\sigma}\frac{d^\nu y}{dt^\nu}-\mathcal{C}\left(\sum_{j=1}^{n}\frac{\partial x_j}{\partial z}L_{a_j}+L_f\right)^\sigma h(x)\right\}$$
$$+W_2(x)r(z)\theta(t)+\epsilon\left(\frac{\partial z_s}{\partial x}\dot{x}+\frac{\partial z_s}{\partial \theta}\dot{\theta}\right).$$
$$y=\mathcal{C}h(x) \tag{B.27}$$

Defining the state vectors $\zeta_\nu=\frac{d^{\nu-1}y}{dt^{\nu-1}}$, $\nu=1,\ldots,\sigma$, the system of Eq. B.27 can be equivalently written in the following form:

$$\dot{\zeta}_1=\zeta_2+\sum_{k=1}^{p}\mathcal{C}L_{\mathcal{L}_{12}^k}h(x)\eta_f$$
$$\vdots$$
$$\dot{\zeta}_{\sigma-1}=\zeta_\sigma+\sum_{k=1}^{p}\mathcal{C}L_{\mathcal{L}_{12}^k}^{\sigma-2}h(x)\eta_f$$
$$\dot{\zeta}_\sigma=-\frac{1}{\gamma_\sigma}\zeta_1-\frac{\gamma_1}{\gamma_\sigma}\zeta_2-\cdots-\frac{\gamma_{\sigma-1}}{\gamma_\sigma}\zeta_\sigma+\frac{1}{\gamma_\sigma}v+\sum_{k=1}^{p}\mathcal{C}L_{\mathcal{L}_{12}^k}^{\sigma-1}h(x)\eta_f$$
$$\frac{\partial x}{\partial t}=\mathcal{L}x+g(x)b(z)\left[\gamma_\sigma\mathcal{C}L_g\left(\sum_{j=1}^{n}\frac{\partial x_j}{\partial z}L_{a_j}+L_f\right)^{\sigma-1}h(x)b(z)\right]^{-1} \tag{B.28}$$
$$\cdot\left\{\bar{\zeta}-\mathcal{C}\left(\sum_{j=1}^{n}\frac{\partial x_j}{\partial z}L_{a_j}+L_f\right)^\sigma h(x)\right\}+\mathcal{L}_{12}\eta_f+W(x)r(z)\theta$$
$$\epsilon\frac{\partial \eta_f}{\partial t}=\mathcal{L}_{22}\eta_f+g_2(x)b(z)\left[\gamma_\sigma\mathcal{C}L_g\left(\sum_{j=1}^{n}\frac{\partial x_j}{\partial z}L_{a_j}+L_f\right)^{\sigma-1}h(x)b(z)\right]^{-1}$$
$$\cdot\left\{\bar{\zeta}-\mathcal{C}\left(\sum_{j=1}^{n}\frac{\partial x_j}{\partial z}L_{a_j}+L_f\right)^\sigma h(x)\right\}$$
$$+W_2(x)r(z)\theta(t)+\epsilon\left(\frac{\partial z_s}{\partial x}\dot{x}+\frac{\partial z_s}{\partial \theta}\dot{\theta}\right).$$

Performing a two-time-scale decomposition, we can show that the fast dynamics of the above system are locally exponentially stable, and the reduced system takes the form:

$$\dot{\zeta}_1^s=\zeta_2^s$$
$$\vdots$$
$$\dot{\zeta}_{\sigma-1}^s=\zeta_\sigma^s$$
$$\dot{\zeta}_\sigma^s=-\frac{1}{\gamma_\sigma}\zeta_1^s-\frac{\gamma_1}{\gamma_\sigma}\zeta_2^s-\cdots-\frac{\gamma_{\sigma-1}}{\gamma_\sigma}\zeta_\sigma^s+\frac{1}{\gamma_\sigma}v$$

$$\frac{\partial x}{\partial t} = \mathcal{L}x + g(x)b(z)\left[\gamma_\sigma \mathcal{C}L_g\left(\sum_{j=1}^{n}\frac{\partial x_j}{\partial z}L_{a_j} + L_f\right)^{\sigma-1} h(x)b(z)\right]^{-1}$$
$$\cdot\left\{\bar{\zeta} - \mathcal{C}\left(\sum_{j=1}^{n}\frac{\partial x_j}{\partial z}L_{a_j} + L_f\right)^{\sigma} h(x)\right\} + W(x)r(z)\theta. \tag{B.29}$$

From theorem 3.2, we have that the state of the above system is bounded. Using the result of theorem 3.4, we have that the state of the closed-loop system of Eq. B.27 is bounded provided that the initial conditions, the uncertainty, the rate of change of uncertainty, and the singular perturbation parameter are sufficiently small. Using the auxiliary variable $\tilde{\zeta}_\nu = \zeta_\nu - \zeta_\nu^s$, $\nu = 1, \ldots, \sigma$, the system of Eq. B.28 can be written as:

$$\dot{\tilde{\zeta}}_1 = \tilde{\zeta}_2 + \sum_{k=1}^{p}\mathcal{C}L_{\mathcal{L}_{12}^k}h(x)\eta_f$$
$$\vdots$$
$$\dot{\tilde{\zeta}}_{\sigma-1} = \tilde{\zeta}_\sigma + \sum_{k=1}^{p}\mathcal{C}L_{\mathcal{L}_{12}^k}^{\sigma-2}h(x)\eta_f$$
$$\dot{\tilde{\zeta}}_\sigma = -\frac{1}{\gamma_\sigma}\tilde{\zeta}_1 - \frac{\gamma_1}{\gamma_\sigma}\tilde{\zeta}_2 - \cdots - \frac{\gamma_{\sigma-1}}{\gamma_\sigma}\tilde{\zeta}_\sigma + \frac{1}{\gamma_\sigma}v + \sum_{k=1}^{p}\mathcal{C}L_{\mathcal{L}_{12}^k}^{\sigma-1}h(x)\eta_f$$
$$\frac{\partial x}{\partial t} = \mathcal{L}x + g(x)b(z)\left[\gamma_\sigma \mathcal{C}L_g\left(\sum_{j=1}^{n}\frac{\partial x_j}{\partial z}L_{a_j} + L_f\right)^{\sigma-1} h(x)b(z)\right]^{-1}$$
$$\times\left\{\bar{\zeta} - \mathcal{C}\left(\sum_{j=1}^{n}\frac{\partial x_j}{\partial z}L_{a_j} + L_f\right)^{\sigma} h(x)\right\} + \mathcal{L}_{12}\eta_f + W(x)r(z)\theta \tag{B.30}$$
$$\epsilon\frac{\partial \eta_f}{\partial t} = \mathcal{L}_{22}\eta_f + g_2(x)b(z)\left[\gamma_\sigma \mathcal{C}L_g\left(\sum_{j=1}^{n}\frac{\partial x_j}{\partial z}L_{a_j} + L_f\right)^{\sigma-1} h(x)b(z)\right]^{-1}$$
$$\times\left\{\bar{\zeta} - \mathcal{C}\left(\sum_{j=1}^{n}\frac{\partial x_j}{\partial z}L_{a_j} + L_f\right)^{\sigma} h(x)\right\} + W_2(x)r(z)\theta(t)$$
$$+\epsilon\left(\frac{\partial z_s}{\partial x}\dot{x} + \frac{\partial z_s}{\partial \theta}\dot{\theta}\right).$$

Using the fact that the state of the above system is bounded and the equality $\zeta_\nu(0) = \zeta_\nu^s(0)$ holds $\forall\, \nu = 1, \ldots, \sigma$, the following bounds can be written for the evolution of the states $\tilde{\zeta} = [\tilde{\zeta}_1^T\ \tilde{\zeta}_2^T\ \ldots\ \tilde{\zeta}_\sigma^T]^T$, η_f of the above system:

$$|\tilde{\zeta}| \leq M_4\|\eta_f\|_2$$
$$\|\eta_f\|_2 \leq K_f\|\eta_{f0}\|_2 e^{-\bar{a}_f\frac{t}{\epsilon}} + \gamma_{\eta_f}\epsilon \tag{B.31}$$

where M_4 is a positive real number. Combining the above inequalities, we have:

$$|\tilde{\zeta}| \leq M_4 K_f \|\eta_{f0}\|_2 e^{-\bar{a}_f \frac{t}{\epsilon}} + M_4 \gamma_{\eta_f} \epsilon. \tag{B.32}$$

Since M_4, K_f, $\|\eta_{f0}\|_2, \gamma_{\eta_f}$ are some finite numbers of order one, and the right-hand side of the above inequality is a continuous function of ϵ which vanishes as $\epsilon \to 0$, we have that there exists an ϵ^* such that if $\epsilon \in (0, \epsilon^*]$, then $M_4 K_f \|\eta_{f0}\|_2 e^{-\bar{a}_f \frac{t}{\epsilon}} + M_4 \gamma_{\eta_f} \epsilon \leq d, \forall\, t > 0$. Thus, $|y^i(t) - y_s^i(t)| = |\tilde{\zeta}_1^i(t)| \leq |\tilde{\zeta}_1(t)| \leq |\tilde{\zeta}(t)| \leq d, \forall\, t > 0$. □

Proof of Theorem 3.7. The detailed presentation of the proof of the theorem is too lengthy and will be omitted for brevity. Instead, we will provide a brief outline of the proof. Initially, it can be shown, following steps analogous to those in the proof of theorem 3.3, that the state of the closed-loop reduced system of Eq. 3.39 is bounded and that there exist positive real numbers K_e, a_e, γ_e such that the following estimate holds for the i-th output error for all $t \geq 0$:

$$|y^i - v^i| \leq K_e |(y^i - v^i)_0| e^{-a_e t} + \gamma_e \phi \tag{B.33}$$

where $(y^i - v^i)_0$ is the output error at time $t = 0$. Picking $\phi^* = \frac{d}{2\gamma_e}$, we have that for $\phi \in (0, \phi^*]$:

$$|y^i - v^i| \leq K_e |(y^i - v^i)_0| e^{-a_e t} + \frac{d}{2}. \tag{B.34}$$

Now, referring to system of Eq. 3.38 we have shown that it satisfies the assumption of theorem 3.4 (the state of the uncertain closed-loop reduced system is bounded and the boundary layer is exponentially stable) and thus, the result of this theorem can be applied. This means that for each positive real number d, there exist positive real numbers $(\bar{K}_e, \bar{a}_e, \bar{\gamma}_e, \tilde{\delta})$, such that if $\phi \in (0, \phi^*]$ there exists an $\epsilon^*(\phi)$ such that, if $\max\{\|x_0\|_2, \|\eta_{f0}\|_2, \|\theta\|, \|\dot{\theta}\|\} \leq \tilde{\delta}$, $\phi \in (0, \phi^*]$ and $\epsilon \in (0, \epsilon^*(\phi)]$, the state of the closed-loop system of Eq. 3.38 is bounded and the following estimate holds for the output error for all $t \geq 0$:

$$|y^i - v^i| \leq \bar{K}_e |(y^i - v^i)_0| e^{-\bar{a}_e t} + \gamma_e \phi + \frac{d}{2}. \tag{B.35}$$

Taking the limit as $t \to \infty$ of the above inequality, we have that if $\max\{\|x_0\|_2, \|\eta_{f0}\|_2, \|\theta\|, \|\dot{\theta}\|\} \leq \tilde{\delta}$, $\phi \in (0, \phi^*]$ and $\epsilon \in (0, \epsilon^*(\phi)]$, then:

$$\lim_{t \to \infty} |y^i - v^i| \leq \gamma_e \phi + \frac{d}{2} \leq d. \tag{B.36}$$

Appendix C

Proofs of Chapter 4

Proof of Proposition 4.1. The proof of the proposition will be obtained in two steps. In the first step, we will show that the system of Eq. 4.34 is exponentially stable, provided that the initial conditions and ϵ are sufficiently small. In the second step, we will use the exponential stability property to prove closeness of solutions (Eq. 4.39).

Exponential stability: First, the system of Eq. 4.34 can be equivalently written as:

$$\begin{aligned} \frac{dx_s}{dt} &= \mathcal{A}_s x_s + f_s(x_s, 0) + [f_s(x_s, x_f) - f_s(x_s, 0)] \\ \epsilon \frac{\partial x_f}{\partial t} &= \mathcal{A}_{f\epsilon} x_f + \epsilon f_f(x_s, x_f). \end{aligned} \tag{C.1}$$

Let μ_1^*, μ_2^* with $\mu_1^* \geq a_4$ be two positive real numbers such that if $|x_s| \leq \mu_1^*$ and $\|x_f\|_2 \leq \mu_2^*$, then there exist positive real numbers (k_1, k_2, k_3) such that

$$\begin{aligned} |f_s(x_s, x_f) - f_s(x_s, 0)| &\leq k_1 \|x_f\|_2 \\ \|f_f(x_s, x_f)\|_2 &\leq k_2 |x_s| + k_3 \|x_f\|_2. \end{aligned} \tag{C.2}$$

Pick $\mu_1 < a_4 < \mu_1^*$ and $\mu_2 < \mu_2^*$. From assumption 4.3 and the converse Lyapunov theorem for finite-dimensional systems [92], we have that there exists a smooth Lyapunov function $V : \mathcal{H}_s \to \mathbb{R}_{\geq 0}$ and a set of positive real numbers $(a_1, a_2, a_3, a_4, a_5)$, such that for all $x_s \in \mathcal{H}_s$ that satisfy $|x| \leq a_4$, the following conditions hold:

$$\begin{aligned} a_1 |x_s|^2 \leq V(x_s) &\leq a_2 |x_s|^2 \\ \dot{V}(x_s) = \frac{\partial V}{\partial x_s} [\mathcal{A}_s x_s + f_s(x_s, 0)] &\leq -a_3 |x_s|^2 \\ \left| \frac{\partial V}{\partial x_s} \right| &\leq a_5 |x_s|. \end{aligned} \tag{C.3}$$

From the global exponential stability property of the fast subsystem of Eq. D.20 (assumption 4.2), and the converse Lyapunov theorem for infinite-dimensional systems [141], we have that there exists a Lyapunov functional $W : \mathcal{H}_f \to \mathbb{R}_{\geq 0}$ and a set of positive real numbers (b_1, b_2, b_3, b_4) such that for all $x_f \in \mathcal{H}_f$ the following conditions hold:

$$\begin{aligned} b_1\|x_f\|_2^2 \le W(x_f) \le b_2\|x_f\|_2^2 \\ \dot{W}(x_f) = \frac{1}{\epsilon}\frac{\partial W}{\partial x_f}\mathcal{A}_{f\epsilon}x_f \le -\frac{b_3}{\epsilon}\|x_f\|_2^2 \\ \left\|\frac{\partial W}{\partial x_f}\right\|_2 \le b_4\|x_f\|_2. \end{aligned} \tag{C.4}$$

Consider now the smooth function $L : \mathcal{H}_s \times \mathcal{H}_f \to \mathbb{R}_{\ge 0}$:

$$L(x_s, x_f) = V(x_s) + W(x_f) \tag{C.5}$$

as Lyapunov function candidate for the system of Eq. C.1. From Eq. C.3 and Eq. C.4, we have that $L(x_s, x_f)$ is positive definite and proper (tends to $+\infty$ as $|x_s| \to \infty$, or $\|x_f\|_2 \to \infty$), with respect to its arguments. Computing the time derivative of L along the trajectories of this system, and using the bounds of Eq. C.3 and Eq. C.4 and the estimates of Eq. C.2, the following expressions can be easily obtained:

$$\begin{aligned} \dot{L}(x_s, x_f) &= \frac{\partial V}{\partial x_s}\dot{x} + \frac{\partial W}{\partial x_f}\dot{x}_f \\ &\le \frac{\partial V}{\partial x_s}[\mathcal{A}_s x_s + f_s(x_s, 0)] + \frac{\partial V}{\partial x_s}[f_s(x_s, x_f) - f_s(x_s, 0)] \\ &\quad + \frac{\partial W}{\partial x_f}\mathcal{A}_f^{\epsilon} x_f + \frac{\partial W}{\partial x_f} f_f(x_s, x_f) \\ &\le -a_3|x_s|^2 + a_5 k_1 |x_s|\|x_f\|_2 - \frac{b_3}{\epsilon}\|x_f\|_2^2 + b_4\|x_f\|_2(k_2|x_s| + k_3\|x_f\|_2) \\ &\le -a_3|x_s|^2 + (a_5k_1 + b_4k_2)|x_s|\|x_f\|_2 - \left(\frac{b_3}{\epsilon} - b_4k_3\right)\|x_f\|_2 \\ &\le -\begin{bmatrix} |x_s| & \|x_f\|_2 \end{bmatrix}\begin{bmatrix} a_3 & -\frac{a_5k_1+b_4k_2}{2} \\ -\frac{a_5k_1+b_4k_2}{2} & \frac{b_3}{\epsilon} - b_4k_3 \end{bmatrix}\begin{bmatrix} |x_s| \\ \|x_f\|_2 \end{bmatrix}. \end{aligned} \tag{C.6}$$

Defining

$$\epsilon_1 = \frac{a_3 b_3}{a_3 b_4 k_3 + \left(\frac{a_5k_1+b_4k_2}{2}\right)^2},$$

we have that if $\epsilon \in (0, \epsilon_1)$, then $\dot{L}(x_s, x_f) < 0$, which from the properties of L directly implies that the state of the system of Eq. C.1 is exponentially stable, that is, there exists a positive real number σ such that:

$$\begin{bmatrix} |x_s| \\ \|x_f\|_2 \end{bmatrix} \le e^{-\sigma t}\begin{bmatrix} \mu_1 \\ \mu_2 \end{bmatrix}. \tag{C.7}$$

Closeness of solutions: First, we define the error coordinate $e_f(\tau) = x_f(\tau) - \bar{x}_f(\tau)$. Differentiating $e_f(\tau)$ with respect to τ, the following dynamical system can be obtained:

$$\frac{\partial e_f}{\partial \tau} = \mathcal{A}_{f\epsilon} e_f + \epsilon f_f(x_s, e_f + \bar{x}_f). \tag{C.8}$$

Referring to the above system with $\epsilon = 0$, we have from the properties of the unbounded operator $\mathcal{A}_{f\epsilon}$ (assumption 4.2) and the converse theorem of [141], that there exists a Lyapunov functional $\bar{W} : \mathcal{H}_f \to \mathbb{R}_{\geq 0}$ and a set of positive real numbers $(\bar{b}_1, \bar{b}_2, \bar{b}_3, \bar{b}_4)$ such that for all $e_f \in \mathcal{H}_f$ the following conditions hold:

$$\begin{aligned} \bar{b}_1 \| e_f(\tau) \|_2^2 \leq \bar{W}(e_f(\tau)) \leq \bar{b}_2 \| e_f(\tau) \|_2^2 \\ \frac{\partial \bar{W}}{\partial \tau} = \frac{\partial \bar{W}}{\partial e_f} \mathcal{A}_f^\epsilon e_f \leq -\bar{b}_3 \| e_f(\tau) \|_2^2 \\ \left\| \frac{\partial \bar{W}}{\partial e_f} \right\|_2 \leq \bar{b}_4 \| e_f(\tau) \|_2. \end{aligned} \tag{C.9}$$

Computing the time derivative of $\bar{W}(e_f)$ along the trajectories of the system of Eq. C.8 and using that $\| f_f(x_s, e_f + \bar{x}_f) \|_2 \leq k_4 \| e_f \|_2 + k_5$, where k_4, k_5 are positive real numbers (which follows from the fact that the states $(x_s, \bar{x}_f)$ are bounded), we have:

$$\begin{aligned} \frac{\partial \bar{W}}{\partial \tau} &\leq -\bar{b}_3 \| e_f \|_2^2 + \epsilon \bar{b}_4 \| e_f \|_2 (k_4 \| e_f \|_2 + k_5) \\ &\leq -(\bar{b}_3 - \epsilon \bar{b}_4 k_4) \| e_f \|_2^2 + \epsilon \bar{b}_4 k_5 \| e_f \|_2. \end{aligned} \tag{C.10}$$

Set $\epsilon_2 = \bar{b}_3 / \bar{b}_4 k_4$ and $\epsilon^* = \min\{\epsilon_1, \epsilon_2\}$. From the above inequality, using the converse Lyapunov theorem [86], we have that if $\epsilon \in (0, \epsilon^*)$, the following bound holds for $\| e_f(\tau) \|_2$ for all $t \in [0, \infty)$:

$$\| e_f(\tau) \|_2 \leq K_3 \| e_f(0) \|_2 e^{-a_3 \frac{t}{\epsilon}} + \epsilon \delta_{e_f} \tag{C.11}$$

where δ_{e_f} is a positive real number. From the above inequality and the fact that $\| e_f(0) \|_2 = 0$, the estimate $x_f(t) = \bar{x}_f(\frac{t}{\epsilon}) + O(\epsilon)$ follows directly.

Defining the error coordinate $e_s(t) = x_s(t) - \bar{x}_s(t)$ and differentiating $e_s(t)$ with respect to time, the following system can be obtained:

$$\frac{de_s}{dt} = \mathcal{A}_s e_s + f_s(\bar{x}_s + e_s, x_f) - f_s(\bar{x}_s). \tag{C.12}$$

The representation of the system of Eq. C.12 in the fast time scale τ takes the form:

$$\frac{de_s}{d\tau} = \epsilon [\mathcal{A}_s e_s + f_s(\bar{x}_s + e_s, x_f) - f_s(\bar{x}_s)] \tag{C.13}$$

where $(e_s, \bar{x}_s)$ can be considered approximately constant, and thus using that $|e_s(0)| = 0$ and continuity of solutions for $e_s(\tau)$, the following bound can be written for $e_s(\tau)$ for all $\tau \in [0, \tau_b]$:

$$|e_s(\tau)| \leq \epsilon k_6 \tau_b, \quad \forall \tau \in [0, \tau_b) \tag{C.14}$$

where k_6 is a positive real number and $\tau_b = \frac{t_b}{\epsilon} = O(1)$, with $t_b = O(\epsilon) > 0$ is the time required for $x_f(t)$ to approach $\bar{x}(t)$, that is, $\|x_f(t)\|_2 \leq k_7\epsilon$ for $t \in [t_b, \infty)$, where k_7 is a positive real number. The system of Eq. C.12 with $\bar{x}_s(t) = x_f(t) \equiv 0$ is exponentially stable (assumption 4.3). Moreover, since $\bar{x}_s(t)$ decays exponentially, the system of Eq. C.12 is also exponentially stable if $x_f(t) \equiv 0$. This implies that for the system:

$$\frac{de_s}{dt} = \mathcal{A}_s e_s + f_s(\bar{x}_s + e_s, 0) - f_s(\bar{x}_s) \tag{C.15}$$

there exists a smooth Lyapunov function $\bar{V} : \mathcal{H}_s \to \mathbb{R}_{\geq 0}$ and a set of positive real numbers $(\bar{a}_1, \bar{a}_2, \bar{a}_3, \bar{a}_4, \bar{a}_5)$, such that for all $e_s \in \mathcal{H}_s$ that satisfy $|e_s| \leq \bar{a}_4$ the following conditions hold:

$$\begin{gathered} \bar{a}_1|e_s|^2 \leq \bar{V}(e_s) \leq \bar{a}_2|e_s|^2 \\ \dot{\bar{V}}(e_s) = \frac{\partial \bar{V}}{\partial e_s}[\mathcal{A}_s e_s + f_s(\bar{x}_s + e_s, 0) - f_s(\tilde{x}_s)] \leq -\bar{a}_3|e_s|^2 \\ \left|\frac{\partial \bar{V}}{\partial e_s}\right| \leq \bar{a}_5|e_s|. \end{gathered} \tag{C.16}$$

Computing the time derivative of $\bar{V}(e_s)$ along the trajectories of the system of Eq. C.12 and using that for $t \in [t_b, \infty)$ $|f_s(\bar{x}_s + e_s, x_f) - f_s(\bar{x}_s + e_s, 0)| \leq k_8\|x_f\|_2 \leq k_7k_8\epsilon$, where k_8 is a positive real number (which follows from the fact that the states $(\bar{x}_s, x_f)$ are bounded), we have for all $t \in [t_b, \infty)$:

$$\dot{\bar{V}}(e_s) \leq -\bar{a}_3|e_s|^2 + k_7k_8\bar{a}_5|e_s|\epsilon. \tag{C.17}$$

From the above inequality, using the converse Lyapunov theorem [92] and that $|e_s(t_b)| = O(\epsilon)$, we have that the following bound holds for $|e_s(t)|$ for all $t \in [t_b, \infty)$:

$$|e_s(t)| \leq \epsilon\delta_{e_s} \tag{C.18}$$

where δ_{e_s} is a positive real number. From inequalities of Eqs. C.14–C.18, the estimate $x_s(t) = \bar{x}_s(t) + O(\epsilon)$ for all $t \geq 0$ follows directly. □

Proof of proposition 4.2. The proof of the proposition will be obtained following a two-step approach similar to the one used in the proof of proposition 4.1.

Exponential stability: This part of the proof of the proposition is completely analogous to the proof of exponential stability in the case of proposition 4.1, and thus, it will be omitted for brevity.

Closeness of solutions: From the first part of the proof, $\epsilon \in (0, \bar{\epsilon}^*)$. Defining the error coordinate $\tilde{e}_s(t) = x_s(t) - \tilde{x}_s(t)$ and differentiating $\tilde{e}_s(t)$ with respect to time, the following system can be obtained:

$$\frac{d\tilde{e}_s}{dt} = \mathcal{A}_s\tilde{e}_s + f_s(\tilde{x}_s + \tilde{e}_s, x_f) - f_s(\tilde{x}_s, \tilde{x}_f) \tag{C.19}$$

where $\tilde{x}_f = \Sigma^0(\tilde{x}_s) + \epsilon\Sigma^1(\tilde{x}_s) + \epsilon^2\Sigma^2(\tilde{x}_s) + \cdots + \epsilon^k\Sigma^k(\tilde{x}_s)$. From assumption 4.3 and the fact that $\tilde{x}_s(t)$ decays exponentially to zero, we have that the system:

$$\frac{d\tilde{e}_s}{dt} = \mathcal{A}_s\tilde{e}_s + f_s(\tilde{x}_s + \tilde{e}_s, \tilde{x}_f) - f_s(\tilde{x}_s, \tilde{x}_f) \tag{C.20}$$

is exponentially stable, which implies that there exists a smooth Lyapunov function $\tilde{V} : \mathcal{H}_s \to \mathbb{R}_{\geq 0}$ and a set of positive real numbers $(\tilde{a}_1, \tilde{a}_2, \tilde{a}_3, \tilde{a}_4, \tilde{a}_5)$ such that for all $\tilde{e}_s \in \mathcal{H}_s$ that satisfy $|\tilde{e}_s| \leq \tilde{a}_4$, the following conditions hold:

$$\tilde{a}_1|\tilde{e}_s|^2 \leq \tilde{V}(\tilde{e}_s) \leq \tilde{a}_2|\tilde{e}_s|^2$$

$$\dot{\tilde{V}}(\tilde{e}_s) = \frac{\partial\tilde{V}}{\partial\tilde{e}_s}[\mathcal{A}_s\tilde{e}_s + f_s(\tilde{x}_s + \tilde{e}_s, \tilde{x}_f) - f_s(\tilde{x}_s, \tilde{x}_f)] \leq -\tilde{a}_3|\tilde{e}_s|^2 \tag{C.21}$$

$$\left|\frac{\partial\tilde{V}}{\partial\tilde{e}_s}\right| \leq \tilde{a}_5|\tilde{e}_s|.$$

Computing the time derivative of $\tilde{V}(\tilde{e}_s)$ along the trajectories of the system of Eq. C.19 and using that for $t \in [0,\infty)$ $|f_s(\tilde{x}_s + \tilde{e}_s, x_f) - f_s(\tilde{x}_s + \tilde{e}_s, \tilde{x}_f)| \leq (\tilde{k}_1\epsilon^{k+1} + \tilde{k}_2e^{\frac{\lambda_1 t}{\epsilon}})$, where $\tilde{k}_1, \tilde{k}_2$ are positive real numbers, we have for all $t \in [0,\infty)$:

$$\dot{\tilde{V}}(\tilde{e}_s) \leq -\tilde{a}_3|\tilde{e}_s|^2 + \left(\tilde{k}_1\epsilon^{k+1} + \tilde{k}_2e^{\frac{\lambda_1 t}{\epsilon}}\right)\tilde{a}_5|\tilde{e}_s|. \tag{C.22}$$

From the above inequality, using the converse Lyapunov theorem [92] and the fact that $|\tilde{e}_s(0)| = 0$, we have that the following bound holds for $|\tilde{e}_s(t)|$ for all $t \in [0,\infty)$:

$$|\tilde{e}_s(t)| \leq \bar{K}e^{\frac{\lambda_1 t}{\epsilon}} + \tilde{K}\epsilon^{k+1} \tag{C.23}$$

where $\bar{K}, \tilde{K}$ are positive real numbers. Since the term $\bar{K}e^{\frac{\lambda_1 t}{\epsilon}}$ vanishes outside the interval $[0, t_b]$ (where t_b is the time required for x_f to approach $\tilde{x}_f$), it follows from Eq. C.23 that for all $t \in [t_b, \infty)$:

$$|\tilde{e}_s(t)| \leq \tilde{K}\epsilon^{k+1}. \tag{C.24}$$

From the above inequality, the estimate $x_s(t) = \tilde{x}_s(t) + O(\epsilon^{k+1})$, for $t \geq t_b$, follows directly. □

Proof of theorem 4.1. Substituting the output feedback controller of Eq. 4.51 into the system of Eq. D.19, we get:

$$\begin{aligned}
\frac{d\eta}{dt} &= \mathcal{A}_s\eta + f_s(\eta, \epsilon\Sigma^1(\eta,u) + \epsilon^2\Sigma^2(\eta,u) + \cdots + \epsilon^k\Sigma^k(\eta,u)) + \mathcal{B}_s\big(p_0(\eta) \\
&\quad + Q_0(x_s)v + \epsilon[p_1(\eta) + Q_1(\eta)v] + \cdots + \epsilon^k[p_k(\eta) + Q_k(\eta)v]\big) \\
&\quad + L\big(q - [\mathcal{Q}\eta + \mathcal{Q}\epsilon\Sigma^1(\eta,u) + \epsilon^2\Sigma^2(\eta,u) + \cdots + \epsilon^k\Sigma^k(\eta,u)]\big) \\
\frac{dx_s}{dt} &= \mathcal{A}_s x_s + \mathcal{B}_s\big(p_0(\eta) + Q_0(x_s)v + \epsilon[p_1(\eta) + Q_1(\eta)v] \\
&\quad + \cdots + \epsilon^k[p_k(\eta) + Q_k(\eta)v]\big) + f_s(x_s, x_f)
\end{aligned} \tag{C.25}$$

$$\epsilon \frac{\partial x_f}{\partial t} = \mathcal{A}_{f\epsilon} x_f + \epsilon \mathcal{B}_f \big(p_0(\eta) + Q_0(x_s) v + \epsilon [p_1(\eta) + Q_1(\eta) v + \cdots + \epsilon^k [p_k(\eta) + Q_k(\eta) v] \big) + \epsilon f_f(x_s, x_f)$$

$$y^i = \mathcal{C}^i x_s + \mathcal{C}^i x_f, \quad i = 1, \ldots, l.$$

Performing a two-time-scale decomposition on the above system, the fast subsystem takes the form:

$$\frac{\partial x_f}{\partial \tau} = \mathcal{A}_{f\epsilon} x_f \tag{C.26}$$

which is exponentially stable. Furthermore, the $O(\epsilon^{k+1})$ approximation of the closed-loop inertial form is given by:

$$\begin{aligned}
\frac{d\eta}{dt} &= \mathcal{A}_s \eta + f_s(\eta, \epsilon \Sigma^1(\eta, u) + \epsilon^2 \Sigma^2(\eta, u) + \cdots + \epsilon^k \Sigma^k(\eta, u)) + \mathcal{B}_s \big(p_0(\eta) \\
&\quad + Q_0(x_s) v + \epsilon [p_1(\eta) + Q_1(\eta) v] + \cdots + \epsilon^k [p_k(\eta) + Q_k(\eta) v] \big) \\
&\quad + L(q - [\mathcal{Q}\eta + \mathcal{Q}\Sigma^0(\eta, u) + \epsilon \Sigma^1(\eta, u) + \epsilon^2 \Sigma^2(\eta, u) \\
&\quad + \cdots + \epsilon^k \Sigma^k(\eta, u)]) \qquad (C.27) \\
\frac{dx_s}{dt} &= \mathcal{A}_s x_s + f_s(x_s, \epsilon \Sigma^1(x_s, u) + \epsilon^2 \Sigma^2(x_s, u) + \cdots + \epsilon^k \Sigma^k(x_s, u)) \\
&\quad + \mathcal{B}_s \big(p_0(\eta) + Q_0(x_s) v + \epsilon [p_1(\eta) + Q_1(\eta) v] \\
&\quad + \cdots + \epsilon^k [p_k(\eta) + Q_k(\eta) v] \big) \\
y_s^i &= \mathcal{C}^i x_s + \mathcal{C}^i [\epsilon \Sigma^1(x_s, u) + \epsilon^2 \Sigma^2(x_s, u) + \cdots + \epsilon^k \Sigma^k(x_s, u)], \\
&\qquad i = 1, \ldots, l.
\end{aligned}$$

Referring to the above closed-loop ODE system, assumption 4.4 yields that it is exponentially stable and the output y_s^i, $i = 1, \ldots, l$, changes in a pre-specified manner. A direct application of the result of proposition 4.2 yields that there exist constants $\tilde{\mu}_1, \tilde{\mu}_2, \tilde{\epsilon}^*$ such that if $|x_s(0)| \leq \tilde{\mu}_1$, $\|x_f(0)\|_2 \leq \tilde{\mu}_2$ and $\epsilon \in (0, \tilde{\epsilon}^*]$, such that the closed-loop infinite-dimensional system is exponentially stable and the relation of Eq. 4.52 holds. □

Appendix D

Proofs of Chapter 5

Definitions

- A function $\gamma : \mathbb{R}_{\geq 0} \rightarrow \mathbb{R}_{\geq 0}$ is said to be of class K if it is continuous, increasing, and is zero at zero. It is of class K_∞, if in addition, $\gamma(s)$ tends to $+\infty$ as s tends to $+\infty$.
- A function $\beta : \mathbb{R}_{\geq 0} \times \mathbb{R}_{\geq 0} \rightarrow \mathbb{R}_{\geq 0}$ is said to be of class KL if, for each fixed t, the function $\beta(\cdot, t)$ is of class K and, for each fixed s, the function $\beta(s, \cdot)$ is nonincreasing and tends to zero at infinity.

Definition [92] *The system in Eq. 5.22 (with $u \equiv 0$) is said to be locally input-to-state stable (ISS) with respect to θ if there exist a function β of class KL, a function γ of class K and a positive real number $\hat{\delta}$ such that for each $x_{s_o} \in \mathbb{R}^n$ and for each measurable, essentially bounded input $\theta(\cdot)$ on $[0, \infty)$ that satisfy $\max\{x_{s_o}, \|\theta\|\} \leq \hat{\delta}$, the solution of Eq. 5.22 with $x_s(0) = x_{s_o}$ exists for each $t \geq 0$ and satisfies*

$$|x_s(t)| \leq \beta(|x_s(0)|, t) + \gamma(\|\theta\|), \quad \forall\, t \geq 0. \tag{D.1}$$

Proof of Theorem 5.1. Under the control law of Eq. 5.28 the closed-loop system takes the form:

$$\begin{aligned}
\frac{dx_s}{dt} &= \mathcal{A}_s x_s + \mathcal{B}_s a_0(x_s, \bar{v}, t) + f_s(x_s, 0) + \mathcal{W}_s(x_s, 0, \theta) + \mathcal{R}_s(x_s, x_f, \theta) \\
\epsilon \frac{\partial x_f}{\partial t} &= \mathcal{A}_{f\epsilon} x_f + \epsilon \mathcal{B}_f a_0(x_s, \bar{v}, t) + \epsilon f_f(x_s, 0) + \epsilon \mathcal{W}_f(x_s, 0, \theta) \\
&\quad + \epsilon \mathcal{R}_f(x_s, x_f, \theta) \\
y &= \mathcal{C} x_s
\end{aligned} \tag{D.2}$$

where $\mathcal{R}_s(x_s, x_f, \theta) = f_s(x_s, x_f) + \mathcal{W}_s(x_s, x_f, \theta) - f_s(x_s, 0) - \mathcal{W}_s(x_s, 0, \theta)$, $\mathcal{R}_f(x_s, x_f, \theta) = f_f(x_s, x_f) + \mathcal{W}_f(x_s, x_f, \theta) - f_f(x_s, 0) - \mathcal{W}_f(x_s, 0, \theta)$ with $\mathcal{R}_s(x_s, 0, \theta) = \mathcal{R}_f(x_s, 0, \theta) = 0$, and

$$\begin{aligned}
a(x_s, \bar{v}, t) := [\mathcal{C}_0(x_s)]^{-1} &\left\{ \sum_{i=1}^{l} \sum_{k=1}^{r_i} \frac{\beta_{ik}}{\beta_{ir_i}} (v_i^{(k)} - L_{F_0}^k h_0^i(x_s)) \right. \\
&+ \sum_{i=1}^{l} \sum_{k=1}^{r_i} \frac{\beta_{ik}}{\beta_{ir_i}} (v_i^{(k-1)} - L_{F_0}^{k-1} h_0^i(x_s)) - \chi[c_0(x_s, t)]
\end{aligned}$$

$$\times \frac{\sum_{i=1}^{l}\sum_{k=1}^{r_i}\frac{\beta_{ik}}{\beta_{ir_i}}\left(L_{F_0}^{k-1}h_0^i(x_s)-v_i^{(k-1)}\right)}{\left|\sum_{i=1}^{l}\sum_{k=1}^{r_i}\frac{\beta_{ik}}{\beta_{ir_i}}\left(L_{F_0}^{k-1}h_0^i(x_s)-v_i^{(k-1)}\right)\right|+\phi}\Bigg\}. \tag{D.3}$$

On the basis of the system of Eq. D.2, it is clear that since the operator $\mathcal{A}_{f\epsilon}$ generates an exponentially stable semigroup, the infinite-dimensional closed-loop fast subsystem is exponentially stable.

Observing the similarity in the structure of the system of Eq. D.2 and the x_s-subsystem of Eq. 5.22, using Assumption 5.1, and introducing the variables $e_k^{(i)} = \zeta_k^{(i)} - v_i^{(k-1)}$, $i = 1, \ldots, l$, $k = 1, \ldots, r_i$, $\tilde{e}_{r_i}^{(i)} = \sum_{i=1}^{l} e_{r_i}^{(i)} + \sum_{i=1}^{l}\sum_{k=1}^{r_i-1}\frac{\beta_{ik}}{\beta_{ir_i}} e_k^{(i)}$, and the notation $\bar{e}^{(i)} = [e_1^{(i)}\ e_2^{(i)}\ \cdots\ e_{r_i-1}^{(i)}]^T$, $\bar{e} = [e^{(1)^T}\ e^{(2)^T}\ \cdots\ e^{(l)^T}]^T$, $\bar{\eta} = [\bar{e}^T\ \eta^T]$, $\tilde{e}_r = [\tilde{e}_{r_i}^{(1)}\ \tilde{e}_{r_i}^{(2)}\ \cdots\ \tilde{e}_{r_i}^{(l)}]^T$, $v_i = [v_i^{(0)}\ v_i^{(1)}\ \cdots\ v_i^{(r_i-1)}]^T$, $\tilde{v} = [v_1^T\ v_2^T\ \cdots\ v_l^T]^T$, the representation of the closed-loop system of Eq. D.2 is:

$$\begin{aligned}
\dot{e}_1^{(1)} &= e_2^{(1)} + \bar{\Psi}_1^{(1)}(\bar{e}, \tilde{e}_r, \tilde{v}, \eta, \theta, x_f)\\
&\ \vdots\\
\dot{e}_{r_1-1}^{(1)} &= -\sum_{i=1}^{l}\sum_{k=1}^{r_i-1}\frac{\beta_{ik}^1}{\beta_{ir_i}^1} e_k^{(i)} + \sum_{i=1}^{l}\tilde{e}_{r_1}^{(1)} + \bar{\Psi}_{r_1}^{(1)}(\bar{e}, \tilde{e}_r, \tilde{v}, \eta, \theta, x_f)\\
\dot{\tilde{e}}_{r_1}^{(1)} &= \bar{\Psi}_{r_1}^{(1)}(\bar{e}, \tilde{e}_r, \tilde{v}, \eta, \theta, x_f) + L_{\mathcal{W}_0} L_{F_0}^{r_1-1} h_0^1(T^{-1}(\bar{e}, \tilde{e}_r, \tilde{v}, \eta, \theta))\\
&\quad + \sum_{i=1}^{l} L_{\mathcal{B}_0^i} L_{F_0}^{r_1-1} h_0^1(T^{-1}(\bar{e}, \tilde{e}_r, \tilde{v}, \eta, \theta)) a^i(\bar{e}, \tilde{e}_r, \tilde{v}, \bar{v}, \eta, t)\\
&\ \vdots\\
\dot{e}_1^{(l)} &= e_2^{(l)} + \bar{\Psi}_1^{(l)}(\bar{e}, \tilde{e}_r, \tilde{v}, \eta, \theta)\\
&\ \vdots\\
\dot{e}_{r_l-1}^{(l)} &= e_{r_l}^{(l)} + \bar{\Psi}_{r_l-1}^{(l)}(\bar{e}, \tilde{e}_r, \tilde{v}, \eta, \theta, x_f)\\
\dot{e}_{r_l}^{(l)} &= \bar{\Psi}_{r_l}^{(l)}(\bar{e}, \tilde{e}_r, \tilde{v}, \eta, \theta, x_f) + L_{\mathcal{W}_0} L_{F_0}^{r_l-1} h_0^l(T^{-1}(\bar{e}, \tilde{e}_r, \tilde{v}, \eta, \theta))\\
&\quad + \sum_{i=1}^{l} L_{\mathcal{B}_0^i} L_{F_0}^{r_l-1} h_0^l(T^{-1}(\bar{e}, \tilde{e}_r, \tilde{v}, \zeta, \theta)) a^i(\bar{e}, \tilde{e}_r, \tilde{v}, \eta, \bar{v}, t)\\
\dot{\eta}_1 &= \Psi_1(\bar{e}, \tilde{e}_r, \tilde{v}, \eta, \theta, \dot{\theta}) + \bar{\Psi}_{\sum_i r_i+1}(\bar{e}, \tilde{e}_r, \tilde{v}, \eta, \theta, x_f)\\
&\ \vdots\\
\dot{\eta}_{m-\sum_i r_i} &= \Psi_{m-\sum_i r_i}(\bar{e}, \tilde{e}_r, \tilde{v}, \eta, \theta, \dot{\theta}) + \bar{\Psi}_m(\bar{e}, \tilde{e}_r, \tilde{v}, \eta, \theta, x_f)\\
\epsilon\frac{\partial x_f}{\partial t} &= \mathcal{A}_{f\epsilon} x_f + \epsilon\mathcal{B}_f a_0(\bar{e}, \tilde{e}_r, \tilde{v}, \eta, \bar{v}, t) + \epsilon f_f(x_s, 0)\\
&\quad + \epsilon\mathcal{W}_f(\bar{e}, \tilde{e}_r, \tilde{v}, \eta, 0, \theta) + \epsilon\mathcal{R}_f(\bar{e}, \tilde{e}_r, \tilde{v}, \eta, x_f, \theta)\\
y^i &= \zeta_1^{(i)}
\end{aligned} \tag{D.4}$$

where $\bar{\Psi}_k^{(i)}$, $i = 1, \ldots, l$, $k = 1, \ldots, r_i$, and $\bar{\Psi}_{\sum_i r_i+1}, \ldots, \bar{\Psi}_m$, are Lipschitz functions of their arguments. In what follows, we derive bounds that capture the evolution of the states $\bar{e}$, $\tilde{e}_r$, $\bar{\eta}$ of the above system. We initially obtain these bounds, when $\epsilon = 0$, and then we show that these bounds continue to hold up to an arbitrarily small offset for $\epsilon > 0$.

First, note that the linear structure of $\bar{e}$ subsystem of the reduced system of Eq. D.4 and the fact that it is exponentially stable when $\tilde{e}_r \equiv 0$ allows using a direct Lyapunov function argument to show that there exist positive real numbers, k_1, a, $\gamma_{\tilde{e}_r}$ such that the following ISS bound holds for the state $\bar{e}$ of the slow subsystem:

$$|\bar{e}(t)| \leq k_1 e^{-at} |\bar{e}(0)| + \gamma_{\tilde{e}_r} \|\tilde{e}_r\|. \tag{D.5}$$

We will now show that the controller of Eq. 5.28 ensures that the state vector $\tilde{e}_r = [\tilde{e}_{r_1}^{(1)} \ \tilde{e}_{r_2}^{(2)} \ \cdots \ \tilde{e}_{r_m}^{(m)}]^T$ of the system of Eq. D.4 with $\epsilon = 0$ possesses an ISS property with respect to $\bar{e}$, $\bar{v}$, η, θ, and, moreover, the gain function saturates at ϕ. To this end, we consider the ϵ-independent system:

$$\begin{aligned}\dot{\tilde{e}}_r = {} & L_{\mathcal{W}_0} L_{F_0}^{r_i-1} h_{i0}(T^{-1}(\bar{e}, \tilde{e}_r, \tilde{v}, \eta, \theta)) + \sum_{i=1}^{l}\sum_{k=1}^{r_i} \frac{\beta_{ik}}{\beta_{ir_i}} e_{k+1}^{(i)} \\ & + \Bigg\{ -\sum_{i=1}^{l}\sum_{k=1}^{r_i} \frac{\beta_{ik}}{\beta_{ir_i}} e_{k+1}^{(i)} - \tilde{e}_r - \chi[c_0((\bar{e}, \tilde{e}_r, \tilde{v}, \eta, \theta), t)] \\ & \times w(T^{-1}(\bar{e}, \tilde{e}_r, \tilde{v}, \eta, \theta)), \phi) \Bigg\}.\end{aligned} \tag{D.6}$$

To establish that the above system is ISS with respect to $\bar{e}$, $\tilde{e}_r$, $\bar{v}$, η, θ, we use the following smooth function $V : \mathbb{R}^m \to \mathbb{R}_{\geq 0}$:

$$V = \frac{1}{2}\tilde{e}_r^2. \tag{D.7}$$

Calculating the time derivative of V along the trajectory of the system of Eq. D.6, we have:

$$\begin{aligned}\dot{V} = \tilde{e}_r^T \Bigg[& L_{\mathcal{W}_0} L_{F_0}^{r_i-1} h_{i0}(T^{-1}(\bar{e}, \tilde{e}_r, \tilde{v}, \eta, \theta)) + \sum_{i=1}^{l}\sum_{k=1}^{r_i} \frac{\beta_{ik}}{\beta_{ir_i}} e_{k+1}^{(i)} \\ & + \Bigg\{ -\sum_{i=1}^{l}\sum_{k=1}^{r_i} \frac{\beta_{ik}}{\beta_{ir_i}} e_{k+1}^{(i)} - \tilde{e}_r - \chi[c_0((\bar{e}, \tilde{e}_r, \tilde{v}, \eta, \theta), t)] \\ & \times w(T^{-1}(\bar{e}, \tilde{e}_r, \tilde{v}, \eta, \theta)), \phi) \Bigg\} \Bigg].\end{aligned} \tag{D.8}$$

Furthermore, it is straightforward to show that the representation of the vector function $w(x, \phi)$ in terms of the vector $\tilde{e}_r$ is given by:

$$w(\tilde{e}_r, \phi) = \frac{\tilde{e}_r}{|\tilde{e}_r| + \phi}. \tag{D.9}$$

Substituting Eqs. D.9, we have:

$$\begin{aligned}\dot{V} &\leq \tilde{e}_r^T \Bigg\{ -\tilde{e}_r + L_{\mathcal{W}_0} L_{F_0}^{r_i - 1} h_{i0}(T^{-1}(\bar{e}, \tilde{e}_r, \tilde{v}, \eta, \theta)) \\ &\quad - \chi [c_0(T^{-1}(\bar{e}, \tilde{e}_r, \tilde{v}, \eta, \theta), t)] \frac{\tilde{e}_r}{|\tilde{e}_r| + \phi} \Bigg\} \\ &\leq \Bigg\{ -\tilde{e}_r^2 - \chi [c_0(T^{-1}(\bar{e}, \tilde{e}_r, \tilde{v}, \eta, \theta), t)] \frac{\tilde{e}_r^2}{|\tilde{e}_r| + \phi} \\ &\quad + |\tilde{e}_r| c_0(T^{-1}(\bar{e}, \tilde{e}_r, \tilde{v}, \eta, \theta), t) \Bigg\} \\ &\leq \Bigg\{ -\tilde{e}_r^2 - (\chi - 1) c_0(T^{-1}(\bar{e}, \tilde{e}_r, \tilde{v}, \eta, \theta), t) \frac{\tilde{e}_r^2}{|\tilde{e}_r| + \phi} \\ &\quad + \phi [c_0(T^{-1}(\bar{e}, \tilde{e}_r, \tilde{v}, \eta, \theta), t)] \frac{\tilde{e}_r}{|\tilde{e}_r| + \phi} \Bigg\}. \end{aligned} \tag{D.10}$$

From the last inequality, it follows directly that if $|\tilde{e}_r| \geq \phi/(\chi - 1)$, the time derivative of the Lyapunov function satisfies $\dot{V} \leq -\tilde{e}_r^2$. This fact implies that the ultimate bound on the state $\tilde{e}_r$ of the system of Eq. D.6 depends only on the parameter ϕ and is independent of the states $\bar{e}, \eta$.

We will now analyze the time derivative of V for $|\tilde{e}_r| < \phi/(\chi - 1)$. For ease of notation, we set $\mathcal{U} = [\tilde{e}_r \ \bar{\eta}^T \ \theta^T \ \bar{v}^T]^T$. Then, Eq. D.10 can be written as:

$$\begin{aligned}\dot{V} &\leq \left\{ -\tilde{e}_r^2 + (\chi - 1)|\tilde{e}_r| [c_0(T^{-1}(\bar{e}, \tilde{e}_r, \tilde{v}, \eta, \theta), t)] \right\} \\ &\leq -\tilde{e}_r^2 + (\chi - 1)|\tilde{e}_r| \rho(|\mathcal{U}|) \end{aligned} \tag{D.11}$$

where ρ is a class K_∞ function. Summarizing, we have that $\dot{V}$ satisfies the following properties:

$$\dot{V} \leq -\frac{|\tilde{e}_r|^2}{2}, \quad |\tilde{e}_r| \geq \min\{\phi, (2(\chi - 1)\rho(|\mathcal{U}|))\} =: \tilde{\gamma}_{\mathcal{U}}(|\mathcal{U}|). \tag{D.12}$$

From the above inequality, we get [92] that the following ISS bound holds for the state $\tilde{e}_r$ of the system of Eq. D.6:

$$\begin{aligned}|\tilde{e}_r(t)| &\leq e^{-0.5t} |\tilde{e}_r(0)| + \tilde{\gamma}_{\mathcal{U}}(\|\mathcal{U}\|) \\ &\leq e^{-0.5t} |\tilde{e}_r(0)| + \phi. \end{aligned} \tag{D.13}$$

Referring to the singularly perturbed system comprised of the states $(\bar{e}, \eta, x_f)$ of the system of Eq. D.4, we have that its fast dynamics are globally exponentially stable and the slow system ($\bar{\eta} = [\bar{e}^T \ \eta^T]^T$) is locally ISS with respect to $\tilde{e}_r, \theta, \dot{\theta}$ [92]. This implies that there exist a function $\beta_{\bar{\eta}}$ of class KL and functions $\bar{\gamma}_{\tilde{e}_r}, \bar{\gamma}_\theta, \bar{\gamma}_{\dot{\theta}}$ of class K such that the following ISS inequality holds for the state $\bar{\eta}$:

$$|\bar{\eta}(t)| \leq \beta_{\bar{\eta}}(|\bar{\eta}(0)|, t) + \bar{\gamma}_{\tilde{e}_r}(\|\tilde{e}_r\|) + \bar{\gamma}_\theta(\|\theta\|) + \bar{\gamma}_{\dot{\theta}}(\|\dot{\theta}\|). \tag{D.14}$$

We will now utilize the result developed in theorem 3.4 that establishes robustness of the ISS property with respect to infinite-dimensional fast dynamics provided that they are stable and sufficiently fast, to show that the ISS inequalities of Eqs. D.13–D.14 continue to hold up to an arbitrarily small offset, for the states $\tilde{e}_r, \bar{\eta}$ of the singularly perturbed system of Eq. D.4. Following theorem 3.4, it can be shown that there exist positive real numbers $(\bar{\delta}, \bar{d}_0, \epsilon^{\tilde{e}_r}(\phi))$ with $\bar{d}_0 = O(\epsilon)$ such that if $\epsilon \in (0, \epsilon^{\tilde{e}_r}(\phi)]$ and $\max\{|\tilde{e}_{\bar{r}}(0)|, \|x_f(0)\|_2, \|\theta\|, \|\bar{v}\|, \|\bar{\eta}\|\} \leq \bar{d}_0$, then

$$|\tilde{e}_r(t)| \leq e^{-0.5t}|\tilde{e}_r(0)| + \phi + \bar{d}_0. \tag{D.15}$$

Furthermore, it can be also shown using the result of theorem 3.4 that the singularly perturbed system comprised of $\bar{e}, \eta, z$, with the same converse function that exists for the reduced system comprised of the states $\bar{e}, \eta$, and its resulting $(\beta_{\bar{\eta}}, \bar{\gamma}_s)$.

Thus, we have that there exist positive real numbers $(\tilde{\delta}, d_{\bar{\eta}}, \epsilon^{\bar{\eta}}(\phi))$ such that if $\epsilon \in (0, \epsilon^{\bar{\eta}}(\phi)]$ and $\max\{|\bar{\eta}(0)|, \|x_f(0)\|_2, \|\theta\|, \|\dot{\theta}\|, \|\bar{v}\|, \|\tilde{e}_r\|\} \leq \tilde{\delta}$, then

$$|\bar{\eta}(t)| \leq \beta_{\bar{\eta}}(|\bar{\eta}(0)|, t) + \bar{\gamma}_{\tilde{e}_r}(\|\tilde{e}_r\|) + \bar{\gamma}_\theta(\|\theta\|) + \bar{\gamma}_{\dot{\theta}}(\|\dot{\theta}\|) + d_{\bar{\eta}}. \tag{D.16}$$

The proof of the theorem can be completed by: (a) analyzing the behavior of the dynamical system comprised of the states $\tilde{e}_r, \bar{\eta}$ of the system of Eq. D.4, for which the inequalities of Eqs. D.15–D.16 hold, using small-gain theorem type arguments (see also [46, 92] for similar results on finite-dimensional systems) to establish boundedness of the state of the closed-loop system, and (b) combining the inequalities of Eqs. D.5–D.15 and using a claim proved in [46] to show that the output of the closed-loop system of Eq. D.2 satisfies the relation of Eq. 5.29, for each $\phi \in (0, \phi^*]$, $\epsilon^*(\phi) \in (0, \epsilon^{\bar{\eta}}(\phi)]$, and $\max\{|x_s(0)|, \|x_f(0)\|_2, \|\theta\|, \|\dot{\theta}\|, \|\bar{v}\|\} \leq \delta$. □

Proof of Theorem 5.3. Substituting the controller of Eq. 5.28 into the parabolic PDE system of Eq. 5.7, we obtain:

$$\begin{aligned} \dot{x} &= \mathcal{A}x + \mathcal{B}a_0(x_s, x_f, \bar{v}, t) + f(x) + \mathcal{W}(x, \theta), \quad x(0) = x_0 \\ y &= \mathcal{C}x, \quad q = \mathcal{Q}x. \end{aligned} \tag{D.17}$$

One can easily verify that assumption 4.1 holds for the above system, and thus, a direct application of Galerkin's method yields the following infinite dimensional system:

$$\begin{aligned} \frac{dx_s}{dt} &= \mathcal{A}_s x_s + \mathcal{B}_s a_0(x_s, x_f, \bar{v}, t) + f_s(x_s, x_f) + \mathcal{W}_s(x_s, x_f, \theta) \\ \frac{\partial x_f}{\partial t} &= \mathcal{A}_f x_f + \mathcal{B}_f a_0(x_s, x_f, \bar{v}, t) + f_f(x_s, x_f) + \mathcal{W}_f(x_s, x_f, \theta) \\ y &= \mathcal{C}x_s + \mathcal{C}x_f, \quad q = \mathcal{Q}x_s + \mathcal{Q}x_f \\ x_s(0) &= P_s x(0) = P_s x_0, \quad x_f(0) = P_f x(0) = P_f x_0. \end{aligned} \tag{D.18}$$

Using that $\epsilon = \frac{|Re\{\lambda_1\}|}{|Re\{\lambda_{m+1}\}|}$, the system of Eq. D.18 can be written in the following form:

$$\begin{aligned}\frac{dx_s}{dt} &= \mathcal{A}_s x_s + \mathcal{B}_s a_0(x_s, x_f, \bar{v}, t) + f_s(x_s, x_f) + \mathcal{W}_s(x_s, x_f, \theta) \\ \epsilon \frac{\partial x_f}{\partial t} &= \mathcal{A}_{f\epsilon} x_f + \epsilon \mathcal{B}_f a_0(x_s, x_f, \bar{v}, t) + \epsilon f_f(x_s, x_f) + \epsilon \mathcal{W}_f(x_s, x_f, \theta)\end{aligned} \tag{D.19}$$

where $\mathcal{A}_{f\epsilon}$ is an unbounded differential operator defined as $\mathcal{A}_{f\epsilon} = \epsilon \mathcal{A}_f$. Since $\epsilon \ll 1$ and the operators $\mathcal{A}_s, \mathcal{A}_{f\epsilon}$ generate semigroups with growth rates that are of the same order of magnitude, the system of Eq. D.19 is in the standard singularly perturbed form (see [95] for a precise definition of standard form), with x_s being the slow states and x_f being the fast states. Introducing the fast time-scale $\tau = \frac{t}{\epsilon}$ and setting $\epsilon = 0$, we obtain the following infinite-dimensional fast subsystem from the system of Eq. D.19:

$$\frac{\partial x_f}{\partial \tau} = \mathcal{A}_{f\epsilon} x_f. \tag{D.20}$$

From the fact that $Re\{\lambda_{m+1}\} < 0$ and the definition of ϵ, we have that the above system is globally exponentially stable. Setting $\epsilon = 0$ in the system of Eq. D.19, we have that $x_f = 0$ and thus, the finite-dimensional slow system takes the form:

$$\begin{aligned}\frac{dx_s}{dt} &= F_0(x_s) + \sum_{i=1}^{l} \mathcal{B}_0^i a_0^i(x_s, 0, \bar{v}, t) + \mathcal{W}_0(x_s, 0, \theta) \\ y^i &= \mathcal{C}^i x_s =: h_0^i(x_s).\end{aligned} \tag{D.21}$$

For the above system we have shown in the proof of theorem 5.1 that there exists a $\phi \in (0, \phi^*]$ such that if $\max\{|x_s(0)|, \|\theta\|, \|\dot{\theta}\|, \|\bar{v}\|\} \leq \delta$, then its state is bounded and its outputs satisfy $\limsup_{t\to\infty} |y^i - v_i| \leq O(\phi)$, $i = 1, \ldots, l$. Finally, since the infinite-dimensional fast subsystem of Eq. D.20 is exponentially stable, we can use standard singular perturbation arguments to obtain that there exists an $\epsilon^*(\phi)$, such that if $\epsilon \in (0, \epsilon^*(\phi)]$, $\max\{|x_s(0)|, \|x_f(0)\|_2, \|\theta\|, \|\dot{\theta}\|, \|\tilde{v}\|\} \leq \delta$, then the state of the closed-loop parabolic PDE system of Eq. D.17 is bounded and that its outputs satisfy the relation of Eq. 5.29. □

Appendix E

Proofs of Chapter 6

Proof of proposition 6.1. The proof of the proposition will be obtained by following a two-step approach. In the first step, we will show that the system of Eq. 6.11 is exponentially stable, provided that the initial conditions and ϵ are sufficiently small. The exponential stability property will be used in the second step to prove closeness of solutions as given in Eq. 6.28.

Exponential stability: The system of Eq. 6.11 can be equivalently written as:

$$\begin{aligned} \frac{dx_s}{dt} &= \mathcal{A}_s(t)x_s + f_s(t, x_s, 0) + [f_s(t, x_s, x_f) - f_s(t, x_s, 0)] \\ \epsilon\frac{\partial x_f}{\partial t} &= \mathcal{A}_{f\epsilon}(t)x_f + \epsilon f_f(t, x_s, x_f). \end{aligned} \tag{E.1}$$

Let μ_1^*, μ_2^* with $\mu_1^* \geq a_5$ and $\mu_2^* \geq b_5$ be two positive real numbers such that if $|x_s| \leq \mu_1^*$ and $\|x_f\|_2 \leq \mu_2^*$, then there exist positive real numbers (k_1, k_2, k_3) such that

$$\begin{aligned} |f_s(t, x_s, x_f) - f_s(t, x_s, 0)| &\leq k_1\|x_f\|_2 \\ \|f_f(t, x_s, x_f)\|_2 &\leq k_2|x_s| + k_3\|x_f\|_2. \end{aligned} \tag{E.2}$$

Pick $\mu_1 < a_5 < \mu_1^*$ and $\mu_2 < b_6 < \mu_2^*$.

Consider the smooth time-varying function $L : \mathcal{H}_s(t) \times \mathcal{H}_f(t) \to \mathbb{R}_{\geq 0}$:

$$L(t, x_s, x_f) = V(t, x_s) + W(t, x_f) \tag{E.3}$$

where $V(t, x_s)$ and $W(t, x_f)$ were defined in assumptions 3 and 4, respectively, as a Lyapunov functional candidate for the system of Eq. E.1. From Eq. 6.25 and Eq. 6.27, we have that $L(t, x_s, x_f)$ is positive definite, proper (tends to $+\infty$ as $|x_s| \to \infty$, or $\|x_f\|_2 \to \infty$) and decrescent, with respect to its arguments. Computing the time derivative of L along the trajectories of the system of Eq. E.1, and using the bounds of Eq. 6.25 and Eq. 6.27 and the estimates of Eq. E.2, the following expressions can be obtained:

$$\begin{aligned} \dot{L}(t, x_s, x_f) &= \frac{\partial V}{\partial t} + \frac{\partial V}{\partial x_s}\dot{x}_s + \frac{\partial W}{\partial t} + \frac{\partial W}{\partial x_f}\dot{x}_f \\ &= \frac{\partial V}{\partial t} + \frac{\partial V}{\partial x_s}[\mathcal{A}_s(t)x_s + f_s(t, x_s, 0)] + \frac{\partial V}{\partial x_s}[f_s(t, x_s, x_f) \\ &\quad - f_s(t, x_s, 0)] + \frac{\partial W}{\partial t} + \frac{1}{\epsilon}\frac{\partial W}{\partial x_f}\mathcal{A}_{f\epsilon}(t)x_f + \frac{\partial W}{\partial x_f}f_f(t, x_s, x_f) \end{aligned}$$

$$
\begin{aligned}
&\leq -a_3|x_s|^2 + a_4k_1|x_s|\,\|x_f\|_2 + b_5\|x_f\|_2^2 - \frac{b_3}{\epsilon}\|x_f\|_2^2 \\
&\quad + b_4\|x_f\|_2(k_2|x_s| + k_3\|x_f\|_2) \\
&\leq -a_3|x_s|^2 + (a_4k_1 + b_4k_2)|x_s|\,\|x_f\|_2 - \left(\frac{b_3}{\epsilon} - b_4k_3 - b_5\right)\|x_f\|_2^2 \\
&\leq -[|x_s| \quad \|x_f\|_2]\begin{bmatrix} a_3 & -\frac{a_4k_1+b_4k_2}{2} \\ -\frac{a_4k_1+b_4k_2}{2} & \frac{b_3}{\epsilon} - b_4k_3 - b_5 \end{bmatrix}\begin{bmatrix} |x_s| \\ \|x_f\|_2 \end{bmatrix}.
\end{aligned} \tag{E.4}
$$

Defining

$$
\epsilon_1 = \frac{(a_3 - \sqrt{\delta})b_3}{(a_3 - \sqrt{\delta})(b_5 + b_4k_3 + \sqrt{\delta}) + \left(\frac{a_4k_1+b_4k_2}{2}\right)^2},
$$

where $\delta \in \mathbb{R}_{>0}$, we have that if $\epsilon \in (0, \epsilon_1]$ then $\dot{L}(x_s, x_f) \leq -\delta(|x_s|^2 + \|x_f\|_2^2)$, which from the properties of L directly implies that the state of the system of Eq. E.1 is exponentially stable, that is, there exists a positive real number σ such that:

$$
\begin{bmatrix} |x_s| \\ \|x_f\|_2 \end{bmatrix} \leq e^{-\sigma t}\begin{bmatrix} |x_s(0)| \\ \|x_f(0)\|_2 \end{bmatrix}. \tag{E.5}
$$

Closeness of solutions: We will initially show the closeness of solution result for the x_f-states, and then for the x_s-states (note that from the first part of the proof $\epsilon \in (0, \epsilon_1]$). To establish the estimate for the x_f-states, we initially prove that the off manifold transients decay quickly (the decay rate will be precisely given below) to the manifold. To this end, we define the error vector $e_f(t) = x_f(t) - \Sigma(t, x_s, 0, \epsilon)$, differentiate $e_f(t)$ with respect to time, and multiply the resulting system with ϵ to obtain the following dynamical system:

$$
\frac{\partial e_f}{\partial \tau} = \mathcal{A}_{f\epsilon}(t)e_f + \epsilon f_f(t, x_s, e_f + \Sigma(t, x_s, 0, \epsilon)) - \epsilon f_f(t, x_s, \Sigma(t, x_s, 0, \epsilon)). \tag{E.6}
$$

Referring to the above system with $\epsilon = 0$, we have from assumption 4 that there exists a Lyapunov functional $\bar{W} : \mathcal{H}_f(t) \to \mathbb{R}_{\geq 0}$ and a set of positive real numbers $(\bar{b}_1, \bar{b}_2, \bar{b}_3, \bar{b}_4, \bar{b}_5, \bar{b}_6)$ such that for all $e_f \in \mathcal{H}_f(t)$ that satisfy $\|e_f\|_2 \leq \bar{b}_6$, the following conditions hold:

$$
\begin{aligned}
\bar{b}_1\|e_f(\tau)\|_2^2 \leq \bar{W}(t, e_f(\tau)) &\leq \bar{b}_2\|e_f(\tau)\|_2^2 \\
\frac{\partial \bar{W}}{\partial e_f}\mathcal{A}_{f\epsilon}(t)e_f &\leq -\bar{b}_3\|e_f(\tau)\|_2^2 \\
\left\|\frac{\partial \bar{W}}{\partial e_f}\right\|_2 &\leq \bar{b}_4\|e_f(\tau)\|_2 \\
\left\|\frac{\partial \bar{W}}{\partial t}\right\|_2 &\leq \bar{b}_5\|e_f(\tau)\|_2^2.
\end{aligned} \tag{E.7}
$$

Computing the time derivative of $\bar{W}(t, e_f)$ along the trajectories of the system of Eq. E.6 and using that $\|f_f(t, x_s, e_f + \Sigma(t, x_s, 0, \epsilon)) - f_f(t, x_s, \Sigma(t, x_s, 0, \epsilon))\|_2 \leq k_5 \|e_f\|_2$, where k_5 is a positive real number (which follows from the fact that the states (x_s, e_f) are bounded), we have:

$$\begin{aligned} \frac{\partial \bar{W}}{\partial \tau} &\leq -\bar{b}_3 \|e_f\|_2^2 + \epsilon \bar{b}_5 \|e_f\|_2^2 + \epsilon \bar{b}_4 \|e_f\|_2 k_5 \|e_f\|_2 \\ &\leq -(\bar{b}_3 - \epsilon(\bar{b}_4 k_5 + \bar{b}_5)) \|e_f\|_2^2. \end{aligned} \tag{E.8}$$

Set $\epsilon_2 = \bar{b}_3 - \bar{\delta}/(\bar{b}_4 k_5 + \bar{b}_5)$ and $\epsilon^* = min\{\epsilon_1, \epsilon_2\}$, where $\bar{\delta} \in \mathbb{R}_{>0}$. From the above inequality we have that if $\epsilon \in (0, \epsilon^*]$, the following bound holds for $\|e_f(\tau)\|_2$ for all $t \in [0, \infty)$:

$$\|e_f(\tau)\|_2 \leq K_3 \|e_f(0)\|_2 e^{-\gamma \frac{t}{\epsilon}} \tag{E.9}$$

where K_3, γ are positive real numbers. From the series expansion of $\Sigma(t, x_s, 0, \epsilon)$ and the definition of the $O(\epsilon^{k+1})$, we also have that:

$$\|\Sigma(t, x_s, 0, \epsilon) - \tilde{x}_f\|_2 \leq \tilde{k}_1 \epsilon^{k+1} \tag{E.10}$$

where $\tilde{x}_f = \Sigma_0(t, \tilde{x}_s, 0) + \epsilon \Sigma_1(t, \tilde{x}_s, 0) + \epsilon^2 \Sigma_2(t, \tilde{x}_s, 0) + \cdots + \epsilon^k \Sigma_k (t, \tilde{x}_s, 0)$ and $\tilde{k}_1$ is a positive number. Combining the bounds of Eqs. E.9–E.10, we can write the following bound:

$$\|x_f - \tilde{x}_f\|_2 \leq \tilde{k}_1 \epsilon^{k+1} + \tilde{k}_2 e^{-\gamma \frac{t}{\epsilon}}, \quad \forall\, t \in [0, \infty) \tag{E.11}$$

where $\tilde{k}_2$ is a positive number. It follows directly that $x_f = \tilde{x}_f + O(\epsilon^{k+1})$, $\forall\, t \in [t_b, \infty)$, where t_b is the time required for x_f to approach the inertial manifold (i.e., $\tilde{k}_2 e^{-\gamma \frac{t}{\epsilon}} = 0$ for $t \geq t_b$).

Defining the error coordinate $e_s(t) = x_s(t) - \tilde{x}_s(t)$ and differentiating $e_s(t)$ with respect to time, the following system can be obtained:

$$\frac{de_s}{dt} = \mathcal{A}_s(t) e_s + f_s(t, \tilde{x}_s + e_s, x_f) - f_s(t, \tilde{x}_s, \tilde{x}_f). \tag{E.12}$$

The representation of the system of Eq. E.12 in the fast time scale τ takes the form:

$$\frac{de_s}{d\tau} = \epsilon[\mathcal{A}_s(t) e_s + f_s(t, \tilde{x}_s + e_s, x_f) - f_s(t, \tilde{x}_s, \tilde{x}_f)] \tag{E.13}$$

where $(e_s, \tilde{x}_s, t)$ can be considered approximately constant, and thus using that $|e_s(0)| = 0$ and the continuity of solutions for $e_s(\tau)$, the following bound holds:

$$|e_s(\tau)| \leq \epsilon k_7 \tau_b, \quad \forall\, \tau \in [0, \tau_b) \tag{E.14}$$

where k_7 is a positive real number and $\tau_b = \frac{t_b}{\epsilon} = O(1)$. Based on the fact that $\tilde{x}_s(t), \tilde{x}_f(t)$ decay exponentially to zero, and from assumption 3, we have that the system:

$$\frac{de_s}{dt} = \mathcal{A}_s(t) e_s + f_s(t, \tilde{x}_s + e_s, \tilde{x}_f) - f_s(t, \tilde{x}_s, \tilde{x}_f) \tag{E.15}$$

is exponentially stable, which implies that there exists a smooth Lyapunov function $\tilde{V}: \mathcal{H}_s \to \mathbb{R}_{\geq 0}$ and a set of positive real numbers ($\tilde{a}_1$, $\tilde{a}_2$, $\tilde{a}_3$, $\tilde{a}_4$, $\tilde{a}_5$), such that for all $e_s \in \mathcal{H}_s$ that satisfy $|e_s| \leq \tilde{a}_5$, the following conditions hold:

$$\begin{aligned} \tilde{a}_1 |e_s|^2 &\leq \tilde{V}(t, e_s) \leq \tilde{a}_2 |e_s|^2 \\ \dot{\tilde{V}}(e_s) &= \frac{\partial \tilde{V}}{\partial t} + \frac{\partial \tilde{V}}{\partial e_s}[\mathcal{A}_s(t)e_s + f_s(t, \tilde{x}_s + e_s, \tilde{x}_f) - f_s(t, \tilde{x}_s, \tilde{x}_f)] \\ &\leq -\tilde{a}_3 |e_s|^2 \\ \left| \frac{\partial \tilde{V}}{\partial e_s} \right| &\leq \tilde{a}_4 |e_s|. \end{aligned} \tag{E.16}$$

From the assumption that $f_s(t, x_s, x_f)$ is Lipschitz continuous with respect to x_s and x_f:

$$|f_s(t, \tilde{x}_s + e_s, x_f) - f_s(t, \tilde{x}_s + e_s, \tilde{x}_f)| \leq \tilde{k} \|x_f - \tilde{x}_f\|_2 \tag{E.17}$$

where $\tilde{k}$ is a positive number. Substituting the inequality of Eq. E.11 to Eq. E.17 the following bound is obtained:

$$|f_s(t, \tilde{x}_s + e_s, x_f) - f_s(t, \tilde{x}_s + e_s, \tilde{x}_f)| \leq \tilde{k}\tilde{k}_1 \epsilon^{k+1} + \tilde{k}\tilde{k}_2 e^{-\gamma \frac{t}{\epsilon}}. \tag{E.18}$$

Computing the time derivative of $\tilde{V}(t, e_s)$ along the trajectories of the system of Eq. E.12 and using that for $t \in [0, \infty)$ the bound described in Eq. E.18 holds, we have for all $t \in [0, \infty)$:

$$\dot{\tilde{V}}(t, e_s) \leq -\tilde{a}_3 |e_s|^2 + (\tilde{k}_1 \epsilon^{k+1} + \tilde{k}_2 e^{-\gamma \frac{t}{\epsilon}}) \tilde{a}_4 |e_s| \tag{E.19}$$

From the above inequality, using corollary 5.3 in [92] and the fact that $|e_s(0)| = 0$, we have that the following bound holds for $|e_s(t)|$ for all $t \in [0, \infty)$:

$$|e_s(t)| \leq \bar{K} e^{-\gamma \frac{t}{\epsilon}} + \tilde{K} \epsilon^{k+1} \tag{E.20}$$

where $\bar{K}$, $\tilde{K}$ are positive real numbers. Since the term $\bar{K} e^{-\gamma \frac{t}{\epsilon}}$ vanishes outside the interval $[0, t_b)$, it follows from Eq. E.20 that for all $t \in [t_b, \infty)$:

$$|e_s(t)| \leq \tilde{K} \epsilon^{k+1}. \tag{E.21}$$

From the above inequality, the estimate $x_s(t) = \tilde{x}_s(t) + O(\epsilon^{k+1})$, for $t \geq t_b$, follows directly. □

Proof of theorem 6.1. Substituting the output feedback controller of Eq. 6.37 into the system of Eq. 6.11, we get:

$$\begin{aligned} \frac{d\eta}{dt} &= \mathcal{A}_s(t)\eta + \mathcal{B}_s(t)\,(p_0(t, \eta) + Q_0(t, \eta)v + \epsilon[p_1(t, \eta) + Q_1(t, \eta)v]) \\ &\quad + f_s(t, \eta, \epsilon \Sigma_1(t, \eta, \bar{u}_0)) + L(\mathcal{Q}(t)x_s + \mathcal{Q}(t)x_f - \mathcal{Q}(t)\eta \\ &\quad - \epsilon \mathcal{Q}(t)\Sigma_1(t, \eta, \bar{u}_0)) \\ \frac{dx_s}{dt} &= \mathcal{A}_s(t)x_s + \mathcal{B}_s(t)\,(p_0(t, \eta) + Q_0(t, \eta)v + \epsilon[p_1(t, \eta) \\ &\quad + Q_1(t, \eta)v]) + f_s(t, x_s, x_f) \end{aligned}$$

$$\begin{aligned}\epsilon\frac{\partial x_f}{\partial t} &= \mathcal{A}_{f\epsilon}(t)x_f + \epsilon\mathcal{B}_f(t)\left(p_0(t,\eta) + Q_0(t,\eta)v + \epsilon[p_1(t,\eta)\right.\\ &\quad \left.+ Q_1(t,\eta)v]\right) + \epsilon f_f(t,x_s,x_f)\\ y^i &= \mathcal{C}^i(t)x_s + \mathcal{C}^i(t)x_f, \quad i = 1,\dots,l.\end{aligned} \tag{E.22}$$

Performing a two-time-scale decomposition on the above system, the fast subsystem takes the form:

$$\frac{\partial x_f}{\partial \tau} = \mathcal{A}_{f\epsilon}(t)x_f \tag{E.23}$$

which is assumed to be exponentially stable (assumption 6.4). The $O(\epsilon^2)$ approximation of the closed-loop inertial form is:

$$\begin{aligned}\frac{d\eta}{dt} &= \mathcal{A}_s(t)\eta + \mathcal{B}_s(t)(p_0(t,\eta) + Q_0(t,\eta)v + \epsilon[p_1(t,\eta) + Q_1(t,\eta)v])\\ &\quad + f_s(t,\eta,\epsilon\Sigma_1(t,\eta,\bar{u}_0)) + L(\mathcal{Q}(t)x_s + \epsilon\mathcal{Q}(t)\Sigma_1(t,x_s,\bar{u}_0) - \mathcal{Q}(t)\eta\\ &\quad - \epsilon\mathcal{Q}(t)\Sigma_1(t,\eta,\bar{u}_0))\\ \frac{dx_s}{dt} &= \mathcal{A}_s(t)x_s + \mathcal{B}_s(t)\left(p_0(t,\eta) + Q_0(t,\eta)v + \epsilon[p_1(t,\eta) + Q_1(t,\eta)v]\right)\\ &\quad + f_s(t,x_s,\epsilon\Sigma_1(t,x_s,\bar{u}_0))\\ y_s^i &= \mathcal{C}_i(t)x_s + \epsilon\mathcal{C}_i(t)\Sigma_1(t,x_s,\bar{u}_0), \quad i = 1,\dots,l.\end{aligned} \tag{E.24}$$

Using the hypothesis $\eta(0) = x_s(0)$, we can write the above system as:

$$\begin{aligned}\frac{dx_s}{dt} &= \mathcal{A}_s(t)x_s + \mathcal{B}_s(t)(p_0(t,\eta) + Q_0(t,\eta)v + \epsilon[p_1(t,x_s)\\ &\quad + Q_1(t,x_s)v]) + f_s(t,x_s,\epsilon\Sigma_1(t,x_s,\bar{u}_0))\\ y_s^i &= \mathcal{C}^i(t)x_s + \epsilon\mathcal{C}^i(t)\Sigma_1(t,x_s,\bar{u}_0), \quad i = 1,\dots,l.\end{aligned} \tag{E.25}$$

Computing the time derivatives of the controlled output y_s^i up to order r_i and substituting into Eq. 6.39, one can show that the input/output response of Eq. 6.39 is enforced in the above closed-loop system. Furthermore, an approach, similar to the one in [104], can be followed to establish that assumptions 1 and 2 of the theorem guarantee that the system of Eq. E.25 is locally exponentially stable. Therefore, we have that system of Eq. E.24 is locally exponentially stable and its outputs y_s^i, $i = 1,\dots,l$, change according to Eq. 6.39. A direct application of the result of proposition 1 then yields that there exist positive real numbers $\tilde{\mu}_1, \tilde{\mu}_2, \tilde{\epsilon}^*$ such that if $|x_s(0)| \le \tilde{\mu}_1$, $\|x_f(0)\|_2 \le \tilde{\mu}_2$ and $\epsilon \in (0, \tilde{\epsilon}^*]$, the closed-loop infinite-dimensional system is exponentially stable and the relation of Eq. 6.39 holds. □

Proof of Theorem 6.2. Substituting the controller of Eq. 6.54 into the parabolic PDE system of Eq. 5.7, we obtain:

$$\begin{aligned}\dot{x} &= \mathcal{A}(t)x + \mathcal{B}(t)a_0(x_s,x_f,\bar{v},t) + f(t,x) + \mathcal{W}(t,x,\theta)\\ y &= \mathcal{C}(t)x, \quad q = \mathcal{Q}(t)x.\end{aligned} \tag{E.26}$$

One can easily verify that assumption 2 holds for the above system, and thus, it can be written in the following form:

$$\begin{aligned} \frac{dx_s}{dt} &= \mathcal{A}_s(t)x_s + \mathcal{B}_s(t)a_0(x_s, x_f, \bar{v}, t) + f_s(t, x_s, x_f) + \mathcal{W}_s(t, x_s, x_f, \theta) \\ \epsilon\frac{\partial x_f}{\partial t} &= \mathcal{A}_{f\epsilon}(t)x_f + \epsilon\mathcal{B}_f(t)a_0(x_s, x_f, \bar{v}, t) + \epsilon f_f(t, x_s, x_f) \\ &\quad + \epsilon\mathcal{W}_f(t, x_s, x_f, \theta). \end{aligned} \tag{E.27}$$

Rewriting the above system in the fast time scale $\tau = \frac{t}{\epsilon}$ and setting $\epsilon = 0$, we obtain the following infinite-dimensional fast subsystem from the system of Eq. E.27:

$$\frac{\partial x_f}{\partial \tau} = \mathcal{A}_{f\epsilon}(t)x_f \tag{E.28}$$

which is globally exponentially stable. Setting $\epsilon = 0$ in the system of Eq. E.27, we have that $x_f = 0$ and thus, the finite-dimensional slow system takes the form:

$$\begin{aligned} \frac{dx_s}{dt} &= F_0(t, x_s) + \sum_{i=1}^{l} \mathcal{B}_0^i(t)a_0^i(x_s, 0, \bar{v}, t) + \mathcal{W}_0(t, x_s, 0, \theta) \\ y^i &= \mathcal{C}^i(t)x_s =: h_0^i(t, x_s). \end{aligned} \tag{E.29}$$

For the above system, there exists a $\phi \in (0, \phi^*]$ such that if $\max\{|x_s(0)|, \|\theta\|, \|\dot{\theta}\|, \|\tilde{v}\|\} \leq \delta$, then its state is bounded and its outputs satisfy $\limsup_{t\to\infty}|y^i - v_i| \leq O(\phi)$, $i = 1, \ldots, l$. Finally, since the infinite-dimensional fast subsystem of Eq. 6.49 is exponentially stable, we can use standard singular perturbation arguments to obtain that there exists an $\epsilon^*(\phi)$, such that if $\epsilon \in (0, \epsilon^*(\phi)]$, $\max\{|x_s(0)|, \|x_f(0)\|_2, \|\theta\|, \|\dot{\theta}\|, \|\tilde{v}\|\} \leq \delta$, then the state of the closed-loop parabolic PDE system of Eq. E.26 is bounded and that its outputs satisfy the relation of Eq. 6.55. □

Appendix F

Karhunen–Loève Expansion

The Karhunen–Loève (K–L) expansion (also known as proper orthogonal decomposition, method of empirical eigenfunctions, principal component analysis, and singular value decomposition) is a procedure used to compute an optimal (in a sense that will become clear in remark F.1 below) basis for a modal decomposition of a PDE system from an appropriately constructed set of data of this system obtained by detailed simulations.

For simplicity of the presentation, we describe the K-L expansion in the context of a PDE system with $n = 1$ and assume that there is available a sufficiently large set of solutions of this system, $\{\bar{v}_\kappa\}$, consisting of K sampled states, $\bar{v}_\kappa(z)$, (which are typically called "snapshots"). The reader may refer to [66, 80] for a detailed presentation and analysis of the K–L expansion and to [75] for approaches for constructing $\{\bar{v}_\kappa\}$. We define the ensemble average of snapshots as $\langle \bar{v}_\kappa \rangle := \frac{1}{K} \sum_{\kappa=1}^{K} \bar{v}_\kappa(z)$ (we note that nonuniform sampling of the snapshots and weighted ensemble average can be also considered; see, for example, [70]). Furthermore, the ensemble average of snapshots $\langle \bar{v}_\kappa \rangle$ is subtracted out from the snapshots, that, is:

$$v_\kappa = \bar{v}_\kappa - \langle \bar{v}_\kappa \rangle \tag{F.1}$$

so that only fluctuations are analyzed. The issue is how to obtain the most typical or characteristic structure (in a sense that will become clear below) $\phi(z)$ among these snapshots $\{v_\kappa\}$. Mathematically, this problem can be posed as the one of obtaining a function $\phi(z)$ that maximizes the following objective function:

$$\text{Maximize} \frac{\langle (\phi, v_\kappa)^2 \rangle}{(\phi, \phi)}$$
$$s.t. \; (\phi, \phi) = 1, \quad \phi \in L^2([\Omega]) \tag{F.2}$$

where Ω is the domain of definition of the process. The constraint $(\phi, \phi) = 1$ is imposed to ensure that the function, $\phi(z)$, computed as a solution of the above maximization problem, is unique. The Lagrangian functional corresponding to this constrained optimization problem is:

$$L = \langle (\phi, v_\kappa)^2 \rangle - \lambda((\phi, \phi) - 1) \tag{F.3}$$

and necessary conditions for extrema is that the functional derivative vanishes for all variations $\phi + \delta\psi \in L^2([\Omega])$, where δ is a real number:

$$\frac{d\,L(\phi + \delta\psi)}{d\delta}(\delta = 0) = 0, \quad (\phi, \phi) = 1. \tag{F.4}$$

Using the definitions of inner product and ensemble average, $\frac{d\,L(\phi+\delta\psi)}{d\delta}(\delta=0)$ can be computed as follows:

$$\begin{aligned}\frac{d\,L(\phi+\delta\psi)}{d\delta}(\delta=0) &= \frac{d}{d\delta}\left[\langle(v_\kappa,\phi+\delta\psi)(\phi+\delta\psi,v_\kappa)\rangle\right.\\ &\quad \left.-\lambda(\phi+\delta\psi,\phi+\delta\psi)\right]_{\delta=0}\\ &= 2\,Re[\langle(v_\kappa,\psi)(\phi,v_\kappa)\rangle-\lambda(\phi,\psi)]\\ &= \left\langle\int_\Omega \psi(z)v_\kappa(z)\,dz\int_\Omega \phi(\bar z)v_\kappa(\bar z)\,d\bar z\right\rangle-\lambda\int_\Omega \phi(\bar z)\psi(\bar z)\,d\bar z\\ &= \int_\Omega\left(\left\{\int_\Omega\langle v_\kappa(z)v_\kappa(\bar z)\rangle\phi(z)\,dz\right\}-\lambda\phi(\bar z)\right)\psi(\bar z)\,d\bar z.\end{aligned} \tag{F.5}$$

Since $\psi(\bar z)$ is an arbitrary function, the necessary conditions for optimality take the form:

$$\int_\Omega \langle v_\kappa(z)v_\kappa(\bar z)\rangle\phi(z)\,dz=\lambda\phi(\bar z),\qquad (\phi,\phi)=1. \tag{F.6}$$

Introducing the two-point correlation function:

$$K(z,\bar z)=\langle v_\kappa(z)v_\kappa(\bar z)\rangle=\frac{1}{K}\sum_{\kappa=1}^{K}v_\kappa(z)v_\kappa(\bar z) \tag{F.7}$$

and the linear operator:

$$R:=\int_\Omega K(z,\bar z)\,d\bar z \tag{F.8}$$

the optimality condition of Eq. F.6 reduces to the following eigenvalue problem of the integral equation:

$$R\phi=\lambda\phi\Rightarrow\int_\Omega K(z,\bar z)\phi(\bar z)\,d\bar z=\lambda\phi(z). \tag{F.9}$$

The computation of the solution of the above integral eigenvalue problem is, in general, a very expensive computational task. To circumvent this problem, Sirovich introduced in 1987 [124, 125] the method of snapshots. The central idea of this technique is to assume that the requisite eigenfunction, $\phi(z)$, can be expressed as a linear combination of the snapshots, that is:

$$\phi(z)=\sum_k c_k v_k(z) \tag{F.10}$$

Substituting the above expression for $\phi(z)$ on Eq. F.9, we obtain the following eigenvalue problem:

$$\int_{\Omega} \frac{1}{K} \sum_{\kappa=1}^{K} v_{\kappa}(z) v_{\kappa}(\bar{z}) \sum_{k=1}^{K} c_k v_k(\bar{z}) d\bar{z} = \lambda \sum_{k=1}^{K} c_k v_k(z). \tag{F.11}$$

Defining:

$$B^{\kappa k} := \frac{1}{K} \int_{\Omega} v_{\kappa}(\bar{z}) v_k(\bar{z}) d\bar{z} \tag{F.12}$$

the eigenvalue problem of Eq. F.11 can be equivalently written as:

$$Bc = \lambda c. \tag{F.13}$$

The solution of the above eigenvalue problem (which can be obtained by utilizing standard methods from matrix theory) yields the eigenvectors $c = [c_1 \ \cdots \ c_K]$ which can be used in Eq. F.10 to construct the eigenfunction $\phi(z)$. From the structure of the matrix B, it follows that it is symmetric and positive semi-definite, and thus, its eigenvalues, λ_{κ}, $\kappa = 1, \ldots, K$, are real and non-negative. Furthermore,

$$\int_{\Omega} \phi_i(z) \phi_j(z) dz = 0, \quad i \neq j. \tag{F.14}$$

Remark F.1 The optimality of the empirical eigenfunctions obtained via K–L expansion can be shown as follows. Consider a snapshot $v_{\kappa}(z)$ of the ensemble of snapshots, v_{κ}, and the set of empirical eigenfunctions obtained by applying K–L expansion to v_{κ}, and let:

$$v_{\kappa}(z) = \sum_{j=1}^{J} \gamma_j \phi_j(z) \tag{F.15}$$

be the decomposition of $v_{\kappa}(z)$ with respect to this basis, where J is any positive integer that satisfies $1 \leq J \leq K$. Assume that the eigenfunctions have been ordered so the corresponding eigenvalues satisfy $\lambda_1 > \lambda_2 > \cdots > \lambda_{j+1}$. Then, it can be shown [80] that if $\{\psi_1, \psi_2, \ldots, \psi_{\kappa}\}$ is some arbitrary set of orthonormal basis functions in which we expand $v_{\kappa}(z)$, then the following result holds:

$$\sum_{j=1}^{J} \langle (\phi_j, v_{\kappa})^2 \rangle = \sum_{j=1}^{J} \lambda_l \geq \sum_{j=1}^{J} \langle (\psi_j, v_{\kappa})^2 \rangle. \tag{F.16}$$

This implies that the projection on the subspace spanned by the empirical eigenfunctions will on average contain the most energy possible compared to all other linear decompositions, for any number of modes J.

References

[1] H. Aling, S. Benerjee, A. K. Bangia, V. Cole, J. Ebert, A. Emani-Naeini, K. F. Jensen, I. G. Kevrekidis, and S. Shvartsman. Nonlinear model reduction for simulation and control of rapid thermal processing. In *Proceedings of American Control Conference*, pages 2233–2238, Albuquerque, NM, 1997.

[2] A. A. Alonso and E. B. Ydstie. Nonlinear control, passivity and the second law of thermodynamics. In *AIChE Annual Meeting, paper 181i, Miami Beach, FL*, 1995.

[3] C. Antoniades and P. D. Christofides. Studies on nonlinear dynamics and control of a tubular reactor with recycle, submitted. *Nonlinear Analysis: Theory, Methods, and Applications*, 2000.

[4] A. Armaou and P. D. Christofides. Crystal temperature regulation in the Czochralski crystal growth process, to appear. *AIChE J.*, 2000.

[5] A. Armaou and P. D. Christofides. Nonlinear feedback control of parabolic PDE systems with time-dependent spatial domains. *J. Math. Anal. Appl.*, 239:124–157, 1999.

[6] A. Armaou and P. D. Christofides. Robust control of parabolic PDE systems with time-dependent spatial domains, to appear. *Automatica*, 2000.

[7] A. Armaou and P. D. Christofides. Wave suppression by nonlinear finite-dimensional control. *Chem. Eng. Sci.*, 55:2627–2640, 2000.

[8] L. J. Atherton, J. J. Derby, and R. A. Brown. Radiative heat exchange in Czochralski crystal growth. *J. Crystal Growth*, 84:57–78, 1987.

[9] J. Baker and P. D. Christofides. Nonlinear control of rapid thermal chemical vapor deposition under uncertainty. *Comp. & Chem. Eng.*, 23 (s):233–236, 1999.

[10] J. Baker and P. D. Christofides. Output feedback control of parabolic PDE systems with nonlinear spatial differential operators. *I & EC Res.*, 38:4372–4380, 1999.

[11] J. BAKER and P. D. CHRISTOFIDES. Finite dimensional approximation and control of nonlinear parabolic PDE systems. *Int. J. Contr.*, 73:439–456, 2000.

[12] M. J. BALAS. Feedback control of linear diffusion processes. *Int. J. Contr.*, 29:523–533, 1979.

[13] M. J. BALAS. Trends in large scale structure control theory: Fondest hopes, wildest dreams. *IEEE Trans. Automat. Contr.*, 27:522–535, 1982.

[14] M. J. BALAS. The Galerkin method and feedback control of linear distributed parameter systems. *J. Math. Anal. Appl.*, 91:527–546, 1983.

[15] M. J. BALAS. Stability of distributed parameter systems with finite-dimensional controller-compensators using singular perturbations. *J. Math. Anal. Appl.*, 99:80–108, 1984.

[16] M. J. BALAS. Finite-dimensional control of distributed parameter systems by Galerkin approximation of infinite dimensional controllers. *J. Math. Anal. Appl.*, 114:17–36, 1986.

[17] M. J. BALAS. Nonlinear finite-dimensional control of a class of nonlinear distributed parameter systems using residual mode filters: A proof of local exponential stability. *J. Math. Anal. Appl.*, 162:63–70, 1991.

[18] M. J. BALAS. Finite-dimensional direct adaptive control for discrete-time infinite-dimensional linear systems. *J. Math. Anal. Appl.*, 196: 153–171, 1995.

[19] A. K. BANGIA, P. F. BATCHO, I. G. KEVREKIDIS, and G. E. KARNIADAKIS. Unsteady 2-D flows in complex geometries: Comparative bifurcation studies with global eigenfunction expansion. *SIAM J. Sci. Comp.*, 18:775–805, 1997.

[20] H. T. BANKS and K. KUNISCH. *Estimation Techniques for Distributed Parameter Systems.* Birkhäuser, Boston, 1989.

[21] H. T. BANKS, R. C. SMITH, and Y. WANG. *Smart Material Structures: Modeling, Estimation, and Control.* John Wiley & Sons, New York, 1996.

[22] A. BENSOUSSAN, G. DA PRATO, M. C. DELFOUR, and S. K. MITTER. *Representation and Control of Infinite Dimensional Systems, Volume I.* Birkhäuser, Boston, 1992.

[23] A. BENSOUSSAN, G. DA PRATO, M.C. DELFOUR, and S. K. MITTER. *Representation and Control of Infinite Dimensional Systems, Volume II.* Birkhäuser, Boston, 1993.

[24] T. BREEDIJK, T. F. EDGAR, and I. TRACHTENBERG. A model predictive controller for multivariable temperature control in rapid thermal processing. In *Proceedings of American Control Conference*, pages 2980–2984, San Francisco, CA, 1993.

[25] H. S. BROWN, I. G. KEVREKIDIS, and M. S. JOLLY. A minimal model for spatio-temporal patterns in thin film flow. In *Pattern and Dynamics in Reactive Media*, pages 11–31, R. Aris, D. G. Aronson, and H. L. Swinney, ed., Springer-Verlag, 1991.

[26] J. A. BURNS and S. KANG. A control problem for Burgers' equation with bounded input/output. *Nonlinear Dynamics*, 2:235–247, 1991.

[27] J. A. BURNS and Y.-R. OU. Feedback control of the driven cavity problem using LQR designs. In *Proceedings of 33rd IEEE Conference on Decision and Control*, pages 289–294, Orlando, FL, 1994.

[28] S. W. BUTLER and T. F. EDGAR. Case studies in equipment modeling and control in the microelectronics industry. *AIChE Symposium Series*, 93:133–144, 1997.

[29] C. A. BYRNES, D. S. GILLIAM, and J. HE. Root-locus and boundary feedback design for a class of distributed parameter systems. *SIAM J. Contr. & Optim.*, 32:1364–1427, 1994.

[30] C. A. BYRNES, D. S. GILLIAM, and V. I. SHUBOV. Global lyapunov stabilization of a nonlinear distributed parameter system. In *Proceedings of 33rd IEEE Conference on Decision and Control*, pages 1769–1774, Orlando, FL, 1994.

[31] C. A. BYRNES, D. S. GILLIAM, and V. I. SHUBOV. On the dynamics of boundary controlled nonlinear distributed parameter systems. In *Proceedings of Symposium on Nonlinear Control Systems Design'95*, pages 913–918, Tahoe City, CA, 1995.

[32] C. I. BYRNES. Adaptive stabilization of infinite dimensional linear systems. In *Proceedings of 26th IEEE Conference on Decision and Control*, pages 1435–1440, Los Angeles, CA, 1987.

[33] J. CARR. *Applications of Center Manifold Theory.* Springer-Verlag, New York, 1981.

[34] S. CHAKRAVARTI, M. MAREK, and W. H. RAY. Reaction-diffusion system with brusselator kinetics: Control of a quasi-periodic route to chaos. *Phys. Rev. E*, 52:2407–2423, 1995.

[35] C. C. CHEN and H. C. CHANG. Accelerated disturbance damping of an unknown distributed system by nonlinear feedback. *AIChE J.*, 38:1461–1476, 1992.

[36] P. D. CHRISTOFIDES. Robust control of parabolic PDE systems. *Chem. Eng. Sci.*, 53:2949–2965, 1998.

[37] P. D. CHRISTOFIDES and J. BAKER. Robust output feedback control of quasi-linear parabolic PDE systems. *Syst. & Contr. Lett.*, 36:307–316, 1999.

[38] P. D. CHRISTOFIDES and P. DAOUTIDIS. Nonlinear control of distributed parameter processes with disturbances. In *Proceedings of Symposium on Nonlinear Control Systems Design'95*, pages 9–15, Tahoe City, CA, 1995.

[39] P. D. CHRISTOFIDES and P. DAOUTIDIS. Feedback control of hyperbolic PDE systems. *AIChE J.*, 42:3063–3086, 1996.

[40] P. D. CHRISTOFIDES and P. DAOUTIDIS. Finite-dimensional control of parabolic PDE systems using approximate inertial manifolds. *J. Math. Anal. Appl.*, 216:398–420, 1997.

[41] P. D. CHRISTOFIDES and P. DAOUTIDIS. Robust control of multivariable two-time-scale nonlinear systems. *J. Proc. Contr.*, 7:313–328, 1997.

[42] P. D. CHRISTOFIDES and P. DAOUTIDIS. Distributed output feedback control of two-time-scale hyperbolic PDE systems. *Int. J. Appl. Math. & Comp. Sci.*, 8(4):101–120, 1998.

[43] P. D. CHRISTOFIDES and P. DAOUTIDIS. Feedback control of nonlinear parabolic PDE systems. In *Proceedings of Nato Conference on Nonlinear Model Based Control*, pages 371–399, NATO ASI Series, Kluwer Academic Publishers, Dordrecht, The Netherlands, 1998.

[44] P. D. CHRISTOFIDES and P. DAOUTIDIS. Robust control of hyperbolic PDE systems. *Chem. Eng. Sci.*, 53:85–105, 1998.

[45] P. D. CHRISTOFIDES and A. R. TEEL. Singular perturbations and input-to-state stability. *IEEE Trans. Autom. Contr.*, 41:1645–1650, 1996.

[46] P. D. CHRISTOFIDES, A. R. TEEL, and P. DAOUTIDIS. Robust semi-global output tracking for nonlinear singularly perturbed systems. *Int. J. Contr.*, 65:639–666, 1996.

[47] D. J. COOPER, W. F. RAMIREZ, and D. E. CLOUGH. Comparison of linear distributed-parameter filters to lumped approximants. *AIChE J.*, 32:186–194, 1986.

[48] M. CORLESS and G. LEITMANN. Continuous state feedback guaranteeing uniform ultimate boundedness for uncertain dynamic systems. *IEEE Trans. Automat. Contr.*, 26:1139–1144, 1981.

[49] R. F. CURTAIN. Finite-dimensional compensator design for parabolic distributed systems with point sensors and boundary input. *IEEE Trans. Automat. Contr.*, 27:98–104, 1982.

[50] R. F. CURTAIN. Disturbance decoupling for distributed systems by boundary control. In *Proceedings of 2nd International Conference on Control Theory for Distributed Parameter Systems and Applications*, pages 109–123, Vorau, Austria, 1984.

[51] R. F. CURTAIN. Invariance concepts in infinite dimensions. *SIAM J. Contr. & Optim.*, 24:1009–1030, 1986.

[52] R. F. CURTAIN and K. GLOVER. Robust stabilization of infinite dimensional systems by finite dimensional controllers. *Syst. & Contr. Lett.*, 7:41–47, 1986.

[53] R. F. CURTAIN and A. J. PRITCHARD. *Infinite Dimensional Linear Systems Theory.* Springer-Verlag, Berlin-Heidelberg, 1978.

[54] R. F. CURTAIN and H. J. ZWART. *An Introduction to Infinite Dimensional Linear Systems.* Springer-Verlag, New York, 1995.

[55] M. A. DEMETRIOU. Model reference adaptive control of slowly time-varying parabolic systems. In *Proceedings of 33rd IEEE Conference on Decision and Control*, pages 775–780, Orlando, FL, 1994.

[56] J. DERBY and R. BROWN. On the dynamics of Czochralski crystal growth. *J. Cryst. Growth*, 83:137–151, 1987.

[57] J. J. DERBY, L. J. ATHERTON, P. D. THOMAS, and R. A. BROWN. Finite-element methods for analysis of the dynamics and control of Czochralski crystal growth. *J. Sci. Comp.*, 2:297–343, 1987.

[58] J. J. DERBY and R. A. BROWN. Thermal-capillary analysis of Czochralski and liquid encapsulated Czochralski crystal growth I. Simulation. *J. Crystal Growth*, 74:605–624, 1986.

[59] J. J. DERBY and R. A. BROWN. Thermal-capillary analysis of Czochralski and liquid encapsulated Czochralski crystal growth II. Processing strategies. *J. Crystal Growth*, 75:227–240, 1986.

[60] J. J. DERBY and R. A. BROWN. On the quasi-steady state assumption in modeling Czochralski crystal growth. *J. Crystal Growth*, 87:251–260, 1988.

[61] D. DOCHAIN, J. P. BABARY, and M. N. TALI-MANAAR. Modeling and adaptive control of nonlinear distributed parameter bioreactors via orthogonal collocation. *Automatica*, 68:873–883, 1992.

[62] C. FOIAS, M. S. JOLLY, I. G. KEVREKIDIS, G. R. SELL, and E. S. TITI. On the computation of inertial manifolds. *Phys. Lett. A*, 131:433–437, 1989.

[63] C. FOIAS, G. R. SELL, and E. S. TITI. Exponential tracking and approximation of inertial manifolds for dissipative equations. *J. Dynamics and Differential Equations*, 1:199–244, 1989.

[64] A. FRIEDMAN. *Partial Differential Equations.* Holt, Rinehart & Winston, New York, 1976.

[65] A. FRIEDMAN. *Partial Differential Equations of Parabolic Type.* R. E. Krieger Pubs, Florida, 1983.

[66] K. FUKUNAGA. *Introduction to statistical pattern recognition.* Academic Press, New York, 1990.

[67] J. P. GAUTHIER and C. Z. XU. H^∞-control of a distributed parameter system with non-minimum phase. *Int. J. Contr.*, 53:45–79, 1989.

[68] D. H. GAY and W. H. RAY. Identification and control of distributed parameter systems by means of the singular value decomposition. *Chem. Eng. Sci.*, 50:1519–1539, 1995.

[69] G. GEORGAKIS, R. ARIS, and N. R. AMUNDSON. Studies in the control of tubular reactors: Part I & II & III. *Chem. Eng. Sci.*, 32:1359–1387, 1977.

[70] M. A. GEVELBER. Dynamics and control of the Czochralski process III. interface dynamics and control requirements. *J. Crystal Growth*, 139:271–285, 1994.

[71] M. A. GEVELBER. Dynamics and control of the Czochralski process IV. control structure design for interface shape control and perfomance evaluation. *J. Crystal Growth*, 139:286–301, 1994.

[72] M. A. GEVELBER and G. STEPHANOPOULOS. Dynamics and control of the Czochralski process I. modeling and dynamic characterization. *J. Crystal Growth*, 84:647–668, 1987.

[73] M. A. GEVELBER, G. STEPHANOPOULOS, and M. J. WARGO. Dynamics and control of the Czochralski process II. objectives and control structure design. *J. Crystal Growth*, 91:199–217, 1988.

[74] M. A. GEVELBER, M. J. WARGO, and G. STEPHANOPOULOS. Advanced control design considerations for the Czochralski process. *J. Crystal Growth*, 85:256–263, 1987.

[75] M. D. GRAHAM and I. G. KEVREKIDIS. Alternative approaches to the Karhunen-Loève decomposition for model reduction and data analysis. *Comp. & Chem. Eng.*, 20:495–506, 1996.

[76] F. K. GREISS and W. H. RAY. Stochastic control of processes having moving boundaries—an experimental study. *Automatica*, 16:157–166, 1980.

[77] P. K. GUNDEPUDI and J. C. FRIEDLY. Velocity control of hyperbolic partial differential equation systems with single characteristic variable. *Chem. Eng. Sci.*, 53:4055–4072, 1998.

[78] E. M. HANCZYC and A. PALAZOGLU. Eigenvalue inclusion for model approximations to distributed parameter systems. *I & EC Res.*, 31:2538–2546, 1992.

[79] E. M. HANCZYC and A. PALAZOGLU. Sliding mode control of nonlinear distributed parameter chemical processes. *I & EC Res.*, 34:557–566, 1995.

[80] P. HOLMES, J. L. LUMLEY, and G. BERKOOZ. *Turbulence, Coherent Structures, Dynamical Systems and Symmetry*. Cambridge University Press, New York, 1996.

[81] K. HONG and J. BENTSMAN. Application of averaging method for integro-differential equations to model reference adaptive control of parabolic systems. *Automatica*, 36:1415–1419, 1994.

[82] D. T. J. HURLE. *Handbook of Crystal Growth Vol. 2: Bulk Crystal Growth. Part B: Growth Mechanisms and Dynamics*. Elsevier Science B.V., Amsterdam, 1994.

[83] D. T. J. HURLE. *Handbook of Crystal Growth. Vol. 2: Bulk Crystal Growth. Part A: Basic Techniques*. Elsevier Science B.V., Amsterdam, 1994.

[84] R. Irizarry-Rivera and W. D. Seider. Model-predictive control of the Czochralski crystallization process. part I: Conduction-dominated melt. *J. Crystal Growth*, 178:593–611, 1997.

[85] R. Irizarry-Rivera and W. D. Seider. Model-predictive control of the Czochralski crystallization process. part II: Reduced-order convection model. *J. Crystal Growth*, 178:612–633, 1997.

[86] A. Isidori. *Nonlinear Control Systems: An Introduction.* Springer-Verlag, Berlin-Heidelberg, second edition, 1989.

[87] C. A. Jacobson and C. N. Nett. Linear state-space systems in infinite-dimensional space: The role and characterization of joint stabilizability/detectability. *IEEE Trans. Autom. Contr.*, 33:541–551, 1988.

[88] Z. P. Jiang, A. R. Teel, and L. Praly. Small-gain theorem for ISS systems and applications. *Math. Control Signals Systems*, 7:95–120, 1995.

[89] D. A. Jones and E. S. Titi. A remark on quasi-stationary approximate inertial manifolds for the Navier-Stokes equations. *SIAM J. Math. Anal.*, 25:894–914, 1994.

[90] A. Kalani and P. D. Christofides. Nonlinear control of spatially-inhomogeneous aerosol processes. *Chem. Eng. Sci.*, 54:2669–2678, 1999.

[91] S. Kang and K. Ito. A feedback control law for systems arising in fluid dynamics. In *Proceedings of 30th IEEE Conference on Decision and Control*, pages 384–385, Tempe, AZ, 1992.

[92] H. K. Khalil. *Nonlinear Systems.* Prentice Hall, New Jersey, second edition, 1996.

[93] W. J. Kiether, M. J. Fordham, S. Yu, A. J. S. Neto, K. A. Conrad, J. R. Hauser, F. Y. Sorrell, and J. J. Wortman. Three-zone rapid thermal processor system. In *Proceedings of 2nd International RTP Conference*, pages 96–101, Albuquerque, NM, 1994.

[94] B. B. King and Y. Qu. Nonlinear dynamic compensator design for flow control in a driven cavity. In *Proceedings of 34th IEEE Conference on Decision and Control*, pages 3741–3746, New Orleans, LA, 1995.

[95] P. V. Kokotovic, H. K. Khalil, and J. O'Reilly. *Singular*

Perturbations in Control: Analysis and Design. Academic Press, London, 1986.

[96] C. KRAVARIS and Y. ARKUN. Geometric nonlinear control—an overview. In *Proceedings of 4th International Conference on Chemical Process Control*, pages 477–515, Y. Arkun and W. H. Ray, eds., Padre Island, TX, 1991.

[97] N. KUNIMATSU and H. SANO. Compensator design of semilinear parabolic systems. *Int. J. Contr.*, 60:243–263, 1994.

[98] I. LASIECKA. Control of systems governed by partial differential equations: A historical perspective. In *Proceedings of 34th IEEE Conference on Decision and Control*, pages 2792–2797, New Orleans, LA, 1995.

[99] L. LASIECKA and R. TRIGGIANI. *Differential and Algebraic Riccati Equations with Applications to Boundary Point Control Problems: Continuous Theory and Approximation Theory*. Lecture Notes in Control and Information Sciences, Vol. 164, Springer-Verlag, Berlin, 1991.

[100] J. L. LIONS. *Optimal Control of Systems Described by Partial Differential Equations*. Springer-Verlag, Berlin, 1971.

[101] C. T. LO. Optimal control of counter-current distributed parameter systems. *Int. J. Contr.*, 18:273–288, 1973.

[102] W. MARQUARDT. Traveling waves in chemical processes. *Int. Chem. Engng.*, 4:585–606, 1990.

[103] M. O. OKUYIGA and W. H. RAY. Modeling and estimation for a moving boundary problem. *International Journal for Numerical Methods In Engineering*, 21:601–616, 1985.

[104] S. PALANKI and C. KRAVARIS. Controller synthesis for time-varying systems by input/output linearization. *Comp. & Chem. Engng.*, 21:891–903, 1997.

[105] A. PALAZOGLU and A. KARAKAS. Control of nonlinear distributed parameter systems using generalized invariants. *Automatica*, 36:697–703, 2000.

[106] A. PALAZOGLU and S. E. OWENS. Robustness analysis of a fixed-bed tubular reactor: Impact of modeling decisions. *Chem. Eng. Comm.*, 47:213–227, 1987.

[107] H. M. PARK and D. H. CHO. The use of the Karhunen-Loève decomposition for the modeling of distributed parameter systems. *Chem. Eng. Sci.*, 51:81–98, 1996.

[108] A. A. PATWARDHAN, G. T. WRIGHT, and T. F. EDGAR. Nonlinear model-predictive control of distributed-parameter systems. *Chem. Eng. Sci.*, 47:721–735, 1992.

[109] A. PAZY. *Semigroups of Linear Operators and Applications to Partial Differential Equations.* Springer-Verlag, New York, 1983.

[110] T. M. PELL and R. ARIS. Control of linearized distributed systems on discrete and corrupted observations. *I & EC Fundamentals*, 9:15–20, 1970.

[111] S. A. POHJOLAINEN. Computation of transmission zeros for distributed parameter systems. *Int. J. Contr.*, 33:199–212, 1981.

[112] A. J. PRITCHARD and D. SALAMON. The linear quadratic control problem for infinite-dimensional systems with unbounded input and output operators. *SIAM Journal of Control and Optimization*, 25:121–144, 1987.

[113] D. RAMKRISHNA and N. R. AMUNDSON. *Linear Operator Methods in Chemical Engineering.* Prentice-Hall, Englewood Cliffs, NJ, 1985.

[114] W. H. RAY. Some recent applications of distributed parameter systems theory—a survey. *Automatica*, 14:281–287, 1978.

[115] W. H. RAY. *Advanced Process Control.* McGraw-Hill, New York, 1981.

[116] W. H. RAY and J. H. SEINFELD. Filtering in distributed parameter systems with moving boundaries. *Automatica*, 11:509–515, 1975.

[117] H. K. RHEE, R. ARIS, and N. R. AMUNDSON. *First-Order Partial Differential Equations: Volumes I & II.* Prentice-Hall, New Jersey, 1986.

[118] D. L. RUSSELL. Controllability and stabilizability theory for linear partial differential equations: Recent progress and open questions. *SIAM Review*, 20:639–739, 1978.

[119] D. M. RUTHVEN and S. SIRCAR. Design of membrane and PSA processes for bulk gas separation. In *Proceedings of 4th International Conference on Foundations of Computer-Aided Process Design*, pages 29–37, Snowmass, CO, 1996.

[120] H. SANO and N. KUNIMATSU. Feedback control of semilinear diffusion systems: Inertial manifolds for closed-loop systems. *IMA J. Math. Contr. Inform.*, 11:75–92, 1994.

[121] H. SANO and N. KUNIMATSU. An application of inertial manifold theory to boundary stabilization of semilinear diffusion systems. *J. Math. Anal. Appl.*, 196:18–42, 1995.

[122] C. D. SCHAPER. Real-time control of rapid thermal processing. In *Proceedings of American Control Conference*, pages 2985–2989, San Francisco, CA, 1993.

[123] H. SIRA-RAMIREZ. Distributed sliding mode control in systems described by quasilinear partial differential equations. *Syst. & Contr. Lett.*, 13:177–181, 1989.

[124] L. SIROVICH. Turbulence and the dynamics of coherent structures: part I: coherent structures. *Quart. Appl. Math.*, XLV:561–571, 1987.

[125] L. SIROVICH. Turbulence and the dynamics of coherent structures: part II: symmetries and transformations. *Quart. Appl. Math.*, XLV:573–582, 1987.

[126] M. W. SMILEY. Global attractors and approximate inertial manifolds for nonautonomous dissipative equations. *Applicable Analysis*, 50:217–241, 1993.

[127] J. SMOLLER. *Shock Waves and Reaction-Diffusion Equations*. Springer-Verlag, Berlin-Heidelberg, 1983.

[128] M. A. SOLIMAN and W. H. RAY. Non-linear filtering for distributed parameter systems having a small parameter. *Int. J. Contr.*, 30:757–771, 1979.

[129] J. P. SORENSEN, S. B. JORGENSEN, and K. CLEMENT. Fixed-bed reactor Kalman filtering and optimal control-I. *Chem. Eng. Sci.*, 35:1223–1230, 1980.

[130] B. E. STANGELAND and A. S. FOSS. Control of a fixed-bed chemical reactor. *I & EC Res.*, 9:38–48, 1970.

[131] J. D. STUBER, T. F. EDGAR, and T. BREEDIJK. Model-based control of rapid thermal processes. In *Proceedings of the Electrochemical Society*, pages 113–147, (Vol. 4), Reno, NV, 1995.

[132] G. SZABÓ. Thermal strain during Czochralski growth. *J. Crystal Growth*, 73:131–141, 1985.

[133] R. TEMAM. *Infinite-Dimensional Dynamical Systems in Mechanics and Physics.* Springer-Verlag, New York, 1988.

[134] A. THEODOROPOULOU, R. A. ADOMAITIS, and E. ZAFIRIOU. Model reduction for optimization of rapid thermal chemical vapor deposition systems. *IEEE Trans. Sem. Manuf.*, 11:85–98, 1998.

[135] A. THEODOROPOULOU, R. A. ADOMAITIS, and E. ZAFIRIOU. Inverse model based real-time control for temperature uniformity of RTCVD. *IEEE Trans. Sem. Manuf.*, 12:87–101, 1999.

[136] A. N. TIKHONOV. On the dependence of the solutions of differential equations on a small parameter. *Mat. Sb.*, 22:193–204, 1948.

[137] E. S. TITI. On approximate inertial manifolds to the Navier-Stokes equations. *J. Math. Anal. Appl.*, 149:540–557, 1990.

[138] N. VAN DEN BOGAERT and F. DUPRET. Dynamic global simulation of the Czochralski process I. principles of the method. *J. Crystal Growth*, 171:65–76, 1997.

[139] N. VAN DEN BOGAERT and F. DUPRET. Dynamic global simulation of the Czochralski process II. analysis of the growth of a germanium crystal. *J. Crystal Growth*, 171:77–93, 1997.

[140] B. VAN KEULEN. *H_∞-Control for Distributed Parameter Systems: A State-Space Approach.* Birkhäuser, Boston, 1993.

[141] P. K. C. WANG. Asymptotic stability of distributed parameter systems with feedback controls. *IEEE Trans. Automat. Contr.*, 11:46–54, 1966.

[142] P. K. C. WANG. On the feedback control of distributed parameter systems. *Int. J. Contr.*, 3:255–273, 1966.

[143] P. K. C. WANG. Control of a distributed parameter system with a free boundary. *Int. J. Contr.*, 5:317–329, 1967.

[144] P. K. C. WANG. Stabilization and control of distributed systems with time-dependent spatial domains. *J. Optim. Theor. & Appl.*, 65:331–362, 1990.

[145] P. K. C. WANG. Feedback control of a heat diffusion system with time-dependent spatial domains. *Optim. Contr. Appl. & Meth.*, 16:305–320, 1995.

[146] J. T. WEN and M. J. BALAS. Robust adaptive control in Hilbert space. *J. Math. Anal. Appl.*, 143:1–26, 1989.

[147] E. B. YDSTIE and A. A. ALONSO. Process systems and passivity via the Clausius–Planck inequality. *Syst. & Contr. Lett.*, 30:253–264, 1997.

[148] E. B. YDSTIE and P. V. KRISHNAN. From thermodynamics to a macroscopic theory for process control. In *AIChE Ann. Mtg., paper 228a, San Francisco, CA*, 1994.

[149] T. K. YU, J. H. SEINFELD, and W. H. RAY. Filtering in nonlinear time-delay systems. *IEEE Trans. Automat. Contr.*, 19:324–333, 1974.

[150] J. ZABCZYK. *Mathematical Control Theory: An Introduction.* Birkhäuser, Boston, 1995.

[151] W. ZHOU, D. E. BORNSIDE, and R. A. BROWN. Dynamic simulation of Czochralski crystal growth using an integrated thermal-capillary model. *J. Crystal Growth*, 137:26–31, 1994.

Index

Systems & Control: Foundations & Applications

Systems & Control: Foundations & Applications publishes research monographs and advanced graduate texts dealing with areas of current research in all areas of systems and control theory and its applications to a wide variety of scientific disciplines.

We encourage the preparation of manuscripts in TEX, preferably in Plain or AMS TEX—LaTeX is also acceptable—for delivery as camera-ready hard copy which leads to rapid publication, or on a diskette that can interface with laser printers or typesetters.

Proposals should be sent directly to the editor or to: Birkhäuser Boston. 675 Massachusetts Avenue, Cambridge, MA 02139, U.S.A.

Estimation Techniques for Distributed Parameter Systems
H.T. Banks and K. Kunisch

Set-Valued Analysis
Jean-Pierre Aubin and Hélène Frankowska

Weak Convergence Methods and Singularly Perturbed Stochastic Control and Filtering Problems
Harold J. Kushner

Methods of Algebraic Geometry in Control Theory: Part I
Scalar Linear Systems and Affine Algebraic Geometry
Peter Falb

H_∞-Optimal Control and Related Minimax Design Problems
Tamer Başar and Pierre Bernhard

Identification and Stochastic Adaptive Control
Han-Fu Chen and Let Guo

Viability Theory
Jean-Pierre Aubin

Representation and Control of Infinite Dimensional Systems, Vol. I
A Bensoussan G. Da Prato, M. C. Delfour, and S. K. Mitter

Representation and Control of Infinite Dimensional Systems, Vol. II
A. Bensoussan, G. Da Prato, M. C. Delfour, and S. K. Mitter

Mathematical Control Theory: An Introduction
Jerzy Zabczyk

H_∞-Control for Distributed Parameter Systems: A State-Space Approach
Bert van Keulen

Disease Dynamics
Alexander Asachenkov, Guri Marchuk, Ronald Mohler, and Serge Zuev

Theory of Chattering Control with Applications to Astronautics, Robotics, Economics, and Engineering
Michail I. Zelikin and Vladimir F. Borisov

Modeling, Analysis and Control of Dynamic Elastic Multi-Link Structures
J. E. Lagnese, Günter Leugering, and E. J. P. G. Schmidt

First Order Representations of Linear Systems
Margreet Kuijper

Hierarchical Decision Making in Stochastic Manufacturing Systems
Suresh P. Sethi and Qing Zhang

Optimal Control Theory for Infinite Dimensional Systems
Xunjing Li and Jiongmin Yong

Generalized Solutions of First-Order PDEs: The Dynamical Optimization Process
Andreí I. Subbotin

Finite Horizon H_∞ and Related Control Problems
M. B. Subrahmanvam

Control Under Lack of Information
A. N. Krasovsku and N. N. Krasovsku

H^∞-Optimal Control and Related Minimax Design Problems: A Dynamic Game Approach
Tamer Başar and Pierre Bernhard

Control of Uncertain Sampled-Data Systems
Geir E. Dullerud

Robust Nonlinear Control Design: State-Space and Lyapunov Techniques
Randy A. Freeman and Petar V. Kokotović

Adaptive Systems: An Introduction
Iven Mareels and Jan Willem Polderman

Sampling in Digital Signal Processing and Control
Arie Feuer and Graham C. Goodwin
Ellipsoidal Calculus for Estimation and Control
Alexander Kurzhanski and István Vályi

Minimum Entropy Control for Time-Varying Systems
Marc A. Peters and Pablo A. Iglesias

Chain-Scattering Approach to H^∞-Control
Hidenori Kimura

Output Regulation of Uncertain Nonlinear Systems
Christopher I. Byrnes, Francesco Delli Priscoli, and Alberto Isidori

High Performance Control
Teng-Tiow Tay, Iven Mareels, and John B. Moore

Optimal Control and Viscosity Solutions of Hamilton-Jacobi-Bellman Equations
Martino Bardi and Italo Capuzzo-Dolcetta

Stochastic Analysis. Control, Optimization and Applications
William M. McEneaney, G. George Yin, and Qing Zhang, Editors

The Dynamics of Control
Fritz Colonius and Wolfgang Kliemann

Optimal Control
Richard Vinter

Advances in Mathematical Systems Theory
Fritz Colonius, Fabian Wirth, Uwe Helmke, and Dieter Prätzel-Wolters

Nonlinear and Robust Control of PDE Systems
Panagiotis D. Christofides